AF714312

Institutional constraints to small farmer development in Southern Africa

Institutional constraints to small farmer development in Southern Africa

edited by:

Ajuruchukwu Obi

Wageningen Academic Publishers

Wageningen Academic Publishers
P.O. Box 220
6700 AE Wageningen
The Netherlands
www.WageningenAcademic.com
copyright@WageningenAcademic.com

ISBN: 978-90-8686-132-3
e-ISBN: 978-90-8686-704-2
DOI: 10.3921/978-90-8686-704-2

First published, 2011

Preface

The Southern Africa region has experienced more than its fair share of problems in recent years. Just when it seemed that the hardships wrought by the devastating cycle of droughts and floods of 2000 to 2002 were a thing of the past, other problems emerged. At one level, there have been the weak and often erratic governance mechanisms and political crises in some countries of the region, leading to severe disruptions in agricultural production to the point that supplies and markets have virtually disappeared. At another level, socio-cultural rigidities have often militated against the adoption of efficient farming practices, resulting in sub-optimal choices that lock smallholders into a low equilibrium trap. In the face of the disappearing supplies and missing markets, these have engendered hyper-inflationary trends of a magnitude unknown anywhere else in the world. But in the midst of all this apparent dreariness, cases are emerging from which immense lessons can be drawn.

This book assembles a collection of research papers based on studies completed in 2008 and 2009 in Southern Africa that examine various dimensions of the problem. Despite its title, the book does not exclusively subscribe to the notion of institutions as constraints in the sense that all they merely perform a restrictive function on economic behaviour. The various contributions to this book cover the spectrum of viewpoints about what institutions constitute and the multiple roles they also play in conditioning economic behaviour by modifying the structure of incentives and enhancing deeper understanding of the workings of the processes that determine economic activity. These perspectives are taken in examining the institutional environment in which small farmers operate in the region and how they are going about either dealing/coping with them (in the case of constraints), or taking advantage of them (in the case of the opportunities afforded by new incentive structures or information).

The papers draw from these diverse and polar experiences and present some theoretical and practical insights that should form the basis for more in-depth, country-level, sector-specific analyses, focusing mainly on citrus, horticultures, cotton and livestock. The thematic issues of income inequality, land reform, value chain governance and chain choice, and natural resource management are covered in this book and expected to hold strong interest for a wide constituency, including researchers, development practitioners, rural animators, and policy makers.

The bulk of the chapters in this book have been based on a study funded under the Southern African Development Community's Fund for Innovative Regional Collaborative Projects (FIRCOP). In late 2003, SADC called for concept notes and received a total of 104 out of which 35 were selected in August 2004. The 35 researchers and groups were contacted to submit full proposals out of which 10 were short-listed for a final selection process that resulted in the approval of 6 projects for funding in 2006. The elaborate review process and the efforts of the anonymous reviewers over an unprecedented period of three years are acknowledged.

Subsequently, the study resulted in a number of dissertations leading to the award of Masters degrees to several students. Eight external examiners and four internal examiners were involved in the formal examination of the original dissertations. In addition, two examiners were involved in the examination of the Swaziland maize market study. The chapter based on that study is the source of three separate papers presented, after double blind reviews, at the International Food and Agri-Business Management Association conference in Chicago in 2005, the annual conference of the Agricultural Economics Association of South Africa in 2005, and the International Association of Agricultural Economists conference in Brisbane, Australia in 2006.

After the chapters were assembled into a manuscript for the publication of the present book, an initial workshop was held at which researchers and practitioners drawn from numerous Universities and research and development organizations were represented. This workshop allowed participants to make major inputs that led to the restructuring of the outline as well as modifications to the content. Two other peer reviewers, namely Professor Noble Jackson Nweze of the Department of Agricultural Economics of the University of Nigeria, and Dr Godfrey Kundhlande of the Department of Agricultural Economics of the University of the Free State, Bloemfontein, South Africa, received the complete manuscript and provided detailed comments which were incorporated during January-June 2010. All these contributions are gratefully acknowledged.

Ajuruchukwu Obi

Table of contents

Part I – Background and issues on the role of institutions in smallholder development in Southern Africa

Part II – Rural livelihood strategies and current circumstances of the small farmer in South Africa

Part IV – Marketing constraints and opportunities in specific commodity lines

Part V – Institutional issues in natural resource management and implications for smallholder development in Southern Africa

Part VI – Alternative trade and support arrangements to enhance livelihoods and welfare of small farmers

Abbreviations and acronyms

A1 FARMERS	small scale farmers who benefitted in Zimbabwe's Fast Track Land Reform Programme (2000-2009)
A2 FARMERS	medium/large-scale commercial farmers who benefitted in Zimbabwe's Fast Track Land Reform Programme (2000-2009)
ADMARC	Agricultural Development and Marketing Corporation
AGOA	Africa Growth and Opportunity Act (of the United States Government)
AGRI-BEE	Black Economic Empowerment in Agriculture
AGRITEX	Agricultural Research and Extension Services (of Zimbabwe).
ANC	African National Congress
ANOVA	Analysis of Variance
ARAP	Accelerated Rainfed Arable Programme (of Botswana)
ARC	Agricultural Research Council
ARD	Agricultural Research for Development
BATAT	Broadening Access to Agriculture Thrust
BEE	Black Economic Empowerment
CAADP	Comprehensive Africa Agricultural Development Programme
CEA	Comparative Economic Advantage
CGIAR	Consultative Group on International Agricultural Research
CIA	Central Intelligence Agency (of the United States of America)
CLARA	Communal Land Rights Act
COMESA	Common Market for Eastern and Southern Africa
COP15	Copenhagen Conference on Climate Change
CPA	Communal Property Association
CPI	Consumer Price Index
CPR	Communal Property Resources
CSC	Cold Storage Company
CTA	The Technical Centre for Agricultural and Rural Cooperation for the Africa, Caribbean and Pacific Group of States and the European Community
DBSA	Development Bank of Southern Africa
DLA/DoA	Department of Land Affairs/Department of Agriculture
DRC	Domestic Resource Costs
ECDC	Eastern Cape Development Corporation
EIU	Economist Intelligence Unit
ESTA	Extension of Security of Tenure Act
FANRPAN	Food, Agriculture and Natural Resources Policy Analysis Network
FAO	Food and Agriculture Organization of the United Nations
FAOSTAT	FAO's Statistical Database
FDI	Foreign Domestic Investment
FIRCOP	Fund for Innovative Regional Collaborative Projects

FTLRP	Fast Track Land Reform Programme
GDP	Gross Domestic Product
GMB	Grain Marketing Board
GOL	Government of Lesotho
GOZ	Government of Zimbabwe
HIPC	Heavily Indebted Poor Countries
HIV/AIDS	Human Immune Deficiency Virus/Acquired Immune Deficiency Syndrome
IAD	Institutional Analysis for Development
ICLARM	International Center for Living Aquatic Resources Management (nowadays WorldFish Center)
ICRA	International Centre for development-oriented Research in Agriculture
IDC	Industrial Development Centre
IES	Income and Expenditure Survey
IFPRI	International Food Policy Research Institute
IMF	International Monetary Fund
IRIN	Integrated Regional Information Networks
LHWP	Lesotho Highlands Water Project
LPM	Landless Peoples' Movement
LRAD	Land Redistribution for Agricultural Development
LSCF	Large Scale Commercial Farms
MDG	Millennium Development Goals
MMIP	Maize Marketing Improvement Programme
MOAC	Ministry of Agriculture and Cooperatives
NAFU	National African Farmers' Union
NAMBOARD	National Agricultural Marketing Board
NATO	North-Atlantic Treaty Organization
NBER	National Bureau of Economic Research
NDA	National Development Agency
NEPAD	New Partnership for Africa's Development
NERPO	National Emergent Red Meat Producers Organization
NEWU	National Early Warning Unit
NFE	Non Farm Employment
NGO	Non-Governmental Organization
NIE	New Institutional Economics
NMC	National Marketing Council
NRM	Natural Resources Management
OLS	Ordinary Least Squares
PAM	Policy Analysis Matrix
RDP	Reconstruction and Development Programme
RSA	Republic of South Africa
SACU	Southern African Customs Union

SADC	Southern African Development Community
SAFEX	South African Futures Exchange
SAP	Structural Adjustment Programme
SLAG	Settlement/Land Acquisition Grant
SNL	Swazi Nations Land
SSA	Sub-Saharan Africa
SSCA	Small Scale Commercial Areas (Zimbabwe)
StatsSA	Statistics South Africa
UNDP	United Nations Development Programme
USAID	United States Agency for International Development
VAT	Value-Added Tax
VOC	Vereenigde Oost-Indische Compagnie
WFP	World Food Programme
ZFU	Zimbabwe Farmers Union

Exchange rates for programme countries

South Africa

US$/R

Year[1]	Rand amount
2003	8.00
2004	6.33
2005	6.62
2006	7.29
2007	7.18
2008	7.90
2009	7.91

[1] Rates as established on 1st January of each year.

Swaziland

1 Swaziland Lilangeni (SZL) = R 0.9714

Lesotho

1 Lesotho Loti = R 1.00

Zimbabwe

US$/ZIM$

Year[1]	ZIM$ amount
2003	57.15
2004	818.52
2005	5,600.1
2006	82,902
2007	258.92
2008	30,475.3
2009	4,066,342
2010	357.75

[1] Rates as established on 1st January of each year.

Part I
Background and issues on the role of institutions in smallholder development in Southern Africa

1. Investigating institutional constraints to smallholder development: the issues and antecedents

Ajuruchukwu Obi and Tebogo Seleka

Abstract

Smallholder farming in Southern Africa has been on the decline in recent years as a result of a combination of institutional, climatic and macro-economic constraints. This has led to some of the worst food and humanitarian crisis in recent years, culminating in drastic food shortages and famines and deepening poverty. Governments and the development community are therefore concerned about the current and potential consequences of this state-of-affairs and are exploring ways and means to reverse the situation. There is increasing awareness about the role of institutions in smallholder development in the region and elsewhere on the continent. But consensus as to what specific institutions are crucial or even how they exert their influence has not yet been achieved despite a considerable amount of theoretical information now available on the subject. This chapter begins the task of defining the broad outlines of a more focused study on the role of institutions in small farmer situations in Southern Africa and proposes an analytical framework consolidated from a wide range of expert views and opinions. Brief descriptions of the diverse studies that fit that framework are provided to attempt to contribute to a debate on the feasible procedures for undertaking a comprehensive analysis of the institutional constraints faced by the small farmer in relation to market access and exploring the competitiveness of the small-scale agriculture in the region.

1.1 Introduction

The initial case for investigating the institutional constraints to smallholder development was pivoted on the reality that smallholder farming in Southern Africa was going through one of its toughest times in history. The scale of the problem was virtually unknown in contemporary experience. The conviction drew largely from literature as well as practical experience with the development terrain in the region as the transition into the new Millennium began. In June, 2001, a joint UN Mission to Lesotho reported a decline in the cereal production in the order of 55% below previous year results and 60% below average results for the 5 years preceding the mission (Joint UN Mission, 2001). Heavy rains and other unfavourable weather conditions coinciding with the planting and harvesting of both winter and summer crops in 2001, severely dislocated the sector's investment profile and created unprecedented deficits as most fields were uncultivated and most crops could not be harvested. By 2002, this situation had placed about 444,800 persons in at least 3 districts on the verge of starvation (The Economist, 2002). In August 2003, similar situations were reported in Tanzania, Malawi, Swaziland, Mozambique, Zimbabwe, etc. (WFP, 2003).

The region's development prospects are also hampered by a number of macroeconomic constraints, including low foreign direct investment (FDI), low savings rates in the region of 17%, and a double-digit inflation in nine member states (Anonymous, 2003). A heavy debt burden, with as many as seven countries participating in the Heavily Indebted Poor Countries (HIPC) Initiative, completes this highly desperate picture (Anonymous, 2003). On top of all these, the HIV/AIDS pandemic began ravaging the region to the point that the human capacity to implement the most basic tasks on farms became compromised. When people cannot even afford the simple tasks of sowing seeds or harvesting crops to feed starving households, nothing can be more desperate. In his book *The End of Poverty: How we can make it happen in our lifetime,* Jeffrey Sachs' presents what he called a 'Global Family Portrait' that sketched the scene in a Malawian village that had lost all its able-bodied young men who, if they had lived, 'could have built small-scale water harvesting units on rooftops...' (Sachs, 2005). This picture obviously has more than a familiar ring to anybody who has been involved in development in the region.

At both regional and continental levels, African leaders then began committing to an expansion of investment in agriculture. The African political leadership meeting in Maputo in its second summit in July 2003, mandated member states to ensure a minimum allocation of 10% to agriculture within five years in order to accelerate growth to at least 6% per annum, which was considered sufficient to effectively address poverty reduction goals on the continent (African Union, 2003). The Comprehensive Africa Agriculture Development program (CAADP) was established at that summit as the means by which the vision of the New Economic Programme for Africa's Development (NEPAD), of redefining African Agriculture by accelerating economic growth, minimizing poverty, and enhancing food security, could be achieved (InWent/IFPRI/NEPAD/CTA, 2003).

The leaders were undoubtedly convinced that potential exists for African countries to expand agricultural output and turn the sector into a profitable one. For instance, in the late 1990s, the United States Agency of International Development (USAID) commissioned several studies to assess the Economic Comparative Advantage (CEA) of Southern African countries in agricultural production. The studies concluded that many of these countries, including Tanzania, Malawi, and Zimbabwe, possess economic comparative advantages in a wide variety of agricultural products whose production would significantly and positively impact on both farmer and national welfare (Nakhumwa *et al.*, 1999; Sukume *et al.*, 2000).

The foregoing clearly suggests that there is strong political commitment for small farmer development at both the national and regional levels. It is also clear that policy makers in the region have come to the realization that alternative policies that ignore small farmer development are not working. In the light of these facts, the question one must ask is why are countries in the region not taking advantage of the available opportunities to expand output and enhance market access (Van Schalkwyk and Jooste, 2002). In the words of the Executive Secretary of the SADC (2003), '...it is a question of management'. But he recognizes that the question goes deeper than that and adds: '...we are looking at irrigation... at other issues, such as agrarian reform, how we can bring extension services to the rural areas, how we can use information technology...' (Anonymous, 2003). A study conducted in Lesotho (Van Schalkwyk *et al.*, 2002) identified numerous technical, managerial, and institutional factors that hampered the success of the government-driven agricultural production programmes in the country.

These views go to the heart of the arguments that proponents of the New Institutional Economics (NIE) have been making for sometime now. As soon as it became clear that Adam Smith's reference to the 'invisible hand' idea never meant that a selfish, mindless, emotionless and value-free mechanism exists for allocating resources to meet the productive ends of society, economists became even more single-minded in advancing the notion of the primacy of institutions. According to North (1992), the efficient markets that are predicted by the neo-classical assumption of instrumental rationality are not achievable in reality because of the numerous shortcomings in the economic system. For instance, the induced innovation model that assumes perfect markets for factors and products, fails to consider situations where prices do not convey all the information needed to decide between alternatives courses of action. The model also assumes that all economic agents in a transaction face the same prices, which would mean that transactions are cost-free and asset distribution would have no effect on efficient allocation of resources, among other conveniences. But nothing can be farther from the truth.

Everybody from North (1992) to Dequech (2002) and Frank (2009), doubts that we are close to a definitive theory of institutions that provides the final answers to these questions. But it is not debatable that transactions involve costs which can often be so high that they constitute veritable obstacles to production and exchange. These transaction costs embrace elements as diverse as the costs of adjustment, information, measuring attributes, and negotiating, monitoring, and enforcing contracts. Human beings are frequently constrained by incomplete information as well as the limited mental capacity for processing whatever information is available (North, 1992). Attempts to overcome these shortcomings impose considerable costs on economic agents. As well, since different economic agents face different prices, there is a tendency for some people to take advantage of others, again due to differential access to information about current and prospective costs and prices, a phenomenon that Akerlof (1970) exploited and elaborated into the sub-discipline of information asymmetries. Information about the possibilities open to human beings to better their lives is never perfect and they have to act often on the basis of hunches and gut feelings which are presented under the more dignified caption of 'perceptions'.

Externalities are another reason that efficient markets do not automatically materialize from the actions of economic agents; benefits and costs of production, marketing and consumption activities accruing to other than those directly responsible for the activities (Frank, 1994). Economists believe that dealing with these problems through collective action is the basis for the evolution of institutions. In turn, institutions contribute to growth in output in agriculture by altering the incentives, rights to use of available productive resources, and knowledge and skills to use what is available or develop new ways and means to attain the goals of economic activity (Eicher and Staatz, 1984; North, 1992). The diverse actions and arrangements that produce the foregoing outcomes, which can broadly be described as coordinating and facilitating actions, can also be included as part of these institutions to obtain what Eicher (1999) describes as 'a good institutional environment'. All these elements, be they formal or informal, which modify the incentive structure, set the rules of engagement in respect to resource use and human interactions, and guarantee some measure of predictability in the system by removing uncertainties, constitute institutions as North (2003) observes. In the way North (2003) has elaborated the concept, institutions embrace the rules and norms that regulate human actions and the arrangements in place to ensure compliance as well as the mechanisms in place to facilitate access to productive resources. A more formal discussion on the link between human actions and institutional development is taken up later in Chapter 2 of this volume. Those mechanisms that facilitate compliance with formal rules and informal norms are what North (2003) identified as enforcement characteristics which are so ubiquitous they embrace almost every aspect of economic, political and social life. Figure 1.1 presents, on the basis of work done by Norton *et al.* (2006), an analysis of these institutional variables implicated in farming systems performance.

In the views of Norton *et al.* (2006), institutions and the human factor interact to determine the farming system. The implication is that these factors are also involved in system performance, as noted by North (1990, 1992, 2003). If formal rules and informal norms recognize and imbue individuals with property rights in some productive resources but they lack sufficient information to efficiently utilize the resources, performance is affected negatively. Information asymmetry in the way George Akerlof (2001) has described it in his Nobel Lecture thus becomes an important institutional factor or enforcement characteristics governing access to markets for producers. This link has also been made by Van Huylenbroeck and Espinel (2007) on the basis of case studies of small livestock producers in Uganda, crop farmers on an irrigation scheme in the Peninsula of Santa Elena (Ecuador), and a biodiversity preservation scheme. So, rules and norms confer rights and individuals are able to take advantage of such rights under appropriate conditions such as affordable prices, access to credit and other facilitating services, effective infrastructure, adequate security, etc. Many or all of these require that appropriate and well-functioning organizations and political systems are in place to provide the requisite governance. It is therefore understandable that institutions can be viewed quite broadly as 'the structure that humans impose on human interactions' (North, 1990). Table 1.1 attempts to condense

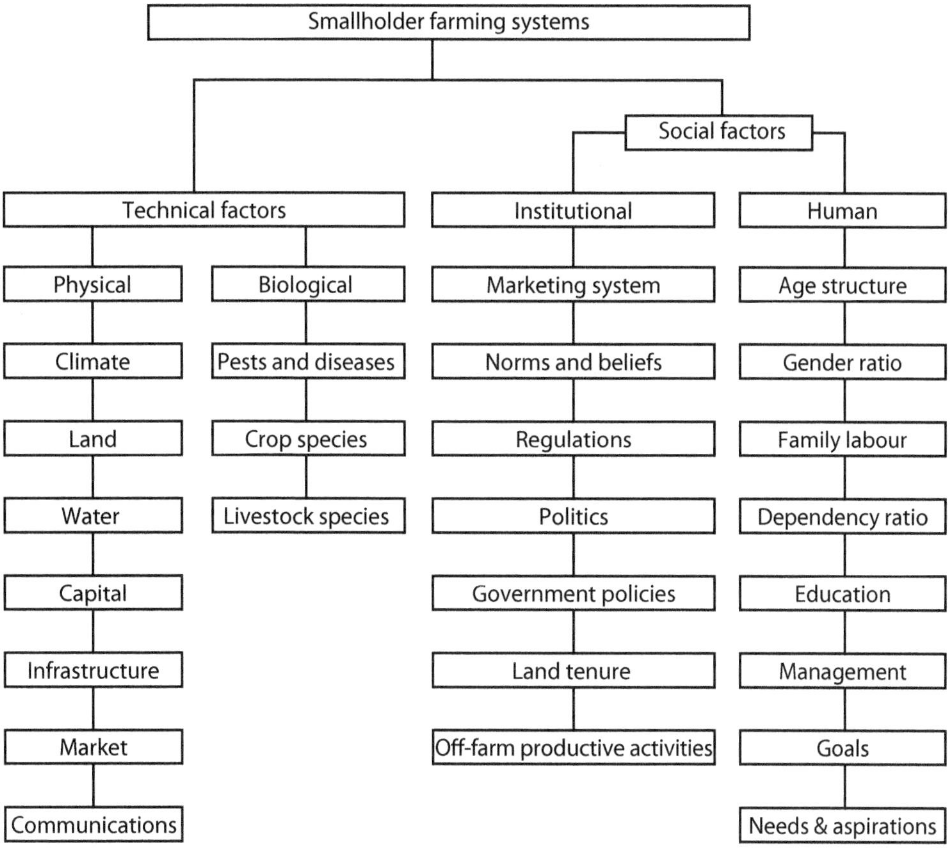

Figure 1.1. Relationships between smallholder farming systems and institutions. Adapted from Norton et al. *(2006).*

these thoughts into a framework that recognizes the ubiquity of institutions and how the diverse elements can be applied for enhanced performance.

Many countries in Africa have since the 1990s been dismantling government controls and converting to market-based food systems, believing that market reforms would enhance farm profitability through their positive effects on prices, investment levels, and commercialization (Jayne *et al.*, 1997). In fact, the need for such agrarian reforms, including commercialization of the smallholder production systems, has received considerable attention from governments and development organizations, including the SADC (FANRPAN, 2003; Anonymous, 2003). But the results of the reform programmes have been mixed and frequently inconsistent with the expected increases in productivity. It is now being realized that the sectoral reform prescriptions have, in many cases, been based upon only superficial knowledge of the prevailing economic institutions and how

Table 1.1. Framework for institutional analysis and policy interventions for agricultural development.[1]

Institutions factor	Implication/outcome	Interventions needed
1. Policy	Lack of operational policy and specific objectives	Balanced operational policy with realistic targets
2. Planning	Lack of decentralized/participatory agricultural planning	Establish participatory agricultural planning systems and procedures
3. Rural infrastructure	Poor water distribution, roads, communication, etc.	Planned piped water, schemes, roads, etc.
4. Inputs/services	Lack of readily available inputs/ services	Establish rural service centers in various districts of the country
5. Marketing/prices	Lack of organized marketing and price incentives	Marketing and pricing policy for major products
6. Credit	Lack of credit facilities	Provide selective controlled credit
7. Research	Lack of local agricultural research	Develop suitable applied research structures
8. Extension	Ineffective and inefficient extension	Reorganize in-time bound Training and Visit System. Then, balanced use of communication channels
9. Land tenure	Lack of security and negotiability of land rights. Uncontrolled communal grazing	Registration and negotiability of arable land rights. Cooperative grazing schedules
10.Development coordination	Uncoordinated rural development approach	Decentralized control and coordination policy at District & regional level
11. Regulations and standards	Lack of clarity; costly compliance requirements	Farmer education programmes. Extension services providing support to producers
12. Cooperation and collection action	Absence of mechanisms for inter-household and inter-institutional cooperation	Civic education on value of cooperation and collective action; demonstration schemes

[1] Consolidated from various workshop/brainstorming sessions with students in editor's development courses, 2008-2010.

they affect economic outcome in particular economies. There is also an emerging general consensus that future productivity growth within the evolving market economies in Africa will require closer attention to the institutional details of the system – i.e. going beyond generalizations that property rights, market rules, and exchange mechanisms need to be identified and worked out, to actually conducting pragmatic applied research on the specific kinds of property rights, rules, and exchange arrangements that would most contribute to economic development under particular circumstances.

The process of globalization has a major impact on the development of institutions and should also be taken into account in research. The globalization process is fuelled by such forces as the simultaneous opening of financial capital markets and the dismantling of closed trade in agricultural commodities, which raises questions about the links between natural resources, government, household and private sector strategies and the economic welfare of a country. If the globalization process and the creation of the correct institutions are to be the catalyst for economic growth and development in emerging and developed markets, then understanding these links is crucial. This calls for a fundamental review of the whole basis for programming of agricultural investment in the region.

In short, a comprehensive assessment of the managerial, technical and institutional basis of farm production is called for, to not only identify the obstacles to agricultural and national economic growth in the region, but also to find out what alternative options are available and how they can be better delivered for greater effectiveness and development of sustainable small-holder farming systems in Southern Africa.

It was therefore proposed to undertake a comprehensive institutional analysis of the agricultural sector of the selected countries such that comparisons of relative performance could be carried out on the basis of a number of criteria, including size, membership of regional economic grouping, notably the Southern African Customs Union (SACU) and the Common Market for Eastern and Southern Africa (COMESA), extent of democratization, spatial relationship with South Africa, etc. For instance, an enclave entity like Lesotho that is completely surrounded by South Africa, and the land-locked countries like Botswana and Swaziland, may experience integration differently from other member states. Partnership agreements being negotiated between some of the countries and the European Union may also produce impacts for those and other countries and these should be examined and understood. Some members of the regional body are also participating in the US Government's Africa Growth and Opportunity Act (AGOA), which will no doubt have implications for the nature and content of intra-SADC and intra-SACU trade and ultimately the economic development of the participating countries. As the SADC is currently implementing a phased program of integration and many members are faced with important decisions concerning their subsequent membership of associated groups and customs unions, a clearer picture of relative costs and benefits of membership of these alternative regional groupings is needed. Funding constraints and administrative decisions regarding the outcome of the project made it impossible to pursue some of the foregoing lines of enquiry, but sufficient insight has no doubt be developed nonetheless and whatever gaps remain can be filled in follow-up investigations in the future.

But what has emerged puts institutions in a somewhat dominant role in terms of the programming of interventions for agricultural development. Among several academics and policy leaders, exasperation with what is described as 'institutional fundamentalism' is building up. They worry that the term 'institutions' is now the 'porridge' concept that is employed to cover up all the confusions we cannot resolve by standard and conventional thought. But we have felt and continue to feel such exasperations with terms such as 'prices', 'markets', 'industrialization' and all the other cure-alls, and this definitely did not render them irrelevant. What is needed is probably some caution and a lot more attention to balance in seeking explanations to the imponderables in African development, of which there are many. Dealing with the development challenges in Africa has all the trappings of caring for a family member whose illness persists even after all known conventional procedures have been exhausted. In some cultures, the care-giver is asked to 'look around' and there is a tendency to hang on to anything that offers a bit of hope.

Given that the smallholder sector seemed most vulnerable, issues around its development continue to dominate discussions and enquiry. In the light of the foregoing, a series of studies were designed in two South African universities, namely the University of the Free State and the University of Fort Hare, with the collective developmental objective to contribute to poverty alleviation in Southern Africa by identifying the roles that institutions can play and what options existed, in addition to the modification of trade patterns, for revamping the smallholder farming sector in the region. On an overall basis, the studies aimed to carry out a comprehensive assessment of the institutional factors that impact on farm production, to not only identify the obstacles to agricultural and national economic growth in the region, but also to find out what alternative options are available and how they can be better delivered for greater effectiveness and development of sustainable smallholder farming systems in Southern Africa. Thus, two obvious immediate objectives were:

- to produce a comprehensive database on the institutional constraints on smallholder farming in Southern Africa, and
- to develop alternative designs that maximize market access for smallholders and lead to measurable improvements in incomes of small scale farmers.

In order to achieve the foregoing immediate objectives, the following specific steps were envisaged:

a. Conduct comprehensive assessment of the existing institutions and organizations that serve the farm sector of the study countries.
b. Recalculate comparative economic advantages for the previously studied countries on the basis of current macroeconomic, production and market data.
c. As a follow-up to the above exercise, conduct more detailed investigation with respect to agricultural products/commodities in which each country has an economic comparative advantage.
d. Comprehensively examine the factors, in terms of assets, institutions, and processes that influence the household's flexibility in responding to alternative production and/or trade regimes.

e. Investigate alternative trading mechanisms to which the farm sector is exposed as a basis for determining the optimal trade patterns for sustainable smallholder development.
f. On the basis of (d) and (e) above, identify measures to link small farmers to the available trade opportunities. In this regard, to come up with alternative designs for addressing access to resources, including land and credit (including through microfinance), knowledge and skills (through an efficient and farmer-relevant extension service).
g. Investigate the supply chain for selected agricultural products within the region and assess extent of vertical coordination among producers and markets and determine measures to improve operational efficiency, profitability and value-adding. In this regard, the issue of informal cross-border operations involving agricultural products was considered crucial for investigation and quantification.

1.2 The problem

The growing number of people living in abject poverty within the region remains a source of serious concern to both governments and development organizations. Theoretically put at that time at about 70% of the population on a regional basis, there is evidence that this is grossly understated for many countries in the region while for others it may be overstated. For instance, rural conditions had been worsening in countries such as Swaziland and Zimbabwe since 2001/2002, when the region as a whole experienced the worst food and humanitarian crises in years. While many of the affected countries had by 2006 embarked on the route to recovery, others had sunk deeper into destitution. The conclusion that was drawn at that time that other factors may have been at play has proved to be quite prophetic in the unfolding drama in Zimbabwe which, even in 2009, was still grappling with the symptoms of an unmistakable government failure. In a country like Swaziland where as many as 60% of the population could be living below the poverty line, the suggestion is that maintenance of stiff controls and regulations on the maize industry may be hurting small farmers as well as the rural poor. Again for Zimbabwe, several factors had been implicated; including the procedures followed in fast-tracking the land reform programme after 1999 and other political factors, although much of the information remained anecdotal in the absence of systematic scientific study.

The situation is not much different in rural Mozambique where severe infrastructure deficiencies hamper farmers' access to markets and opportunities to profitably market produce or obtain inputs. While Botswana as an economy continues to show promise in terms of efficient economic and public policy management, there seems to be a disconnect in terms of the pace of rural transformation, with many rural areas being still steeped in poverty and untouched by the significant progress at the macroeconomic level. The South African situation is now well-known in respect to the serious dichotomies between rural and urban areas and among races despite considerable progress on the political front and the sustenance of a modern industrial sector. Many parts of South Africa, particularly the

Eastern Cape, North West and Limpopo Provinces, still exhibit severe rural poverty levels reminiscent of the old apartheid days and not much different from what obtains elsewhere on the African continent.

The situation for the entire region is being worsened by the impact of HIV/AIDS which is leading to unrelenting depletion of the human resource capacity in the region, more so in the farm sector. The results of these studies were thus expected to contribute to policy formulation on strategies to improve smallholder participation in the development process through output expansion and enhanced market access. Implementation of such strategies would contribute to poverty alleviation by providing more income to farmers. This should have a positive effect on the economy as a whole, with special considerations being the empowerment of women and other vulnerable groups.

Among the numerous problems of traditional agriculture, the prevalence of low-level technology ranks very high. According to Spencer (1994), farmers in Africa have not widely adopted the Green Revolution technologies that brought phenomenal improvements in yield in America and Asia. As Lele (1984) noted, whatever increases occurred in the production of the major cereals and root and tuber crops in Africa up to the early 1980s were as a result of area expansion rather than improvements in the productivity per unit of input, or yield, which results from technical change. In fact, agricultural production in general has been on the decline for nearly three decades. It is estimated that agricultural production per capita declined by some 22% between 1971 and 1984. By all accounts, sub-Saharan Africa is the only region in the world where per capita agricultural growth has been declining over the past four decades. In both Asia and Latin America, agricultural growth per capita began a phenomenal growth from the mid-1970s and has remained positive since then.

This situation has continued to be a source of serious concern among policy makers and researchers. Coincidentally, during the same period, African governments began to cut back on expenditures devoted to agricultural technology development and transfer. According to a review conducted in the late 1990s to assess the impacts and lessons in agricultural technology development and transfer in Africa, the spending on agricultural technology development and transfer had actually declined by as much as 37% over the 20 year-period between 1971 and 1991 (Oehmke *et al.*, 1997). Since up to one-third of the growth in agricultural productivity can be explained by past investments in agricultural research (Oehmke *et al.*, 1997), there is no question that the observed decline in agricultural production per capita can be attributed to underinvestment in research and development.

Reviewing the performance of traditional arable farming in the context of the Accelerated Rainfed Arable Programme (ARAP) in Botswana, Seleka (1999) mentioned low adoption of improved technologies as one of the 'sub-sectoral factors' implicated in the poor performance of the country's arable agriculture. A key challenge identified in the effort to develop agricultural production in the communal areas of South Africa is the use of low level technology among the farmers in the former 'independent homelands' of the country where inadequate infrastructure, knowledge and skills constrain the use of improved inputs and methods in farming (Van Schalkwyk *et al.*, 2004). A similar point has been made for several sub-Saharan African countries, notably Swaziland (Magagula and Faki, 1999; Mashinini, 2004; Mkhabela and Mashinini, 2005). The situation has persisted despite the large amount of research efforts and results currently available and the indisputable evidence of the important role of improved technologies in agricultural development.

As is already clear from the foregoing, smallholder farmer problems are both production and marketing related, and the situation in Southern Africa is no different from what obtains in the rest of Africa. In relation to markets, two equally serious problems can be identified. One is market failure in which the markets that exist are so severely weakened that they are unable to perform the signalling function of markets in a way that serves as positive incentives to producers. The reason for this could be that the state system considers that the markets need to be assisted to perform their role, leading to various degrees of intervention by the state. In such a situation, prices do not bear a clear relationship to the value of goods and services because they either carry heavy subsidies or have been inflated by taxes. This is a serious problem wherever it exists.

The other problem is that the markets simply are not there. The development literature has generally attributed this phenomenon of missing markets to things like externalities, coordination failure, technological development, transactions costs, and failure of trust and information (Arrow, 1969; Hart, 1980; Makowski, 1980; Durlauf, 1992; Makowski and Ostroy, 1995; Heller, 1997, among others). A practical situation that has elements of the concepts outlined above is probably one in which physical product has dwindled to levels that are insufficient to motivate exchange or there is no cash in the economy and producers are unsure that their output will be paid for, leading to a termination of market participation. Are there cases that can be used to illustrate the concepts in the case of Southern Africa? A country like Zimbabwe has experienced both market failure and missing markets and details of how these situations develop and manifest can be examined to see what lessons can be drawn for policy.

Naturally, concerns about market development are an integral part of the efforts to develop smallholder farmers. When the international development community, notably the World Bank, decided on Structural Adjustment Programmes (SAP) in the 1980s, the idea was to tackle the problem of market failure. The market liberalization prescriptions were aimed to achieve this goal as explained in the previous section. But when markets are not there at all, it becomes a lot more complicated. There are many reasons for this, and economists as temporally distant as Hodder and Ukwu (1969), Obi (1984) and Gabre-Madhin (2006) have long recognized that much of the difficulty in addressing market development in Africa stems from the fact that traditional markets are intimately linked with human interactions and influenced by culture, values, history and social forces. One often-missed fact is that markets in Africa are not only about prices and getting them right as required by the market liberalization strategies of the international development institutions. Markets are points of social and cultural interactions in many societies; men go to markets to look for future brides and village governments punish wrong-doings by exposing culprits to the marketplace, thus making markets 'multifunctional institutions associated with several non-economic aspects of (local) cultures' (Hodder and Ukwu, 1969). As human cognitive ability to deal with these issues is often limited, resolution requires immense investment of time and effort and innovative analytical procedures, hence the focus on institutional analysis.

In South Africa, agricultural produce from smallholder farmers is often lost after production due to spoilage and inability to access the markets. This is mainly because most smallholder and emerging farmers are faced with a range of technical and institutional challenges that affect market access and compromise the profitability of farming at that level. There are also concerns about natural resource management especially in parts of the country where customary tenure systems still prevail and conflicts arise about the optimal allocation of common property resources. Whereas the marketing infrastructure is poorly developed, smallholder and emerging farmers lack supportive organizations that represent and serve them. These factors reduce smallholder and emerging farmers' incentives to participate in formal markets. A reduction in formal market participation, in turn, makes it difficult for these farmers to shift into commercial farming and thus, a reduction in economic development. With these facts in mind, this study analyzes the extent to which technical and institutional factors influence the marketing channel choices among emerging and smallholder farmers in the Kat River Valley.

Literature on marketing constraints and market opportunities for indigenous cattle is still limited since over the past years studies conducted were based on mixed and exotic cattle breeds. In South Africa, off-take rates in the small-scale cattle sector is much lower than in the commercial sector, i.e. an off-take rate of between 5-10% compared to 25% in the commercial sector (Montshwe, 2006). This situation creates serious obstacles to developing an effective cattle marketing system that targets the smallholders in the communal areas of the country that suffered the most during the period of apartheid policies in the country. After the implementation of Nguni cattle projects in the Eastern Cape Province by the University of Fort Hare, market off-take is still very low in communities that benefited from the project as evidenced by the deterioration of grazing lands in the province due to overstocking.

In addition, the majority of the beneficiaries sell cattle by-products with no value added to them and they get relatively lower prices that make production unsustainable and lead to a view of agriculture as a business with low returns. As a result, the youths also view agriculture as a non-profitable business. In extreme cases, some farmers do not even see the value of other cattle by-products such as manure. It is, therefore, crucial to understand why these farmers do sell their cattle, see value of other cattle by-products and do not add value to cattle by-products. In other words, there is need for research and development strategies aimed at identifying marketing constraints that the communal farmers experience and possibility of a niche market for Nguni cattle products.

Despite efforts to make poverty history in several regions, there is no agreement on which income-generating activities should be promoted and/or how development agencies should allocate available limited resources to reduce poverty. Policymakers view the rural economy as a sector driven almost entirely by agriculture, with almost all poverty reduction initiatives biased towards agriculture (Kirsten and Moldenhauer, 2001; Machethe, 2004). They advocate that agriculture is the critical source of livelihood of rural societies and the best hope for reduction of poverty in rural Africa. Provision of land and promotion of agriculture are thus seen as the pathway out of poverty and inequality in rural South Africa (Lipton, 1996). Such a policy position obviously calls for a bold land reform programme that squarely addresses South Africa's history of land-related racial exclusion.

Maize and other related coarse grains constitute half of the global cereal production and trade (Falcon and Naylor, 1998). Maize alone constitutes about 60% of coarse grain production globally and is the basic staple in much of sub-Saharan Africa, Central America, the Andean regions and many of the poorest parts of Asia (Falcon and Naylor, 1998). Maize also provides fodder for livestock in many developing nations and is the basic energy source for the livestock sector of middle and higher income countries. It is, therefore, not surprising that the long-term demand prospects for maize appear stronger than for most other crops. However, barley, sorghum and millet are preferred and grown in drier ecosystems and have a high degree of substitutability among all the grains in feed rations (Falcon and Naylor, 1998).

Maize is the staple food of Swaziland and the main crop grown by the vast majority (90%) of small, primarily subsistence farmers on the Swazi Nation Land (SNL) (Magagula and Faki, 1999; Mkhabela and Mashinini, 2005). The Swazi Nation Land is communal land held in trust and allocated by the traditional authority. The other land tenure system in the country embraces the Title Deed Land (TDL) that includes commercial farms, estates and ranches that are held under freehold or various concession agreements. Agriculture on the TDL is mainly commercial oriented.

Rice and wheat are consumed in Swaziland in increasing quantities especially in the urban areas and together they now account for 25% of the cereal consumption in the country (MOAC, 2003). However, neither rice nor wheat is grown in Swaziland in significant quantities and at present the economic prospects for sizeable domestic production is questionable.

Following multi-party elections in 1994, the incoming ANC government launched a multi-pronged land reform programme that promised to return land to persons dispossessed by apartheid land laws, redistribute existing land to enhance greater access of the black population to land, and change customary tenure arrangements in communal areas (Bryceson, 2002). Unfortunately, facts reveal that the pace of land redistribution has been slow and its quantum is inadequate to the extent that the target of completing the transfer of 30% of land to Black people by 2014 is under threat. The reform has been delayed by protracted court and administrative issues (Xingwana, 2008). More importantly, farm activities have been operated in several rural communities of South Africa but poverty still persists. Climatic variability, inadequate infrastructure, extension services and markets pose serious challenges to plans for agricultural development for previously disadvantaged black rural communities (Obi, 1984). The high hopes of reducing poverty seem elusive as the predicament of poor households has not improved significantly, despite promotion of agriculture. This poses a general pessimism about agriculture's potential to reduce poverty in rural South Africa. Thus, it is widely assumed that increased participation in rural non-farm development is critical for the country's growth, raising living standards and reducing poverty.

Nonetheless, studies on rural non-farm activities reflected their varied effects on rural livelihoods, poverty level and income growth. Kirsten (1995) has recorded quite a number of non-farm activities in rural South Africa but analysis of the 2001 census data reveals that high levels of poverty still persist in the same rural areas (StatsSA, 2001). In Latin America, rural non-farm activities have been actively promoted in response to the strong advocacy work of policymakers although this does not seem to have relieved the serious poverty situation (Avarez and Naglar, 1995; Davis and Pearce, 2001). In Bangladesh, studies by Nargis and Hossain (2006) found that income growth during the period 1988-2004 was due to successful participation in non-farm activities. Generally, in all the afore-mentioned countries and the world as a whole; little progress has been made in reaching the MDG of reducing by half the number of the poor. Therefore, it is unclear which forms of capital should be made available to the rural population and which activities should be pursued by rural households that have the potential to reduce poverty and improve rural livelihoods. Optimal public policy can be achieved by adopting the right mix of activities to facilitate more effective programming and sequencing of government support. It is also important to know what is responsible for the trend in non-farm income diversification, activity choice and income received from each activity practised. Besides, studies on non-farm activities in South Africa have remained a by-product of poverty studies (Machethe, 2004). Without a more explicit focus on these activities, it will not be possible to more precisely determine their relative importance and the specific ways in which they can be deployed as a veritable vehicle for poverty reduction and linking of the rural areas into the mainstream of the country's economy and finally put to rest the aberrant notion of a 'second economy' in South Africa.

1.3 Plan of the book

The rest of the book is devoted to a critical evaluation of a broad range of issues relevant to the rural economies of the countries within the Southern African region and the role of institutional innovation in smallholder agriculture. The second part of this book consists of two chapters which take up an important issue that is often ignored in the standard literature on the subject, and that is, to what extent rural households actually depend on the agricultural sector for their livelihoods. Very often discussions about rural livelihoods make the assumption that agriculture is the sole income-generating activity and that there are no, or limited opportunities, outside that sector. These questions are addressed in this chapter, drawing from the results of a study designed to explore the baseline situation in the rural areas around which more focused studies and intervention can be planned. Chapter 2 reviews the literature in respect to the key issues around rural household sources of income and livelihoods strategies, and broadly examines the institutional constraints in the separate commodity contexts that constituted the focus of the studies covered in the book.

Chapters 3 and 4 address the subject matter of rural livelihoods and income inequalities, making use of the standard measures of the Gini-coefficients and Lorenz curve to draw some conclusions about the progress thus far in dealing with this matter in South Africa. Support is sought from the literature to generalize these results for the region.

The third part of the book examines the Socio-economic and institutional factors implicated in commodity marketing among small farmers. This part consists of four chapters, namely Chapters 5, 6, 7 and 8. Chapter 5 examines the situation of small-scale farmers growing a wide range of tree and field crops in the Kat River Valley of the former Ciskei homeland area of South Africa. The range of challenges faced by these farmers in accessing alternative market channels are analysed by means of descriptive statistics and multinomial regression procedures to conclude that market information, awareness about standards and grades, availability of contractual agreements, and group dynamics rank among the most important factors to consider in designing interventions to improve market access for this group of farmers.

Chapter 6 weighs in on this subject in respect to the circumstances of smallholders in communities that are participating in the promotion of the indigenous Nguni cattle under an initiative started by the University of Fort Hare several years ago. The questions of how these farmers are making decisions about livestock off-take and the obstacles they encounter in operating in formal and informal cattle markets are addressed.

Chapter 7 looks at the experiences of maize producers under different regulatory frameworks in Swaziland and the implications of regulation for profitability of maize production and the welfare of maize producers. It is concluded that maize production in Swaziland is a loss-making endeavour that is supported solely by the cultural value of maize and that much of the problem arises from the pricing mechanisms adopted by the state maize regulatory apparatus and the heavy influence of culture and traditions in resource allocation, especially in respect of land, for maize production in the land-locked Kingdom.

Chapter 8 presents the results of a study conducted to determine the constraints to the development of the horticulture sub-sector in Lesotho. The chapter attempts to identify the factors that are crucial for the functioning of the marketing system for horticultural produce are also examined and provide a basis for making policy recommendations on how to improve the competiveness of the horticulture sub-sector in Lesotho.

The fourth part of the book looks at natural resources in terms of the politics around them which mediate how they are distributed, managed and controlled. The first chapter in this part is Chapter 9 which addresses the South African land question in totality, with the aim of capturing the principal debates around why land became such a contentious issue in South Africa and became the pivot around which the negotiated settlement of the long-standing political crisis of South Africa was built. There is almost a mindset that if only everyone who needs land had one there would be no problem in South Africa. Granted that this is a deliberate over-simplification, there is no question that no resource has had as much power and influence as land in the overall scheme of things in the country and a thorough review was therefore inevitable, more so as the rural crisis of unemployment, underemployment, destitution and hunger threatens to take a turn for the worst several years after enthronement of democratic rule in the country.

Chapter 10 then looks more specifically at the conflicts and confusions over definition, responsibilities, management procedures and enforcement mechanisms in natural resource management in three specific community-based case studies in the Eastern Cape Province of South Africa. A key finding is that lack of government support to communities has resulted in the poor management of resources. The consequence of this has been deterioration of the rangelands and impoverishment of the communities. The indication is that transaction costs and good governance structures are important in the common property institution and high levels of trust and cooperation between members are necessary to reduce these transaction costs. Furthermore, collective action on communal lands enables communities to share ideas on how best to manage their resources to ensure sustainability.

The final part of this book is Part 5 which consists of two chapters in which an attempt is made to contextualize the effects of recent desperate measures to deal with the food crisis and then summarize the evidence in respect to the links between livelihoods, institutions and the small farmer. A natural and convenient example of such desperate policy making process and its implications is found in the fast track land reform programme in Zimbabwe. How this programme has affected the markets and marketing of key commodities is examined in Chapter 11. Chapter 12 summarizes the evidence and proffers thoughts on the links among livelihoods, institutions and the small farmer in Southern Africa. Some methodological notes are included in the appendix to this book.

References

Akerlof, G.A. (1970). The market for lemons. Quality uncertainty with the market mechanism. Quarterly Journal of Economics 84, 488-500.

Akerlof, G.A. (2001). The Nobel Prizes 2001. T. Frangsmyr (ed.) Nobel Foundation, Stockholm, Sweden.

Anonymous (2003). Interview with the Executive Secretary of SADC. Courier 2003.

Arrow, K.J.(1969). The organisation of economic activity: issues pertinent to the choice of market vs non-market allocation. In: Joint committee (eds.) The analysis of public expenditure: the PPB system. Government Printing Office, Washington, DC, USA, pp. 47-64.

Avarez, R.M. and J. Naglar (1995). The spatial model and specification of choice models. Working paper. California Institute of Technology, Pasadena, CA, USA.

Bryceson, D.F. (2002). Multiplex livelihoods in rural Africa: recasting the terms and conditions of gainful employment. Journal of Modern African Studies 40, 1-28.

Davis, R. Jr. and D. Pearce (2001). The non-agricultural rural sector in Central and Eastern Europe. Natural Resources Institute Report No. 2630. Available at: http://www.nri.org/rnfe/pub/papers/2630.pdf. Accessed on 27 May 2007.

Dequech, D. (2002). The demarcation between the 'old' and the 'new' institutional economics: recent complications. Journal of Economic Issues 36, 565-572.

Durlauf, S.N. (1992). Nonergodic economic growth. Review of Economic Studies 60, 349-366.

Eicher, C.K. (1999). Institutions and the African farmer. Issues in Agriculture No. 14, Consultative Group on International Agricultural Research (CGIAR), Washington, DC, USA.

Eicher, C.K. and J.M. Staatz (1984). Agricultural development in the Third World. Baltimore, MA, USA: Johns Hopkins University Press.

Falcon, W.P. and R.L. Naylor (1998). The maize transition in Asia: unlocking the controversy. American Journal of Agricultural Economics 80, 960-968.

FANRPAN (2003). Equipping SADC member states for effective participation in multi-lateral agricultural trade negotiations. FANRPAN policy discussion paper No. Pan 2/03, Harare, Food Agriculture and Natural Resources Policy Analysis Network.

Frank, R.H. (1994). Microeconomics and behaviour. New York, NY, USA: Mcgraw-Hill, Inc.

Frank, R.H. (2009). Economist naturalist's guide: common sense principles for troubled times. Playaway: New York, NY, USA.

Gabre-Madhin, E. (2006). Making markets work and work for the poor. Washington, DC, USA: International Food Policy Research Institute (IFPRI).

Hart, O. (1980). Some remarks on missing markets. University of Cambridge (Mimeo).

Heller, W.P. (1997). Equilibrium market formation causes missing markets. Discussion paper 93-07R, Department of Economics, University of California, San Diego.

Hodder, B.W. and I. Ukwu (1969). Markets in West Africa. Ibadan, University Press.

INWENT/IFPRI/NEPAD/CTA (2003). Pretoria statement on the future of African agriculture. International Conference on Successes in African Agriculture – Building for the Future, Pretoria, 1-3 December, 2003.

Jayne, T., J. Shaffer, J. Staatz and T. Reardon (1997). Improving the impact of market reform on agricultural productivity in Africa: how institutional design makes a difference. Michigan State University International Development Working Paper No. 66, East Lansing, Michigan State University.

Joint UN Mission (2001). Special report. FAO/WFP Crop and Food Supply Assessment Mission to Lesotho 5-16 May, 2001.

Kirsten, J. (1995). Rural non-farm enterprises: a vehicle for rural development in South Africa? Agrekon 34, 198-244.

Kirsten, J. and W. Moldenhauer (2001). Measurement and analysis of rural household income in a rural household income in a dualistic economy: the case of South Africa. University of Pretoria, South Africa.

Lele, U. (1984). Rural Africa: modernization, equity, and long-term development. In: Eicher, C. and J.M. Staatz (eds.) Agricultural development in the Third World. Baltimore, USA: The Johns Hopkins University Press.

Lipton, M. (1996). Rural reforms and rural livelihoods: The context of international experience. In: M Lipton, M de Klerk, and M Lipton (eds.) Land, labour and livelihoods in rural South Africa. Volume one: Western Cape. Durban, South Africa: Indicator Press.

Machethe, C.L. (2004). Agriculture and poverty in South Africa: can agriculture reduce poverty? Available at: http://cfapp1-docs-public.undp.org/eo/evaldocs1/sfcle/ eodoc357114047.pdf.

Magagula, G.T. and H.M. Faki (1999). Comparative economic advantage of alternative agricultural production options in Swaziland. United States Agency for International Development, Technical Paper, Washington, DC, USA.

Makowski, L. (1980). Perfect competition, the profit criterion and the organization of economic activity. Journal of Economic Theory 22, 222-242.

Makowski, L. and J. Ostroy (1995). Appropriation and efficiency: a revision of the first theorem of welfare of economics. American Economic Review 85, 808-827.

Mashinini, N.N. (2004). Welfare effects of the swaziland maize market system. Faculty Research Paper. Faculty of Natural and Agricultural Sciences, Bloemfontein, University of the Free State, South Africa.

Mkhabela, M.S. and N.N. Mashinini (2005). Early maize yield forecasting in the four agro-ecological regions of Swaziland using NDVI data derived from NOAA's AVHRR. Agricultural and Forest Meteorology 129, 1-9.

MOAC (2003). Ministry of Agriculture and Cooperatives publications. Available at: www.gov.sz.

Montshwe D.B. (2006). Factors affecting participation in mainstream cattle markets by small-scale cattle farmers in South Africa. MSc Thesis, University of Free State, South Africa.

Nakhumwa, T.O, D.H. Ng'ong'ola, I.J. Minde, V. Lungu and H.E. Mapemba (1999). Comparative economic advantage in agricultural trade and production in Malawi. USAID Publication Series Technical Paper No. 93, United States Agency for International Development (USAID).

Nargis, N. and M. Hossain (2006). Income dynamics and pathways out of poverty in Bangladesh, 1988-2004. Agricultural Economics 35, 425-435.
North, D.C. (1990). Institutions, institutional change and economic performance. Cambridge, UK: Cambridge University Press.
North, D.C. (1992). Institutions and economic theory. American Economist 36, 3-6.
North, D.C. (2003). Effects of institutions on market performance. Lecture given at Washington University, DC, USA.
North, D.C. (2003). The role of institutions in economic development. Discussion Paper Series No 2003.2. Geneva, Switzerland: United Nations Economic Commission for Europe.
Norton, G.W., J. Alwang, and W.A. Masters (2006). The economics of agricultural development: world food systems and resource use. Oxon, UK: Routledge Press.
Obi, A. (1984). Marketing of sheep and goats: analysis of supply and price in selected markets in Oyo State, Nigeria. Unpublished MSc Dissertation, Department of Agricultural Economics & Extension, University of Nigeria, Nsukka.
Oehmke, J.F., P. Anandajasekeram and W.A. Masters (1997). Agricultural technology development and transfer in Africa: impacts achieved and lessons learned. United States Agency for International Development. Technical Paper No. 77, Washington, DC, USA.
Sachs, J. (2005). The end of poverty: how we can make it happen in our lifetime. Penguin Books: London, UK.
SADC (2003). Press briefing: end-of-year briefing on regional developments by sadc executive secretary Dr. Prega Ramasamy', 9 December, 2003.
Seleka, T.B. (1999). The performance of Botswana's traditional arable agriculture: growth rates and the impact of the accelerated rainfed arable programme (ARAP). Agricultural Economics 20, 121-133.
Spencer, D.S.C. (1994). Infrastructure and technology constraints to agricultural development in the humid and subhumid tropics of Africa. International Food Policy Research Institute, EPTD Discussion Paper No. 3, Washington, DC, USA.
Statistics South Africa (Statsa) (2001). Available at: http://www.statssa.gov.za/.
Sukume, C., E. Makudze, R.M. Chimedza and N. Zitsana (2000). Comparative advantage of crop production in Zimbabwe. USAID Publication Series Technical Paper No. 99, Harare, Zimbabwe.
The Economist (2002). Southern Africa's food shortage: with the wolf at the door. The Economist Newspapers, Vol. 363, No. 8275, 1 June, 2002.
Van Huylenbroeck and R.L. Espinel (2007). Importance of institutions and governance structures for market access and protection of property rights of small farmers in developing countries. In: E. Bulte and R. Ruben (eds.) Development economics between markets and institutions-incentives for growth, food security and sustainable use of the environment, Mansholt Publications Series Vol. 4, Wageningen Academic Publishers, Wageningen, the Netherlands, pp. 327-344.
Van Schalkwyk, H.D. and A. Jooste (2002) The external agricultural trade environment of Lesotho with special reference to trade agreements. Maseru, Lesotho, Ministry of Industry, Trade and Industry.

Van Schalkwyk, H.D., E. Nel, and A. Obi (2004). Factors inhibiting agricultural development in the former homelands of South Africa. Bloemfontein, University of the Free State (mimeo).
Van Schalkwyk, H.D., M.L. Mhalanga, and A. Jooste (2002). The enhancement of rural marketing in Lesotho. Maseru, Lesotho, Ministry of Industry, Trade and Marketing.
WFP (2003). ODJ Regional Consolidated Report for Southern Africa. Issue No 2, 31 August, 2003.
Xingwana, L. (2008). Challenges and opportunities for land and agrarian reform: towards 2025. Speech by the Honourable Minister for Agriculture and Land Affairs at the Agri-Consultation Protea Ranch Hotel, Polokwane, 30 July.

Part II
Rural livelihood strategies and current circumstances of the small farmer in South Africa

2. Rural household sources of income, livelihoods strategies and institutional constraints in different commodity contexts

Simbarashe Ndhleve, Bridget Jari, Lovemore Musemwa and Ajuruchukwu Obi

Abstract

A substantial amount of research has been carried out globally and in South Africa on the socioeconomic circumstances of rural households and the strategies they adopt to deal with their daily realities of poverty, unemployment, food shortage, among others. Increasingly, the links are being made between these issues and the institutional environment in which smallholders operate. A useful analysis of the role of institutions in smallholder development must begin with an understanding of the existing livelihoods and patterns of socioeconomic participation. In Southern Africa, it is still being debated how roles for rural incomes and employment are split between farm and non-farm activities. Since this has important implications for the focus of public policy and for the pattern and extent of institutional development for poverty alleviation, it is important to examine this element and gain an understanding of the current status as well as the trends. This chapter reviews the broad development literature on income earning strategies of rural households, the motivations for diversification into different activities and the determinants of participation in these activities. The chapter further presents a review of the literature on the institutional constraints faced by smallholder and emerging farmers in general and the smallholders and communities participating in the production and marketing of the indigenous Nguni cattle in the Eastern Cape Province of South Africa, given its importance in the rural economy.

2.1 Introduction and problem context

Among the numerous problems of traditional agriculture, the prevalence of low-level technology ranks very high. According to Spencer (1994), farmers in Africa have not widely adopted the Green Revolution technologies that brought phenomenal improvements in yield in Latin America and Asia. Nweke *et al.* (2002) re-affirmed these observations and showed that the low adoption rates for improved technologies remain a serious obstacle to fighting hunger on the continent. Back in the 1980s, analysts, notably Eicher (1982), and Lele (1984) noted that whatever increases occurred in the production of the major cereals and root and tuber crops in Africa up to the early 1980s were as a result of area expansion rather than improvements in the productivity per unit of input, or yield, which results from technical change. In fact, agricultural production in general has been on the decline for nearly three decades. It is estimated that agricultural production per capita declined by some 22% between 1971 and 1984. According to Southgate *et al.* (2007), agricultural production per capita in Africa may have declined by as much as 6% in 1981, and by about

10% in 2001. On the other hand, per capital food production in Asia and South America has been going up, by between 14-15% in 1981 and by between 44-74% in the early 2000s (Southgate *et al.*, 2007; Southgate, 2009). In both Asia and Latin America, agricultural production per capita began its phenomenal growth from the mid-1970s, suggesting that this was directly linked to the Green Revolution initiated in the 1960s. As observed by Nweke *et al.* (2002) and Lele (2010), agricultural research requires long-term investments and takes at least 10 years to begin to show results. These gains in Asia and Latin have been sustained and have remained positive since then.

This situation has continued to be a source of serious concern among policy makers and researchers. Coincidentally, during the same period, African governments began to cut back on expenditures devoted to agricultural technology development and transfer. According to a review conducted in the late 1990s to assess the impacts and lessons in agricultural technology development and transfer in Africa, the spending on agricultural technology development and transfer had actually declined by as much as 37% over the 20 year-period between 1971 and 1991 (Oehmke *et al.*, 1997). Since up to one-third of the growth in agricultural productivity can be explained by past investments in agricultural research (Oehmke *et al.*, 1997), there is no question that some of the observed decline in agricultural production per capita can be attributed to underinvestment in research and development. This is therefore only one part of the picture. According to the brief posted by the African Green Revolution network (Toenniessen *et al.*, 2008), the following may account for the poor performance of African agriculture:
- population increase outstripping agricultural productivity;
- changes in consumption patterns;
- nature's curtailment;
- political neglect.

Echoing earlier opinions on this subject, notably Nweke (1978), USDA (1981), Eicher (1982), and Eicher and Baker (1982), the Alliance for a Green Revolution in Africa (AGRA) (Toenniessen *et al.*, 2008) has identified the root causes of rural poverty and hunger in sub-Saharan Africa as revolving around eight key points, namely:
1. Persistent use of traditional varieties of crops in place of the improved varieties that have been developed and introduced into the farming systems of these countries.
2. Deterioration of soil fertility due to poor soil management practices.
3. Fragmentation of land that results in plot sizes that are too small to support meaningful economic empowerment.
4. Insufficient access to adequate water resources to support irrigated agriculture.
5. Excessive loss of crops and livestock due to post-production losses as a result of pest and disease infestation.
6. Poor land management and administration procedures that lead to skewed distribution of land.

7. Insufficient access to profitable markets to sell farm output.
8. Infrastructure deficiencies in both institutional and physical terms (Toenniessen *et al.*, 2008).

The foregoing views invariably blame human actions of one type or the other for the unrelenting hunger and poverty in Sub-Saharan Africa. In some cases, the problems have arisen from improper actions, namely those that lead to exploitation of scarce resources and result in inefficiencies. In other cases, actions that would have led to desirable outcomes are not taken at all. Again, some of the problems have resulted from failures to adequately regulate human actions which lead to outcomes that could have been avoided. A number of institutional and structural theorists, including Barley and Tolbert (1997) and Myint (2003), have shown that 'institutions and actions are inextricably linked'. The consensus of these theorists is that it is human action that creates, maintains and alters institutions. To that extent, we can safely say that all the identified causes of hunger and poverty in Africa are traceable to institutional capacity to the extent that they lead to political neglect, representing varying degrees of government failure which may easily be the most potent of these causes given that it affects all the other elements. As Anne Krueger noted several years ago, 'in many developing countries, government policies have been highly distortive and harmful to economic growth' (Krueger, 1990). In the context of an assessment of institutional constraints to smallholder development on the continent, the role of government will obviously rank very high.

Reviewing the performance of traditional arable farming in the context of the Accelerated Rainfed Arable Programme (ARAP) in Botswana, Seleka (1999) mentioned low adoption of improved technologies as one of the 'sub-sectoral factors' implicated in the poor performance of the country's arable agriculture. A key challenge identified in the effort to develop agricultural production in the communal areas of South Africa is the use of low level technology among the farmers in the former 'independent homelands' of the country where inadequate infrastructure, knowledge and skills constrain the use of improved inputs and methods in farming (Van Schalkwyk *et al.*, 2004). A similar point has been made for several sub-Saharan African countries, including Swaziland (Magagula and Faki, 1999; Mashinini, 2004; and Mkhabela and Mashinini, 2005). The situation has persisted despite the large amount of research efforts and results currently available and the indisputable evidence of the important role of improved technologies in agricultural development.

2.2 Household sources of income

Income diversification into non-farm activities has come to be recognised as typical practice among rural households (Barrett *et al.,* 2001). Analyses of rural livelihoods in Africa, Asia, and Latin America, show that rural households derive a significant proportion of their livelihoods from non-farm employment (NFE) (Adams, 1999; Barrett *et al.,* 2001; Escobal, 2001; Fraser *et al.,* 2003 and Otsuka and Yamano, 2006). The rural non-farm sector has developed as several farm households becoming more actively involved in non-farm activities in developing countries (Barrett *et al.,* 2001). Non-farm activities imply a set of activities carried out in the rural areas that are not agricultural (Barrett *et al.,* 2001).

Attempts have been made by researchers to classify economic activities carried out by rural populations. In South Africa, Machethe (2004) came up with three classes of rural economic activities, namely: (a) smallholder (subsistence or semi-subsistence) farming consisting of self-employed farmers producing staple foods and selling surplus; (b) commercial farming comprising medium and large-scale farmers and providing employment to a significant number of the landless; and (c) the rural non-farm employment. Perret *et al.* (2005) made almost similar classifications based on studies in the former homeland areas of Limpopo and North West Provinces. Income generating activities in these Provinces manifest themselves in the form of either farm or non-farm activities. The Eastern Cape Province is not different in this regard; Fraser *et al.* (2003) recorded a number of commercial and trading enterprises in the Province. Rural households in the Province derive income from inheritances and transfers, provision of personal and community services, and value adding activities such as food processing, construction and manufacturing (Fraser *et al.,* 2003).

The above sources of income are not exclusive to rural South Africa. Studies by Sanchez (2005) in Bolivia and Barrett *et al.* (2001) in rural areas elsewhere in Africa show diverse sources of income in rural areas. Furthermore, analysis of rural livelihoods by Otsuka and Yamano (2006) in Asia and East Africa confirm income diversification as a typical practice in most rural areas. Figure 2.1 below adapted from Davis and Pearce (2001) shows the sources of income of farm households in Latin America.

As shown by in the diagram by Davis and Pearce (2001), sources of income in rural areas can be classified into three categories, namely on-farm income, off-farm income and transfers. Non-farm income being all the income associated with wage work or self-employment. Own farm income implies income from own agricultural activities.

Despite the wide range of activities pointed out above, rural households' transition into high return, non-farm activities is not without constraints. Barrett *et al.* (2001) wrote that not all households enjoy equal access to high-return, non-farm activities. Selected households with the required capacities are able to diversify into high return, non-farm sector (Deininger and Olinto, 2001). Household activity choice is largely dependent on endowments in terms of assets, access to credit and the required skills. For policy implication, Barrett *et al.* (2001) wrote that reforms that fail to address these constraints leave the less fortunate sub-populations trapped in low-return, high-risk livelihood strategies based largely on unskilled labour and part-time, self-employed farming. Policies addressing the underlying constraints to activity choice facilitate the access of the poor populations to livelihood strategies that are relatively more lucrative, less risky, or both (Barrett *et al.,* 2001).

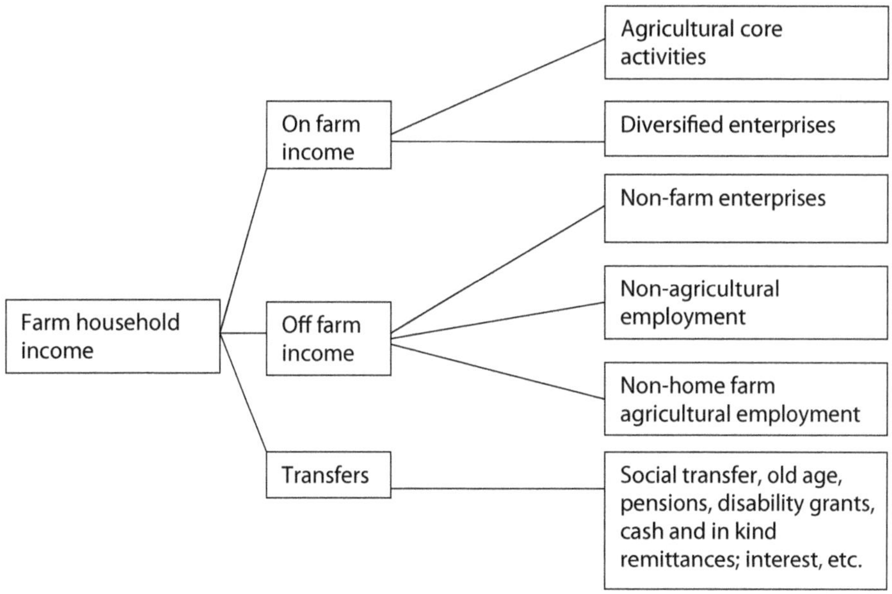

Figure 2.1. Sources of income (Davis and Pearce, 2001).

2.3 Rural households' choice of activities

Desperate economic realities in many African countries oblige households and individuals to seek solutions to the circumstances in which they find themselves (Nel and Binns, 2000). Day-by-day resolution of tensions due to low income among rural households generates differing degrees of change in a variety of ways. Income portfolio diversification is a typical practice among rural dwellers and a lot more households are attracted into this livelihood strategy. According to Davis and Bezemer (2004), 'in studying households' diversification strategies it is important to account for the fact that the motivations, means and outcomes of diversifying are heterogeneous'. Households and individuals fall into two distinct groups on the basis of the motives behind their livelihood diversification *viz*; 'demand pull' and 'distress push' diversification (Davis and Bezemer, 2004; Barrett *et al.,* 2001; Davis, 2003). Demand pull diversification is a response to emerging opportunities in the non-farm sector such as those brought about by market or technological opportunities. On the other hand, distress push diversification is driven by little or no opportunities on-farm, e.g. drought or market constraints (Davis, 2003). The literature has for long emphasized the relative importance of 'pull' and 'push' factors for households' involvement in non-farm activities (Barrett *et al.,* 2001; Davis, 2003; Bezemer and Davis, 2004). The key features of distress-push and demand-pull diversification are outlined in Table 2.1.

The manifestations and modus operandi of the factors listed in the table are elaborated in the next sections below.

Table 2.1. Key features of distress-push and demand-pull diversification.

Push factors	Pull factors
Population growth	Availability of a wide range of resources
Inadequate access to fertile land	Higher returns on labor in non-farm activities
Low farm productivity	Higher returns on investment in rural non-farm activities
Low returns to farming	Lower risk of non-farm activities compared to farm activities
Lack of access to farm input markets	Generation of cash to meet household objectives
Deterioration of the natural resource base	
Temporary events and shocks	
Lack of access to rural financial markets	

2.3.1 Demand-pull factors

The existence of a wide range of resources (labour, assets, education, etc.) that are not suitable for traditional farming practices drives households' efforts to explore alternative opportunities for utilising these resources (Davis, 2003). Diversification is undertaken to use these resources in a manner that is sustainable on the one hand, while also expanding total household incomes (Matsumoto *et al.,* 2006). Diversification in this perspective is characterized by response to evolving market or technological opportunities, which offer the potential for increasing labour productivity and household incomes (Barrett *et al.,* 2001).

Households are 'pulled' into non-farm activities as a means of obtaining more income and improving their current living conditions (Davis and Pearce, 2001). They might be attracted into non-farm activities because of higher returns on factors of production (labour and capital) in these activities relative to agriculture (Barrett *et al.,* 2001). Potentially higher returns to labour that could be obtained from working off the farm would lure households into diversifying. Studies conducted by Estudillo *et al.* (2006) in the rural Philippines suggest a structural shift as households move away from farm-based sources of income and specifically away from agriculture into non-farm activities between 1985 and 2004. Such a shift may have resulted from an increase in the relative profitability of rural non-farm employment (NFE) opportunities vis-à-vis farming. A study by Fraser *et al.* (2003) recorded an increase in agro-processing activities and service-providing institutions in rural South Africa and this shows that the rural non-farm sector is growing. The development of the rural non-farm and also the urban labour markets has raised the wage rates in these two sectors, thus inducing households to reallocate their labour resources away from own farm production (Estudillo *et al.,* 2006).

Agricultural development is widely recognised in many countries of Africa, Asia and other continents as behind reduction in poverty and household income growth (Cherduchuchai and Otsuka, 2006). The literature recognizes the impact of the Green Revolution represented by the development and adoption of fertilizers and high-yielding modern varieties on the change in household income and poverty. Despite these crucial developments, the importance of income from farming is declining partly due to price variability in the output markets for agricultural produce and escalating prices of inputs. All this is happening at the time when the non-farm sector is developing; wages are increasing with better protection of workers through the unions. All these render the non-farm sector less risky relative to own farming. Factors such as lower risk non-farm activities will tend to 'pull' households into these activities. In this way, households will be undertaking diversification into non-farm activities as a safety net (Sanchez, 2005). This is because of the higher predictability of income flows in most non-farm sources relative to farm sources of income (Barrett *et al.*, 2001).

2.3.2 Distress-push factors

Distress-push diversification typically occurs in an environment of risk, market imperfections, and disguised agricultural unemployment. This is typically triggered by economic misfortune, which sets the household on a downward income spiral (Davis, 2003). According to Barrett *et al.* (2001), distress-push diversification emerges naturally due to diminishing or time varying returns to household productive assets (land or labour), from market failures, from *ex ante* risk management and from *ex post* coping with adverse shocks.

Individual factors of production face diminishing returns to scale (Parkin, 2008). Households endowed with much labour relative to land will, in the absence of well-functioning land markets, typically apply some labour to their own farm, and hire out excess labour for off-farm wage employment. This explains what Barrett *et al.* (2001) pointed out as desperation-led-diversification. Diversification might be derived by the existence of incomplete markets for land, labour, credit and insurance. Where markets often do not operate in a competitive or efficient manner, personal and institutional constraints can play an important role in determining participation in non-farm activities.

Households diversify as a way of achieving self-sufficiency if located in remote areas where physical access to markets is costly and causes factor and product failures (Nel and Binns, 2000). Limited access to market forces individuals and households to develop local coping strategies which facilitate self-reliance. Barrett *et al.* (2001) also indicate that diverse work portfolios arise because neither subsistence production (agricultural commodity production), nor non-agricultural activities, guarantee security of livelihoods. In such an environment, the search for the right balance of activities becomes necessary thus resulting in households indulging in various activities. This results mostly in rural-urban migration; a typical practice in South Africa as noted by Davis (2003).

Although the above discussion portrays diversification into non-farm activities as widespread, Barrett *et al.* (2001) noted that not all households enjoy equal access to attractive non-farm income opportunities. Selected households with human or physical capital diversify out of agriculture into non-farm income sources as a refuge from low return semi-subsistence farming, which has little prospect for economic advancement (Deininger and Olinto, 2001). Only households which can meet the investment requirements for entry into remunerative non-farm activities are able to diversify. Those who own sufficiently high levels of assets, are able to access credit or possess the required skills will be able to make full use of opportunities for increased returns to labour provided by rural non-farm income sources.

2.4 Determinants of household activity choice and diversification

De-Janvry and Sadoulet (2001), in their study in Mexico, pointed out that gaining a good understanding of the origins of non-farm activities and determinants of households' participation in these activities, coupled with levels of income achieved, is essential for the design of effective policies for rural development. Several studies suggest that rural households' diversification into non-farm activities is influenced by individual characteristics, household characteristics and community characteristics (Davis, 2003; Bryceson, 2002; Sanchez, 2005; Matsumoto *et al.*, 2006). An attempt is made below to examine each of these factors and analyse among all the variables in each of the three components that affect the household's successful participation in diversified income generating activities.

2.4.1 Individual characteristics

Individual characteristics expected to influence participation in various income generating activities and income include age, gender, marital status and level of education. Age influences the way the individual values the future, choice of livelihood strategies and the propensity to invest (Schwarze, 2004). Age may increase the likelihood to participate in agriculture relative to migratory activities especially for elderly heads of household (Matsumoto *et al.*, 2006). The older one gets, the more likely it is to select more sedentary or home-based activities (Matsumoto *et al.*, 2006).

A gender perspective adds significant insight into rural poverty and livelihood issues (Davis, 2003). During this research, gender has emerged as an important factor influencing participation patterns in rural non-farm income activities. Regarding gender and marital status, Matsumoto *et al.* (2006) in their studies in Ethiopia, Kenya and Uganda, found that larger proportions of individuals in non-farm activities are male and single. Compared to women, male individuals are more willing to explore opportunities, and single people are more mobile than married people, thus exhibiting greater propensity for migrating for employment.

Better-paid jobs and most non-farm activities require formal schooling, usually to completion of primary school or beyond and also tend to favour men over women (Davis, 2003). Positive correlation between education and participation in rural non-farm activities is often reported. This is supported by empirical findings by Bryceson (2002) in both Malawi and Ethiopia, Sanchez (2005) in Bolivia, and Matsumoto *et al.* (2006) in Kenya. They showed that the level of education indicated in years of formal education, enhances household members' capability to exploit better income generating opportunities. Education is linked to information acquisition, skills development and human resource development (Schwarze, 2005).

2.4.2 Household variables

In rural or traditional societies the household is the primary decision making unit (Hebinck and Lent, 2007). According to Matsumoto *et al.* (2006), households' access to land, asset endowments, demographic composition and transfers determines the capability to participate in non-farm activities.

Land and other assets are the key requirements for entry into most local non-farm activities in rural areas. Economic theory of investment supports the view that household endowment in assets places them in a relatively better position to respond to incentives (Parkin, 2008). This implies capacities in terms of human capital, physical capital, social capital and organizational capital (Sanchez, 2005). Household diversification into non-farm income varies in extent and nature based on relative household wealth (Escobal, 2001). Wealth-differentiated barriers to entry into non-farm activities, as noted by Holden *et al.* (2004), have been noted in Burkina Faso, Côte d'Ivoire, Ethiopia, Kenya, Rwanda, South Africa and Tanzania.

A household may have the incentive to participate in non-farm employment, say because of higher returns to labour, but if the capacities are not in place (such as capital), the household will not be able to take advantage of them (Barrett *et al.*, 2001). Therefore participation in non-farm activities is directly related to households' asset endowments. Although having various assets could be a determinant of participation, it could also be a result of additional incomes from non-farm activities (Barrett *et al.*, 2001).

Land-constrained households resort to local non-farm activities and wage employment to earn a living (Matsumoto *et al.*, 2006). If a household has limited access to land, the only way to earn income is through adoption of non-farm activities. Thus, it is presumed that land size and participation in non-farm activities are negatively related.

Transfers or remittances negatively affect participation in rural farm or non-farm activities. Chaplin *et al.* (2000) found negative relationship between transfers and participation in off-farm employment in Poland and Czech Republic. Transfers act as insurance, reducing variability in total income and, therefore, decrease the income risk factor, i.e. reduce push and pull (Barrett *et al.*, 2001; Davis, 2003). In the absence of transfers, diversification into non-farm activities is widely understood as a form of self-insurance. Barrett *et al.* (2001) reported similar results in Africa and Asia. Conversely, households' access to transfers can facilitate participation in non-farm activities. If it coincides with adequate household food production, transfers can provide the required capital to venture into non-farm activities.

2.4.3 Community variables

According to Matsumoto *et al.* (2006), variables which entail infrastructure, community average land productivity, the distance to the nearest market, and the distance to the government support agencies are the major determinants of participation in non-farm activities among households.

Basic problems with infrastructure, scarce information and poor utility services militate against the successful development of the rural non-farm sources of income in many parts of Africa (Perret, 2002). A case study in Poland (Chaplin *et al.*, 2000) show that average distance to public transport negatively affects the diversification into non-farm activities. Infrastructure with special focus on transport and communication are important factors in the development of viable non-farm activities. An improvement in rural infrastructure increases the chances of adoption of non-farm activities by rural dwellers. High land productivity encourages individuals to adopt farming instead of non-farm activities. Land productivity has a negative impact on the participation in local non-farm activities (Matsumoto *et al.*, 2006). People from low-potential agricultural areas participate in non-farm activities to a greater extent than people from high potential agricultural areas. High population density in most rural areas of South Africa has led to diversification to non-farm activities as a way of reducing pressure on limited land (Machethe, 2004).

2.5 Household income by source

Despite the rural economy being assumed to be naturally agriculture based, a large number of rural households derive most of their income and employment from the non-farm sector (Barrett *et al.*, 2001). Agriculture alone cannot provide sufficient livelihood opportunities for the rural population (Davis and Pearce, 2001; Mehta, 2002; Davis, 2003). In South Africa, StatsSA (2000) reports that only 4% of poor households have agriculture as their primary source of income, 57% depends on income from wages or salaries, 14% from social grants and 10% from remittances. Using a sample of households from two villages in the Eastern Cape, Guquka and Koloni, Fraser *et al.* (2003) found that agricultural activities provide 15.4% and 12.3% of total household income for Guquka and Koloni, respectively, with the rest of the income being derived from non-farm sources and transfers. Kirsten and Moldenhauer (2001) report almost similar results, finding that almost 76.8% of total income received by rural households in the former homelands in 1997 was from non-farm sources with farm income contributing only 23.2% of total income (Kirsten and Moldenhauer, 2001).

This situation is not peculiar to South Africa. Selected studies in other countries in Africa reflect a similar pattern. Based on a nationally-representative household budget survey in Egypt, Adams (1999) found that agricultural income accounts for less than one-quarter (26.6%) of total rural household income with non-farm income accounting for 42.2%. The above statistics are quite revealing about typical African rural economy. Barrett *et al.* (2001) also found an average of 42.2% for Africa as a whole, 42.2% of household income in rural areas is from non-farm sources. Similar results were also found in Bangladesh by Nargis and Hossain (2006). Income patterns recorded in Bangladesh were not only among the landless; landholders equally depend on non-farm income to supplement their agricultural earnings.

Table 2.2 provides regional data on the share of non-farm income to total household income. The table presents a summary of the proportion of income derived from non-farm activities by households from the three continents, Africa, Asia and Latin America.

Besides providing a higher share of rural households' income, non-farm activities are sources of rural income growth. Findings by Fraser *et al.* (2003) suggest that income derived from non-farm sources by rural communities represents a substantial and sometimes growing share. In their studies on growth in rural households' income in Bangladesh over the period of 1988-2004, Nargis and Hossain (2006) attributed most of the growth in income to widespread diversification into non-farm activities. Income increase during the period was a factor of occupational shift into the non-farm sector with trade, business and service provision topping the list. Rural income pattern is therefore not static but growing in favour of non-farm employment.

Table 2.2. Proportion of income from non-farm activities according to region.

Area	Proportion of income from non-farm activities	Sources
Africa	40-45%	Holden *et al.*, 2004; Barrett *et al.*, 2001; AllAfrica.com, 2007; Escobal, 2001
Asia	30-40%	Onchan, 2001; De Janvry *et al.*, 2005; Escobal, 2001
Latin America	40-50%	Escobal, 2001; Sanchez, 2005; De Janvry and Sadoulet, 2001; Deininger and Olinto, 2001

Following the above findings, a conclusion can be drawn that the composition of rural income is undergoing changes. The share of non-farm income is growing. This implies that promotion of non-farm activities is essential for rural households' income growth. Explanations for high and increasing portion of non-farm income in rural areas have been sought with researchers giving varied reasons. In the case of Bangladesh, Nargis and Hossain (2006) found that a high and increasing portion of rural non-farm income could be explained by rapidly decreasing land size of farm households, as well as production and marketing constraints in subsistence agriculture. This makes subsistence agriculture an unsustainable income generating strategy, pushing individuals into non-farm activities to reduce risks and in search of stable sources of income.

2.6 Income distribution

Piesse *et al.* (1999) suggest that non-farm incomes address inequality in income distribution across rural societies. With corresponding views, Reardon *et al.* (2006) note that non-farm sources of income will only address inequality if: (1) the income created by such sources is large enough to influence the rural income distribution; (2) that non-farm income is unequally distributed (an income source that is perfectly equally distributed, by definition, cannot alter the distribution of total income); and (3) that this unequally distributed income source favours the poor.

Several arguments have been posited in the literature on the effect of different sources of income on overall income inequality. Holden *et al.* (2004) found that involvement in non-farm activities increases income inequality in most rural areas. Holden *et al.* (2004) argued that asset poverty, a characteristic feature of the poor, appeared to inhibit entry into remunerative non-farm earnings, e.g. own-business. Furthermore, Adams (1999) argued that ample land access may tend to keep the poor in low returns subsistence agriculture and to 'pull' only richer households into the non-farm sector. Empirical evidence from various parts of Africa indicates that often only rich households are able to engage in non-farm activities, such that non-farm enterprises development increases inequality and have only a limited impact on poverty (Van de Berg and Kumbi, 2006). The poor are trapped in a vicious circle of poverty, from which they can only escape through acquisition of selected assets.

In another study on inequality carried out De-Janvry *et al.* (2005) in China using detailed household survey data from Hubei province established that the Gini index of the observed income is lower than that of predicted income in the absence of non-farm activities. This implies that participation in non-farm activities reduces income inequality. With the aid of statistical tools the studies found that in the absence of non-farm incomes, the Gini index of total household income would increase by 36.8% in that region. In this view, income earned in the rural non-farm sector represents the agent of positive change towards addressing income inequality rather than income earned from traditional agricultural sector.

Evidence from Zimbabwe is however mixed: using a sample of rural households from two communal areas (Chiweshe and Gokwe), Piesse *et al.* (1999) found that non-farm income sources have an equalizing effect in Chiweshe, whereas they deepen inequality in Gokwe. These contradictory results are not surprising. As mentioned by Van de Berg and Kumbi (2006), such contradictions are credited to differences in development of factor markets and institutions and the biophysical environment which varies across regions. Conflicting results were credited to differences in geographical location of the two areas and the main activities carried out in each. Gokwe is located away from a major urban centre and has comparative advantage in agriculture, while Chiweshe is located a few kilometres away from Harare. These locational factors give Chiweshe a competitive advantage in non-farm activities because of easy access to market for the outout and ready labour market.

Income equalising effect of various activities depends on the economic environment and geographical location of the area under study. It does appear that there is some evidence to suggest that in more remote areas, with traditional, mainly subsistence agriculture, the agrarian power structures result in a situation where those who have better farm incomes are also in a better position to exploit non-farm income opportunities. Conversely, where there is a more developed infrastructure and urban proximity, the commercialisation of agriculture may result in less equal farm incomes, but gives greater opportunities for non-farm employment and thus more equalising non-farm income.

2.7 Institutional factors in agricultural marketing

In this section, the institutional factors, which influence smallholder farmers' decision to market their produce, are reviewed. Institutional aspects in marketing and economic development include transaction costs, market information flows and the institutional environment. Smallholder farmers in less developed rural economies lack adequate market information and contractual arrangements, lack lobbies in the legal environment and are not easily receptive to changes (Delgado, 1999; Kherallah and Kirsten, 2001). These factors tend to result in high transaction costs and, hence, difficulties in formal market participation.

According to Kherallah and Kirsten (2001), market institutions are the underlying determinants of economic performance since they shape the organisation of market transactions. Moreover, institutions provide for more certainty in human interaction. Ruijs (2002) argues that institutions, which facilitate transactions, are necessary for sustainable trade. Thus, the institutional development of a society has a substantial influence on transaction costs. For instance, where trade laws (institution) are well implemented, legally enforced agreements may result. Under such circumstances, traders can easily get information on demand, market conditions and prices, leading to lower trade uncertainties (Dorward and Kydd, 2005). Consequently, transaction costs will decrease considerably, encouraging continuous transactions.

2.7.1 Informal rules

Informal institutions refer to non-legal rules that are enforced by peers and these include norms, traditions, customs, value systems, religions and sociological trends (Kherallah and Kirsten, 2001). Informal rules are usually taken as exogenous factors because they change slowly and may retain agreements or habits for a long time, even if they have become less suitable. North (1990) suggested that the governing structure is overwhelmingly defined by informal rules, because once they are established, they constrain individual actors. Informal rules are important, particularly amongst smallholder farmers in developing countries, because many exchange relationships are based on ethnical or kinship ties. For instance, smallholder farmers offer services to relatives even if it would be more efficient not to offer the services.

2.7.2 Formal rules

Formal rules refer to legal rules such as laws, contracts, constitutions, political systems and markets. The formal rules are usually enforced by the government. Ensminger (1992) pointed out that since bargaining typically takes place 'in the shadow of the law,' formal rules are important. However, North (1990) is of the view that formal rules only make up a small part of the constraints that shape choices. North (1990) explained that the law can only shape the outcome of private bargaining by serving as a backup mechanism for resolving disputes that cannot be resolved privately.

Both formal and informal institutions are essential as societies are generally governed by both. It is also important to note that both the formal and informal rules do not only facilitate economic growth, but they can also impede it.

2.8 The New Institutional Economics (NIE)

The New Institutional Economics (NIE) builds on Coase's 1937 article entitled 'The nature of the firm,' but the term 'new institutional economics' was originated by Williamson (1975). Williamson came up with the term in an effort to differentiate NIE from old institutional thoughts. The old institutional school argued that institutions were a key factor in explaining and influencing economic behaviour, but it had no theory and little analytical rigour. On the other hand, neoclassical economics ignored the role of institutions (Klein, 1999). The NIE can be regarded as a bridge between the two since it acknowledges the importance of institutions and at the same time argues that these institutions can be analyzed within the framework of neoclassical economics (Kherallah and Kirsten, 2001). Thus, the NIE tries to provide an economics with both theory and institutions.

Citing Benham and Benham (1998), Makhura (2001) highlighted that NIE postulates that economic activity is accompanied by multiple costs termed transaction costs and these determine economic performance. The essence of NIE is that the success of marketing and economic activity is dependent upon the institutions that facilitate transactions. The NIE paradigm is well suited to evaluate the organisation of individual transactions (Nabli and Nugent, 1989). As such, the paradigm seems ideal in explaining the market participation behaviour of smallholder farmers.

2.8.1 Markets and institutions

Markets can be grouped into informal and formal. In the agricultural context, Kherallah and Minot (2001) explained that informal markets embrace unofficial transactions between farmers and from farmers directly to consumers. On the other hand, formal markets have clearly defined grades, quality standards and safety regulations and prices are formally set. Smallholder farmers find it difficult to penetrate the formal markets, due to high transaction costs, high risks, missing markets and lack of collective action (Mangisoni, 2006). In summary, institutional aspects in marketing include transaction costs, market information flows and the institutional environment.

2.8.2 Transaction costs in smallholder farming

Transaction costs are observable and non-observable costs associated with enforcing and transferring property rights from one person to another (Eggertson, 1990). These include the costs of searching for a trading partner with whom to exchange with, the costs of screening partners, of bargaining, monitoring, enforcement and, eventually, transferring the product to its destination (Jaffee and Morton, 1995; Hobbs, 1997). Delgado (1999) identified high transaction costs as the embodiment of market access barriers among resource poor smallholders. These high transaction costs result from individual produce transportation and selling, difficulties in getting trading partners and poor bargaining power (Delgado, 1999). When transaction costs are high, smallholder farmers may cease produce marketing. In other words, with high transaction costs, markets fail in their role of allocating scarce resources to alternative ends. For South Africa, Makhura (2001) explained that high transaction costs prevail among the smallholder farmers.

2.8.3 Market information

Market information is vital to market participation behaviour of smallholder farmers. Market information allows farmers to take informed marketing decisions that are related to supplying necessary goods, searching for potential buyers, negotiating, enforcing contracts and monitoring. Necessary information includes information on consumer preferences, quantity demanded, prices, produce quality, market requirements and opportunities (Ruijs,

2002). Of equal importance is the source of market information because it determines accuracy of the information.

According to Montshwe (2006), smallholder farmers have difficulties in accessing market information, exposing them to a marketing disadvantage. Smallholder farmers normally rely on informal networks (traders, friends and relatives) for market information due to weak public information systems (FAO, 2004). However, such individuals may not have up to date and reliable market information, making the usefulness of the information doubtful. Additionally, farmers relying on informal networks for market information are at risk of getting biased information due to opportunistic behaviour of the more informed group. For instance, Mangisoni (2006) explained that smallholders usually accept low prices for their crops when the broker informs them that their produce is of poor quality. Smallholder farmers accept these low prices mainly because they are unable to negotiate from a well-informed position.

2.8.4 Grades and standards

Consumers demand high quality for the goods they buy. In addition, they will not buy food products unless there is a guarantee that they are safe to eat (Kherallah and Kirsten, 2001). In other words, consumers make purchasing decisions depending on packaging, consistency as well as uniformity of goods.

Most smallholder crops have no clearly defined grades and standards and, therefore, cannot meet the consumers' demands (Reardon and Barrett, 2000). Produce from smallholder farmers do not meet certain market grades and standards because the farmers lack the knowledge and resources to ascertain such requirements. In addition, institutions for determining market standards and grades tend to be poorly developed in smallholder farmers environments. Due to uncertainty on the reliability and quality of their goods, they usually cannot get contracts to supply formal intermediaries such as shops and processors (Benfica *et al.*, 2002). This indicates that only well organized farmers can benefit from trade liberalization by adopting strict quality control measures and obtaining the necessary certification for their goods.

2.8.5 Organization in markets

Smallholder farmers tend not to be organized in the markets as they usually sell their few agricultural produce surpluses individually and directly to the consumers without linking with to other market actors (Key and Runsten, 1999). In other words, smallholder farmers lack collective action in markets. Individual marketing of small quantities of produce weakens the smallholder farmers' bargaining positions and often exposes them to price exploitation by traders. They also do not benefit from economies of scale (Kherallah and Minot, 2001).

In a globalised world, there is increasing vertical integration and alliance formation in the agricultural marketing channels and markets, in an effort to meet consumer needs. Such alliances include contract farming, cooperatives and farmer organizations. Agribusiness firms favour contracts with medium to large-scale farmers, such that individual smallholder farmers cannot be part of these contracting arrangements (Key and Runsten, 1999; Kherallah and Kirsten, 2001). Lack of facilitation in the formation of producers associations or other partnership arrangements makes it more difficult for smallholder producers to participate in formal markets. The greater the degree of organisation in the market, the smaller the transaction costs are likely to be and the easier it is to benefit from the exchange opportunity (Frank and Henderson, 1992). Unfortunately, lack of collective action among smallholder farmers denies them entry into formal market channels.

2.8.6 Legal environment

Legal institutions influence the activities performed on the market and the costs of exchange. Minot and Goletti (1997) affirm that the formal institutional development of a society has a considerable influence on transaction costs. Thus, if trade laws are transparent then agreements can be legally enforced, leading to information accessibility and lower costs. In other words, effective legal institutions may improve the organisation of the marketing channels and decrease marketing costs.

In many developing countries, laws are not always executed and enforced correctly, bribery and cheating are often not penalised, courts are out of reach for the majority of the population, and market rules are often not transparent to the producers and traders (Ruijs, 2002). In addition, formal contract enforcement mechanisms are weak (Fafchamps, 1996). It is even worse for the smallholder farmers because they lack lobbies in the legal environment. As a result, rural trade prospers where trust has been developed based on repeated transactions or informal relationships (Randela, 2005). Thus, an unfavourable legal environment creates a significant barrier to entry into formal food trade and limits participation by smallholders in the modern marketing system.

2.9 Marketing challenges faced by the Nguni project beneficiaries

Productivity of cattle in communal areas is affected by diseases and parasites, lack of feed resources and poor rangeland management (Chimonyo *et al.*, 2000; Bester *et al.*, 2003; Montshwe, 2006; Musemwa *et al.*, 2007). However, for the Nguni farmers, production-related challenges are likely to be minimal due to Nguni's resistance to tick-borne diseases and ability to survive under harsh environmental conditions. Therefore, market-related constraints such as lack of information, high transaction costs and poor infrastructure are more likely to be important concerns for Nguni cattle producers.

2.9.1 Infrastructure

Lack of marketing facilities imposes a serious constraint on the marketing of livestock (Mahabile *et al.*, 2002). Most of the beneficiaries are located in areas remote from major markets, where there is a serious lack of both physical and institutional infrastructure (NDA, 2005). This partly explains the poor livestock supplies to formal market outlets by small-scale farmers (USAID, 2003). In communities that have marketing facilities, they are either in poor state or non-functional because farmers do not have funds to maintain them (Frisch, 1999). The most important physical infrastructural weaknesses for the communal Nguni cattle producers are related to transport and holding facilities (Bailey *et al.*, 1999).

In South Africa, lack of marketing facilities such as sale pens and loading ramps are some of the numerous factors that impose a serious constraint on small-scale farmers' ability to market their cattle (NERPO, 2004). On the contrary, Fidzani (1993) reported that poor infrastructure does not influence livestock marketing since in most cases buyers provide their own loading and transport services. Comparatively, NERPO (2004), states that apart from the distance to formal markets, the poor state of road networks in South African communal areas imposes a serious constraint. It affects the farmers' ability to attract many buyers in their areas since bad road network systems are associated with very high transport costs. There is need for community members and all stakeholders to collaborate in constructing and maintaining community infrastructures. The involvement of community members can instil some sense of ownership and responsibility and enable them to maintain their infrastructure.

2.9.2 High transactional costs

Transactional costs are barriers to the efficient participation of farmers in different markets. Producers will not use a particular channel when value of using that channel is outweighed by the costs of using it. The remote location of most communal cattle producers, coupled with poor road networks, result in high transactional costs (especially transport costs) reducing the price that traders are prepared to pay for the cattle (Musemwa *et al.*, 2007). Makhura (2001), Mahabile *et al.* (2002) and Nkhori (2004) noted that even if farmers are in areas with good road linkages, the distance from the markets tends to influence transaction costs. The further away the farmers are from markets, the higher the transport costs they incur. In addition, farmers' incur extra transport costs to obtain transporting and selling permits from the police station and veterinary offices, respectively. It is a statutory requirement that when purchasing or selling cattle, they must have valid identification certificates and transporting permits (NDA, 2005). These restrict farmers' participation in distant markets.

2.9.3 Lack of information

None or poor provision of agricultural information is a key factor that has greatly limited agricultural development in developing countries (Bailey *et al.*, 1999). The farmers' information needs are those that enable him them to make rational, relevant decisions and strengthen their negotiating ability during transactions with buyers and consequently prevent possible exploitation by better informed buyers (Coetzee *et al.*, 2004). Information needs for communal farmers range from information on prevailing production techniques and market conditions, type of product demanded, quality, quantity, price and market opportunities (Bailey *et al.*, 1999).

According to Montshwe (2006), lack of timeous and reliable information is severe, particularly in the communal areas. Although considerable progress has been observed in the provision of communication systems such as telephone and cellular phone network facilities, communal farmers still remain uninformed in terms of new production techniques, market prices, trends and auction sale dates. Radio and personal communication are still used as main source of information. However, access by smallholder farmers to radios, televisions and internet is still limited. In most cases, information is broadcasted and written in Afrikaans and English. This makes the information irrelevant to the majority of communal farmers who in most cases only understand their local languages (Xhosa, Sotho and Zulu). The poor transfer of knowledge, skills and information is further manifested by limited interaction of the farmers with extension officers due to poor road networks and resources (Coetzee *et al.*, 2004). Training and education will further improve the capacity of the farmers and allow them to make informed decisions.

2.9.4 Diseases

Diseases are a major constraint to both livestock production and marketing in the tropics (Devendra *et al.*, 2000). Animal health issues are barriers to trade in livestock and their products, whilst specific diseases decrease production and increase morbidity and mortality (Düvel and Stephanus, 2000; Mwacharo and Drucker, 2005; Chawatama *et al.*, 2005). These diseases include anthrax, foot and mouth, black-leg and contagious abortion. The Mail and Guardian (2007) reported that the South African government has confirmed that until further notice, no veterinary import permits will be issued for cloven-hoofed animals and products derived there-of originating from the United Kingdom due to the current outbreak of the diseases in the United Kingdom. The outbreaks of such diseases in South Africa can be a threat to the communal cattle producers, who do not have medicine and proper disease control infrastructure. Furthermore, movement of cattle and their by-products are difficult to monitor in the communal areas.

2.9.5 Other marketing challenges

Other marketing challenges include lack of marketable livestock numbers and poor condition of livestock. According to Stevens and Jabara (1988), livestock numbers in communal areas are generally low per producer and the average weight of animals are generally lower compared to those of the commercial farming sector. The lack of marketable livestock numbers is also as a result of livestock theft. Excellent prices offered for Nguni cattle are likely to increase theft cases from this breed (Van den Bos, 2004; Dzimba and Matooane, 2005). Furthermore, farmers often have inadequate or no insurance coverage on livestock (Smith, 2002). Lack of marketable livestock numbers and poor condition of livestock therefore results in buyers not coming to purchase livestock since they will face very high transactional costs (Makhura, 2001).

Poor condition of livestock results in farmers getting low farm gate prices especially during dry spells (Makhura, 2001). More often, it results in farmers refusing to sell their livestock. Livestock auctioneers and speculators often raise concerns that they cannot pay competitive prices for animals that are in poor condition or not ready for the market (Nkhori, 2004). In addition to this, Nkhori (2004) also highlighted that the poor condition of livestock is important, but the age of animals affect prices. The animals are often too old when farmers do sell and this equally contributes to poor prices.

2.10 Existing market opportunities for Nguni cattle

A number of cattle market outlets are available to the beneficiaries of the Nguni cattle project; however access to formal markets is limited by a number of factors, chiefly of which are the distance from the market and inadequate infrastructure. There are many marketing channels that the beneficiaries can use when selling their cattle. These include private sales, auctions, butcheries and abattoirs.

2.10.1 Private sales and informal markets

The shortest, simplest, and the most popular option, especially amongst smallholder livestock owners, is private sales directly to the ultimate consumer (Nkosi and Kirsten, 1993). Private sale occupies an important position in the livestock marketing arena of the emerging sector. Private sales include individuals buying livestock for different reasons which include slaughter, investment or for socio-cultural functions such as funerals, weddings, customary and religious celebrations (USAID, 2003).

Due to the important functions performed by livestock in African societies, there exists a market amongst individual households (Nkhori, 2004). Private selling is a common practice to communal farmers as they are in a position to determine prices for their animals.

In addition, farmers incur less low marketing costs. Private sales are therefore, the cheapest and most probably, the simplest form of market outlet.

Nkhori (2004) reported that on-farm or direct sales to the consumer offer the greatest profit margin on live animals for the producer because all middlemen and their fees are eliminated. It offers a year-round marketing outlet; however the demand is irregular with high demand during certain times of the year, like festive seasons and Easter holiday. Most of the cattle traded in these informal markets are primarily old oxen destined for service as draught animals and ultimately for slaughter (Swallow and Brokken, 1987).

2.10.2 Auctions

Livestock auction markets are established places of business where livestock are assembled at regular intervals and sold by public bidding to the buyer who offers the highest price per head (Nkosi and Kirsten, 1993). These markets are public markets open to all buyers and sellers. As indicated by the NDA (2005), buyers include individuals buying for household use, butchers, commercial farmers and speculators. Nguni cattle can also be sold at better prices as breeding stock to breeders, commercial farmers and other communal farmers. The number of cattle sold through auctions varies considerably between locations. This influences the number of prospective buyers which in turn may affect the prices paid for cattle at a particular market (Benson *et al.*, 2001). In the case of the Nguni, the Nguni Breeders' Association do advertise these auctions, the prices paid in such auctions are very high compared to convectional auctions were all breeds of cattle are sold (Nkhori, 2004).

2.10.3 Butcheries

Another available option to communal farmers is to sell cattle directly to the butchers. Butcheries provide basic marketing services for farmers, particularly communal farmers, who are unable to market their cattle efficiently and profitably through other existing formal channels. Butchers enhance the marketability of livestock by acting as buyers in their own right and by acting as buyers at auctions. Nkhori (2004) found that good prices and farmers having a strong bargaining power in determining the prices of their stock are the main reasons for some farmers' satisfaction with sales to butchers. For Nguni cattle producers, they can sell their cattle to butcheries that do sell natural meat; hence there is need for the beneficiaries to develop some contract with big butcheries like Woolworths. Organic beef costs 8-12 US$/kg retail, compared with at least half that amount for non-organic beef (ECDC, 2003b).

2.10.4 Abattoirs

According to the NDA (2005), the abattoir is the least used marketing channel by communal farmers because of factors which include distance from the producers, payments delays, high risk factor of animals being condemned on the basis of health status, and many charges involved in using this channel. It is not economical to sell one or two animals as transport costs will not be justified. Group marketing can assist farmers to enjoy economies of scale when using this channel. However, group marketing is not always possible since farmers sell their animals at different times. Abattoirs pay farmers according to age, weight and grade of the animal (Nkhori, 2004). This grading system does not consider the breed and feeding practices. However, abattoirs tend to sell natural beef at high prices at both local and international markets than genetically modified beef and this result in them getting higher than normal returns at farmers' expense. The ability to sell stock at market-related prices would translate small scale farmers' cattle base into a capital base and improved livelihoods.

2.11 Potential markets for other Nguni cattle products

Since the Nguni is a multi-purpose animal, marketing should take a holistic approach, and promote development of other Nguni cattle products such as beef, milk, skins and hides, draught power and manure. Neglecting other uses of cattle can reduce household food security and exacerbate poverty (Shackleton *et al.,* 1999).

2.11.1 Nguni beef

Due to the dynamic nature of the world, naturally produced foods are currently in vogue. Beef that contains a high fat content is increasingly frowned upon by consumers. Beef from animals reared with artificial hormones have become unpopular. Natural beef from Nguni cattle is already being 'grown' in the Eastern Cape Province for instance in Sterksroom and exportation have already started (ECDC, 2003a). The pesticide and herbicide free natural grasslands of the province in South Africa provide the opportunity to develop high quality beef for both domestic and international natural beef markets (ECDC, 2003b).

Nguni cattle producers have the potential to dominate natural beef market due to the adaptability of the Nguni breed to the local environment. As far as local market is concerned, local supplies are unable to cope with high demand for naturally produced beef, the current biggest market for natural beef in South Africa is Woolworths (Raats *et al.,* 2004). Hotel and other butcheries, especially those that are located in low density suburbs have a high potential of being possible markets for naturally produced beef.

2.11.2 Milk

A Nguni cow produces 2-4 kg of milk per day on average, compared to established breeds which produce 10-20 kg under improved conditions (Bester *et al.*, 2003). The low milk quantity is further manifested by few lactating cows that the beneficiaries normally have at the same time. The available markets for Nguni milk in the Eastern Cape Province are mainly neighbours, street milk vendors and small local dairy shops. However, for the farmers to be able to meet the required supply by these dairy shops and for street vendors, they have to do cooperative marketing and this involves putting their milk together so that they can reduce transactional cost and meet the required volume (ECDC, 2003a). Nguni cattle producers can add value to their milk by processing and marketing their own products such as farm bottled pasteurised milk, powdered milk, butter, cheese, yoghurt, ice-cream, chocolates and sweets. In the long-term, organic certification and group marketing can result in higher premium prices and profits for the smallholder milk producers (Mapiye *et al.*, 2007). Labelling such products could further enhance value and could even lead to potential lucrative contracts with foreign investors.

2.11.3 Skins and hides

In the past, most of the hides and skins were exported by many companies in South Africa (ECDC, 2003b). However, the communal farmers failed to penetrate this market because of lack of information and large volumes required. For the few well informed communal farmers that were able to penetrate this market they only managed to obtain meagre returns due to high transaction costs (many middlemen were involved). As pointed out by Nkhori (2004), the higher the transaction costs the lower the profits and this reduces the chances of farmers participating in a given market. However, nowadays a number of leather industries have been set up throughout the country and these are currently experiencing shortage of domestic hides, especially of higher quality. These industries form a local market for hides from Nguni cattle, which, according to SAADA (2005), are of high quality.

In the Eastern Cape, the Eastern Cape Development Co-operation (ECDC) is partnering with Triple Trust, the Eastern Cape Tourism Board and with local farmers and businesses to train small-scale tanners (ECDC, 2003b). The aim of the project is to process skins and hides to produce hand-crafted 'organic' leather products to sell to the tourist market. The project has potential to provide products and a market for the hides that are currently either used locally or sold to brokers without any value being added (ECDC, 2003b). The high demand for quality automotive leather and the need to import finished leather for auto seat manufacture provides an ongoing opportunity. For example, the Daimler Chrysler committed itself to the project and will use 40,000 Nguni hides in exported Mercedes vehicles (Raats *et al.*, 2004). In addition to this, the South Africa Antique Dealers Association (SAADA) (2005) highlighted that the decor and fashion industries are taking advantage of the beauty and quality of the Nguni skin. Top furniture and interior designers are also known to be excited about high quality of Nguni cowhide due to its multiple colours and toughness.

2.11.4 Draught power

Nguni cattle are good draught animals; they are used to provide traction power during cultivation or transportation of goods (Bester *et al.*, 2003). Since some communal farmers do not own cattle, and most of them cannot afford to purchase or hire tractors for tillage (Chimonyo *et al.*, 1999), Nguni cattle can be a source income to farmers through provision of draught power to neighbours. This can only be achieved if Nguni cattle farmers can market this service to their neighbours by informing them at community gatherings about the service they can offer, especially those without cattle. If the animals can be well trained, they can be used to pull wagons and this form of transport commands significant tourist attraction. The Nguni cattle owners can advertise this service in tourist magazines or provide wagons at tourist centres near their communities.

2.11.5 Manure

Manure plays a key role in crop production by providing essential crop nutrients (Chimonyo *et al.*, 2000). In addition, the use of fertilizers is now becoming unattractive due to their effect on the environment. This creates a big demand for manure as source of organic nutrients for food and fodder crops. Nguni cattle producers can sell manure to their neighbours. Cattle owners can also combine their manure and sell them to large companies that manufacture organic fertilizers. Cow dung can be a source of income in some communal areas where fire wood and electricity are not available.

References

Adams, R.H. Jr. (1999). Nonfarm income, inequality and land in rural Egypt. Draft. World Bank Washington, DC, USA.

AllAfrica.com. (2007). Africa: more rural families earning money from non-farm work. Available at: http://allafrica.com/stories/200706050803.html.

Bailey, D., C.B. Barrett, P.D. Little and F. Chabari (1999). Livestock markets and risk management among East African pastoralists: A review and research agenda. Utah University, USA.

Barley, S.R. and P.S. Tolbert (1997). Institutionalization and structuration: studying the links between action and institution. Organization Studies 18, 93-117.

Barrett, C.B., T. Reardon and P. Webb (2001). Nonfarm income diversification and household livelihood strategies in rural africa: concepts, dynamics, and policy implications. Cornell University, Ithaca, USA.

Benfica, R., D. Tschirley and L. Sambo (2002). Agro-industry and smallholder agriculture: Institutional arrangements and rural poverty reduction in Mozambique. Journal of Modern African Studies 39, 333-358.

Benham, A. and L. Benham (1998). Measuring the costs of exchange. Paper prepared for the 2nd Annual Meeting of the International Society for New Institutional Economics. September 17-19, 1998.

Benson, G., D. Miller and R. Lichtenwalner (2001). Beef cattle marketing in North Carolina. Available at: http://www2.ncsu.edu/unity/lockers/project/arepublication/AREno32.pdf.

Bester, J., I.E. Matjuda, J.M. Rust and H.J. Fourie (2003). The Nguni: case study. In: FAO Community-based management of animal genetic resources. Rome: UNDP, GTZ, CTA, FAO. pp. 45-68.

Bryceson, D.F. (2002). Multiplex livelihoods in rural Africa: recasting the terms and conditions of gainful employment. Journal of Modern African Studies 40, 1-28.

Chaplin, H., S. Davidova and M. Gorton (2000). Analysis of diversification of farm enterprises and household income in CEECs: definitions and propositions. Idara Working Paper 2/3, Wye College, London, UK.

Chawatama, S., C. Mutisi and A.C. Mupawaenda (2005). The socio-economic status of smallholder livestock production in Zimbabwe: a diagnostic study. Livestock Research for Rural Development 17 (12). Available at: http://www.cipav.org.co/lrrd/lrrd17/12/chaw17143.htm.

Cherdchuchai, S. and K. Otsuka (2006). Rural income dynamics and poverty reduction in Thai villages from 1987 to 2004. Agricultural Economics 35, 409-423.

Chimonyo M, N.T. Kusina, H. Hamudikuwanda, O. Nyoni and I. Ncube (2000). Effects of dietary supplementation and work stress on ovarian activity in non-lactating Mashona cows in a smallholder farming area of Zimbabwe. Animal Science 70, 317-323.

Coetzee, L., B.D. Montshwe and A. Jooste (2004). Livestock marketing: constraints, challenges and implications for extension services. Paper presented at the 38th Conference of the South African Society for Agricultural Extension. May 6, 2004.

Davis, R. Jr. (2003). The rural non-farm economy, livelihoods and their diversification: Issues and options. Report No: 2753. Natural Resources Institute, Chatham, UK.

Davis, R. Jr. and D. Bezemer (2004). The development of the rural non-farm economy in developing countries and transition economies: key emerging and conceptual issues. Natural Resources Institute, Chatham, UK.

Davis, R. Jr. and D. Pearce (2001). The non-agricultural rural sector in Central and Eastern Europe. Report No. 2630. Natural Resources Institute, Chatham, UK. Available at: http://www.nri.org/rnfe/pub/papers/2630.pdf.

De Janvry, A. and E. Sadoulet (2001). Income strategies among rural households in Mexico: the role of off-farm activities in poverty reduction. World Development 29, 467-480.

Deininger, K. and P. Olinto (2001). Rural employment and income diversification in Colombia. World Development 29, 455-465.

De-Janvry, A., E. Sadoulet and N. Zhu (2005). The role of non-farm incomes in reducing rural poverty and inequality in China. Johns Hopkins University Press, Baltimore, USA.

Delgado, C. (1999). Sources of growth in smallholder agriculture in sub-Saharan Africa: the role of vertical integration of smallholders with processors and marketers of high-value added items. Agrekon 38 (Special Issue), 165-189.

Devendra C, D. Thomas, M. Jabbar and E. Zerbini (2000). Improvement of livestock production in crop-animal systems in agro-ecological zones of South Asia. ILRI, Nairobi, Kenya. pp. 37-38.

Dorward, A. and J. Kydd (2005). Making agricultural market systems work for the poor: Promoting effective, efficient and accessible coordination and exchange. Imperial College, London, UK.

Düvel G. H. and A. L. Stephanus (2000). Production constraints and perceived marketing problems of stock farmers in some districts of the Northern communal areas of Namibia. South African Journal of Agricultural Extension, 29, 89-104.

Dzimba, J. and M. Matooane (2005). The impact of stock theft on human security: Strategies for combating stock theft in Lesotho. In: Muloongo, K., Kibasomba, R. and Kariri, J.N. (eds.) The many faces of human security: case studies of seven countries in Southern Africa. Institute for Security Studies, Pretoria, South Africa, pp. 277-281.

Eggertson, T. (1990). Economic behaviour and institutions: Cambridge Survey of Economic Literature. Cambridge University Press, Cambridge.

Eicher, C.K. (1982). Facing up to Africa's food crisis. Foreign Affairs 61, 151-174.

Eicher, C.K. and D.C. Baker (1982). Research on agricultural development in Sub-Saharan Africa: a critical survey. East Lansing, Michigan State University, USA. pp. 1-14.

Ensminger, J. (1992). Making a market: the institutional transformation of an African society. Cambridge University Press, New York, USA.

Escobal, J. (2001). The determinants of nonfarm income diversification in rural Peru. World Development 29, 497-508.

Estudillo, J.P., Y. Sawada, and K. Otsuka (2006). The Green Revolution, development of labour markets, and poverty reduction in the Philippines, 1985-2004. Agricultural Economics 35, Supplement 3, 399-406.

Fafchamps, M. (1996). The enforcement of commercial contracts in Ghana. World Development 24, 427-448.

FAO (2004). Changing patterns of agricultural trade. The evolution of trade in primary and processed agricultural products. Available at: http://www.fao.org/docrep/007/y54.

Fidzani, N.H. (1993). Understanding cattle offtake rates in Botswana. PhD Thesis, Boston University, USA.

Frank, S.D. and D.R. Henderson (1992). Transaction costs as determinants of vertical coordination in the US food industries. American Journal of Agricultural Economics 74, 941-950.

Fraser, G., N. Monde, and W., Van Averbeke (2003). Food Security in South Africa: A case study of rural livelihoods in the Eastern Cape Province. In: Lieb Nieuwoudt and Jan Groenewald (eds.), The challenge of change: agriculture, land and the South African cconomy University of Natal Press: Pietermaritzburg, South Africa, pp. 171-183.

Frisch, J.E. (1999). Towards a permanent solution for controlling cattle ticks. International Journal of Parasitology 29, 57-71.

Gabre-Madhin, E. (2001). Understanding how markets work: transaction costs and institutions in the Ethiopian grain market. IFPRI Research Report. International Food Policy Research Institute, Washington, DC, USA.

Hobbs, J.E. (1997). Measuring the importance of transaction costs in cattle marketing. American Journal of Agricultural Economics 79, 1083-1095.

Holden, S., B. Shiferawi, and J. Pender (2004). Non-farm income, household welfare, and sustainable land management in a less-favoured area in the Ethiopian highlands. Discussion Paper. Department of Economics and Resource Management, Agricultural University of Norway.

Jaffee, S. and J. Morton (1995). Marketing Africa's high-value foods: comparative experiences of an emergent private sector. Kendall/Hunt Publishing: Iowa, USA.

Key, N. and D. Runsten (1999). Contract farming, smallholders, and rural development in Latin America: the organization of agro-processing firms and the scale of outgrower production. World Development 27, 381-401.

Kherallah, M. and J. Kirsten (2001). The new institutional economics: applications for agricultural policy research in developing countries. Markets and Structural Studies Division, Discussion Paper No.41, International Food Policy Research Institute, Washington, DC, USA.

Kherallah, M. and N. Minot (2001). Impact of agricultural market reforms on smallholder farmers in Benin and Malawi. IFPRI Collaborative Research Project. International Food Policy Research Institute, Washington, DC, USA.

Klein, P.G. (1999). New institutional economics. American Journal of Agricultural Economics 30, 456-463.

Krueger, A.O. (1990). Government failures in development. NBER Working Paper Series No 3340. National Bureau of Economic Research: Cambridge, UK.

Lele, U. (1984). Rural Africa: modernization, equity, and long-term development. In: Eicher, C.K. and J.M. Staatz (eds.), Agricultural development in the Third World. The Johns Hopkins University Press: Baltimore, USA.

Lele, U. (2010). Food security for a billion poor. Science 327, 1554.

Machethe, C.L. (2004). Agriculture and poverty in South Africa: can agriculture reduce poverty? Available at: http://cfapp1-docs-public.undp.org/eo/evaldocs1/sfcle/ eodoc357114047.pdf.

Magagula, G.T. and H.M. Faki. (1999). Comparative economic advantage of alternative agricultural production options in Swaziland. United States Agency for International Development, Technical Paper. Washington, DC, USA.

Mahabile, M., M. Lyne and A. Panin (2002). Factors affecting the productivity of communal and private livestock farmers in Southern Botswana: A descriptive analysis of sample survey results. Agrekon 41, 326-338.

Mail and Guardian (2007). SA halts UK meat imports after disease outbreak. 6 August 2007.

Makhura, M.T. (2001). Overcoming transaction costs barriers to market participation of smallholder farmers in the Northern Province of South Africa. PhD Thesis, University of Pretoria, RSA.

Mangisoni, J. (2006). Markets, institutions and agricultural performance in Africa. ATPS special paper series No. 27, the African Technology Policy Studies Network, Nairobi, Kenya.

Mapiye, C., M. Chimonyo, V. Muchenje, K. Dzama, M.C. Marufu and J.G. Raats (2007). Potential for value-addition of Nguni cattle products in the communal areas of South Africa: a review. African Journal of Agricultural Research 2, 488-495.

Mashinini, N.N. (2004). Welfare effects of the Swaziland maize market system. Faculty Research Paper. Faculty of Natural and Agricultural Sciences, Bloemfontein, University of the Free State, RSA.

Matsumoto, T., Y. Kijima, and T. Yamano (2006). The role of local nonfarm activities and migration in reducing poverty: evidence from Ethiopia, Kenya and Uganda. Agricultural Economics 35, 449-458.

Mehta, G.S. (2002). Non-farm economy and rural development. Giri Institute of Development Studies. Lucky Now. Available at: http://developmentfirst.org/india/planning_commission/special_study_reports/nonfarm_eco_ruraldev.pdf.

Minot, N. and F. Goletti (1997). Impact of rice export policy on domestic prices and food security. markets, trade, and institutions. International Food Policy Research Institute (IFPRI).

Mkhabela, M.S. and N.N. Mashinini (2005). Early maize yield forecasting in the four agro-ecological regions of Swaziland Using NDVI data derived from NOAA's AVHRR. Agricultural and Forest Meteorology 129, 1-9.

Montshwe, D.B. (2006). Factors affecting participation in mainstream cattle markets by small-scale cattle farmers in South Africa. MSc Thesis, University of Free State, RSA.

Musemwa, L, C. Chagwiza, W. Sikuka, G. Fraser, M. Chimonyo and N. Mzileni (2007). Analysis of cattle marketing channels used by small scale farmers in the Eastern Cape Province, South Africa. Livestock Research for Rural Development 19 (9), Article #131.

Myint, T. (2003). Understanding 'problem of fit' between institutions and environment: lessons from the Rhine River Basin. Paper presented at the LARS2 Conference, The Second International Symposium on the Management of Large Rivers for Fisheries: Sustaining livelihoods and biodiversity in the new millennium, February 11, 2003, Phnom Penh, Cambodia. Available at: http://ssrn.com/abstract=1555422.

Nabli, M.K. and J.B. Nugent (1989). The new institutional economics and its applicability to development. World Development 17, 1333-1347.

Nargis, N and M. Hossain (2006). Income dynamics and pathways out of poverty in Bangladesh, 1988-2004. Agricultural Economics 35, 425-435.

Nel, E.L. and J.A. Binns (2000). Rural self-reliance strategies in South Africa: community initiatives in the former homelands. Journal of Rural Studies 19, 367-377.

NERPO (2004). Marketing infrastructure development. Available at: http://www.nerpo.org.za.

Nkhori, P.A. (2004). The impact of transaction costs on the choice of cattle markets in Mahalapye district, Botswana. MSc Dissertation, University of Pretoria, RSA.

Nkosi, S.A. and J.F. Kirsten (1993). The marketing of livestock in South Africa's developing areas: A case study of the role of speculators, auctioneers, butchers and private buyers in Lebowa. Agrekon 32, 230-237.

North, D. (1990). Institutions, institutional change and economic performance. Cambridge University Press, Cambridge, UK.

Nweke, F.I. (1978). Agricultural credit in Ghana: priorites and needs for domestic food production. Canadian Journal of Agricultural Economics 26, 38-46.

Nweke, F.I., D.S.C. Spencer, and J.K. Lynam (2002). The cassava transformation: Africa's best-kept secret. Michigan State University Press: East Lansing, MI, USA

Oehmke, J.F., P. Anandajasekeram, and W.A. Masters (1997). Agricultural technology development and transfer in Africa: impacts achieved and lessons learned. United States Agency for International Development. Technical Paper No. 77, Washington, DC, USA.

Onchan, T. (2001). Non-farm employment opportunities in rural areas in Asia. (ICD-SE2-01) Report of the Asian Productivity Organization (APO) Seminar on Non-Farm Employment Opportunities in Rural Areas, Philippines, 24-29 September 2001. Tokyo (Japan), Asian Productivity Organization.

Otsuka, K. and T. Yamano (2006). Introduction to the special issue on the role of non-farm income in poverty reduction: evidence from Asia and East Africa. Agricultural Economics 35 (Supplement), 393-397.

Parkin, M. (2008). Economics. Addison-Wesley: London, UK.

Perret, S. (2002). Livelihood strategies in rural Transkei (Eastern Cape Province): how does wool production fit in? University of Pretoria, Pretoria (mimeo), RSA.

Perret, S., W. Anseeuwand and N. Mathebula (2005). Poverty and livelihoods in rural South Africa. Investigating diversity and dynamics of livelihoods. Case studies in Limpopo. Unpublished Project report num.05/01, Kellogg's Foundation / University of Pretoria, RSA.

Raats, J.G., A.M. Magadlela, G.C.G. Fraser and A. Hugo (2004). Re-introducing Nguni nucleus herds in 100 communal villages of the Eastern Cape Province. A proposed co-operative project between the University of Fort Hare, Agricultural and Development Research Institute (ARDRI) and the East Cape Department of Agriculture and the Kellogg Foundation, Unpublished report.

Randela, R. (2005). Integration of emerging cotton farmers into the commercial agricultural economy. Unpublished PhD thesis, University of the Free State, Bloemfontein, RSA.

Reardon, T. and C. Barrett (2000). Agro-industrialization, globalization, and international development: An overview of issues, patterns, and determinants. Journal of Agricultural Economics 23, 195 -205.

Reardon, T., J. Berdegue, C.B. Barrett and K. Stamoulis (2006). Household income diversification into rural nonfarm activities. In: S. Haggblade, P. Hazell and T. Reardon (eds.) Transforming the Rural Nonfarm Economy. Baltimore, USA: Johns Hopkins University Press.

Ruijs, A. (2002). Cereal trade in developing countries: a stochastic equilibrium analysis of market liberalisation and institutional changes in Burkina Faso. Labyrinth Publication: the Netherlands.

SAADA (2005). Beauty is Nguni skin deep. Available at: http://www.streetwires.co.za/pressroom1.htm.

Sanchez, V. (2005). The determinants of rural non-farm employment and incomes in Bolivia. Masters of Agricultural Economics Thesis, Michigan State University, MI, USA.

Schwarze, S. (2004). Determinants of income generation activities of rural households in Central Sulawesi, Indonesia. Institute of Rural Development, Georg-August University Göttingen, Germany.

Schwarze, S. (2005). Determinants of Income generating activities of rural households: a quantitative study in the vicinity of the Lore Lindu National Park in Central Sulawesi Indonesia. PhD Thesis. Institute of Rural Development, University of Goettingen, Germany.

Seleka, T.B. (1999). The performance of Botswana's traditional arable agriculture: growth rates and the impact of the accelerated rainfed arable programme (ARAP). Agricultural Economics 20, 121-133.

Shackleton, C.M., S.E. Shackleton, T.R. Netshiluvhi, F.R. Mathabela and C. Phiri (1999). The direct use value of goods and services attributed to cattle and goats in the Sand River Catchment, Bushbuckridge, CSIR-Environmentek Report No. ENV-P-C 99003, CSIR, Pretoria, RSA.

Smith, D. (2002). Crime in rural areas on increase. Available at: http://deltafarmpress.com/mag/farming_crime_rural_areas\.

Southgate, D. (2009). Population growth, increases in agricultural production and trends in food prices. The Electronic Journal of Sustainable Development 1 (3), 29-35.

Southgate, D., D. Graham, and D. Southgate (2007). The world food economy. Malden: Blackwell Publishing.

Spencer, D.S.C. (1994), 'infrastructure and technology constraints to agricultural development in the humid and sub-humid tropics of Africa', EPTD Discussion Paper No. 3, International Food Policy Research Institute (IFPRI), Washington, DC, USA.

Statistics of South Africa (StatsSA) (2000). Household survey statistical release P0317.

Stevens, R.D. and C.L. Jabara (1988). Agricultural development principles: economic theory and empirical evidence. The John Hopkins University Press: Baltimore, USA.

Swallow, B.M. and R.P. Brokken (1987). Cattle marketing policy in Lesotho. Available at: http://www.fao.org/Wairdocs/ILRI/x5508E/x5508e00.htm#Contents.

Toenniessen, G., A. Adesina, and J. DeVries (2008). Building an Alliance for a Green Revolution in Africa. Annals of the New York Academy of Sciences 1136, 233-242.

USDA (1981). Food problems and prospects in Sub-Saharan Africa: the decade of the 1980s. Washington, DC, USA. United States Department of Agriculture. Foreign Agricultural Research Report No. 166.

Van De Berg, M. and G.E. Kumbi (2006). Poverty and the rural non farm income in Oromia, Ethiopia. Agricultural Economics 35, 469-475.

Van Schalkwyk, H.D., E. Nel, and A. Obi (2004). Factors inhibiting agricultural development in the former homelands of South Africa. Bloemfontein, University of the Free State (mimeo).

Williamson, O.E. (1998). The institutions of governance. American Economic Review 88, 75-79.

3. Determinants of household activity choice, rural income strategies and diversification

Simbarashe Ndhleve and Ajuruchukwu Obi

Abstract

South Africa's post-apartheid economy continues to invest in agricultural development but poverty remains endemic among households with agriculture as their main source of livelihood. There are now doubts that agriculture is the dominant income-generating activity. This chapter analyses the potential livelihoods in rural areas of South Africa. Using household and community level data, the chapter develops simultaneous equations to estimate the determinants of both poverty and income from activities practised in the study area. It is expected that these would enable policymakers to identify the weak links in rural poverty reduction efforts and determine whether agriculture should remain a priority for government investment. Despite low returns per capita, own agriculture and wage employment were shown to be the main sources of income and strongly influenced by level of education, labour availability, social capital, physical capital and geographical location. To improve rural income and reduce poverty, urgent improvements in education, capital resources, infrastructure and communication are necessary.

3.1 Introduction and problem context

Development strategies articulated by the South African government in the democratic era have been oriented towards improving the lot of the historically disadvantaged majority of the black population through the development of agriculture. The post-apartheid economy continues to invest in agricultural development but poverty remains high among households with agriculture as their main source of livelihood. About 72% of the population of South Africa is poor, with the bulk of these persons residing in the Eastern Cape Province which ranks as the second poorest province in the country, harbouring some 6.2 million inhabitants out of whom 70.7% were classified as poor (Perret, 2002). With such disturbing numbers, the country's ability to meet the Millennium Development Goal (MDG) of halving poverty by 2015 is in serious doubt (World Bank, 2006). Reducing poverty by promoting small scale agriculture is seemingly hard in rural South Africa. The most disappointing aspect of post-apartheid economic performance is the widespread poverty and the widening of inequalities.

Agriculture must be the key driver of rural income growth but this is not the case in rural South Africa. According to 1995 and 2000 income and expenditure surveys' figures, South Africa's rural share of income poverty declined by approximately 5% but this decline was largely due to the unprecedented expansion of social grants expenditure (Katungi *et al.*, 2007). Many advocates that own agriculture continues to be the main livelihood for the poor in South Africa but small scale agriculture is in a moribund state. The debate on small scale agricultural development and rural poverty in South Africa remains an unsettled issue due to continuous poverty among the rural population. A greater proportion of households practicing agriculture is reeling with poverty.

Many still argue that agriculture has the potential to reduce poverty in rural South Africa (Byerlee *et al.*, 2005, Katungi *et al.,* 2007). Its characteristic features like the concentration of the poor in the sector, its growth linkages to other sectors and the positive externalities from assuring food security and reducing prices makes it an important driver of poverty. Increasing agricultural income through investment in small scale agriculture like infrastructural development, research and development, land reform and land redistributions have until recently been used to promote rural income growth in South Africa. But South Africa's small scale agriculture remains passive, so that many doubt whether this sector alone will be sufficient to address rural poverty.

World Bank (2006) data suggest that, before the current global food and humanitarian crises, some progress was being made towards reducing poverty on an overall basis. For instance, it was reported that global poverty rates fell from 33% in 1981 to 21% in 2001 (World Bank, 2006). However, the fact that at the dawn of the 21st century as much as 1.2 billion people around the world still live in abject poverty means that much still needs to be done to achieve the MDGs (World Bank, 2006). Evidence from elsewhere in Sub-Saharan Africa shows slow progress towards halving poverty by 2015 (Matsumoto *et al.*, 2006). Further, the fact that according to the World Bank (2006), about 75% of the poor live in rural areas, calls for deeper enquiry into the dynamics of rural livelihoods and poverty. A logical starting point of such enquiry is gaining an understanding of the factors driving whatever positive changes might have taken place. In that regard, it is important to ascertain whether whatever positive trends there have been in poverty reduction are due to income growth derived from small scale agriculture as the only rural activity, or whether they have been due to other sources of income.

This chapter presents an inventory of rural income-generating activities and examines income-enhancing opportunities in rural areas in order to propose a roadmap out of poverty in rural South Africa. Specifically, this objective will be achieved by presenting an analysis of rural income levels and poverty status in Eastern Cape Province of South Africa. It intends to examine the nature and dimensions of income poverty in the Province and how they relate to household activities. Specifically, this study aims to:

- identify and catalogue the various sources of income and livelihoods in the study area;
- estimate the poverty situation in the study communities in Eastern Cape Province;

- examine the determinants of income from different rural households sources of income in the study area; and
- analyse the determinants of poverty in Eastern Cape Province. Based on the research findings, the paper identifies and recommends strategies for improving the poverty situation.

3.2 Rural sources of income

Despite the rural economy being assumed to be naturally agriculture based, a large number of rural households derive most of their income and employment from non-farm related activities (Barrett *et al.*, 2001). In South Africa, StatsSA (2000) reports that only 4% of poor households have agriculture as their primary source of income, 57% depends on income from wages or salaries, 14% from social grants and 10% from remittances. Agriculture alone cannot provide sufficient livelihood opportunities for the rural population (Davis and Pearce, 2001; Mehta, 2002; Davis, 2003).

This situation is not peculiar to South Africa. Selected studies in other countries in Africa reflect a similar pattern. Based on a nationally-representative household budget survey in Egypt, Adams (1999) found that agricultural income accounts for less than one-quarter (26.6%) of total rural household income with non-farm income accounting for 42.2%. The above statistics are quite revealing about typical African rural economy. Barrett *et al.* (2001) also found an average of 42.2% for Africa as a whole as the proportion of household income in rural areas coming from non-farm sources. Similar results were also found in Bangladesh by Nargis and Hossain (2006). Income patterns recorded in Bangladesh were not only among the landless; landholders equally depend on non-farm income to supplement their agricultural earnings.

On the basis of the foregoing findings, a conclusion can be drawn that besides agricultural income, the rural share of non-farm income is high. This implies that promoting both small scale agriculture and non-farm activities is essential for rural households' income growth. Explanations for the seemingly high proportion of non-farm income in rural areas have been sought with researchers giving varied reasons.

In the case of Bangladesh, Nargis and Hossain (2006) found that a high and increasing portion of rural non-farm income could be explained by decreasing agricultural income due to decreasing land size of farm households, as well as production and marketing constraints in subsistence agriculture. This makes subsistence agriculture an unsustainable income generating strategy, pushing individuals into income diversification searching for stable sources of income. Although agriculture remains the backbone of most rural livelihoods, this shows that the notion of South Africa's rural economies as purely agricultural is changing.

3.3 Model structure and data

This study employs a model that enables us to calculate the determinants of income from different income generating activities practiced by households in the study area. Classification of sources of income in this study follows almost similar classification as that adopted by Fraser *et al.* (2003) in Guquka and Koloni villages. These villages are in the same district with the present study area. The following classification of activities was used:

1. Own agriculture income (income earned from agriculture-crop, poultry and livestock).
2. Own business income (income from activities such as industry, transportation, construction, and services).
3. Remittances (money or goods received from migrant household member by the household).
4. Wage income (income earned from formal or informal wage employment, including salaries, allowances, bonuses, and other kinds of remuneration).
5. Inheritance and grants – incomes without *quid pro quo*, such as pensions, transfers, grants/subsidies, rents, and financial income.

With the view of tracking the determinants of income received by each household in 2008, separate simultaneous equations were modeled for each activity. The equations were developed to estimate the effects of household members' characteristics, household variables and community level variables on household income. The estimates of income by activity are derived using the Probit model, and represented as:

$$Y^j = \beta_0 + \beta_i X_i + \mu_i \qquad (1)$$

where:
Y^j = the dependent variable representing income from each income category;
β_0 = the constant term;
β_i = the vector of coefficients;
X_i = the vector of explanatory variables;
μ_i = the error term.

Increases in households income can only reduce poverty if it benefits the poor sect of the society. An attempt was made to model poverty directly as function of household characteristics, household assets and community characteristics (Fan *et al.*, 2002). In this case household poverty is a binary variable representing either household whose annual per capita income falls below the poverty line or above. Using ordinary least squares (OLS) will result in biased estimates. Therefore, a Probit model is used to estimate the poverty determination equation. The Probit model was necessary to avoid selection bias in estimating income distribution within the communities (Yúnez-Naude and Taylor, 2001).

$$P_i = \beta_0 + \beta_i X_i + \mu_i \quad (2)$$

where:
P_i = the dependent variable representing household poverty status;
β_0 = the constant term;
β_i = the vector of coefficients;
X_i = the vector of explanatory variables;
μ_i = the error term.

The above econometric model was used and treated against the potential variables, which are assumed to affect household poverty status. Potential variables which may influence household's poverty status were acquired from literature (Barrett *et al.*, 2001; Fan *et al.*, 2002; Schwarze, 2004). Variables included in the model were age, education, household size, gender, labour availability, possession of physical assets, social capital, dependency ratio, land size, and geographical location. βs are the parameters to be estimated giving the marginal effects of each independent variable on poverty.

The study area, Stutterheim, is situated in the Eastern Cape province of South Africa, located along the N6 road to East London (Figure 3.1). Approximately sixty kilometres from King Williams Town, the area was formerly part of the Ciskei Bantustan under the

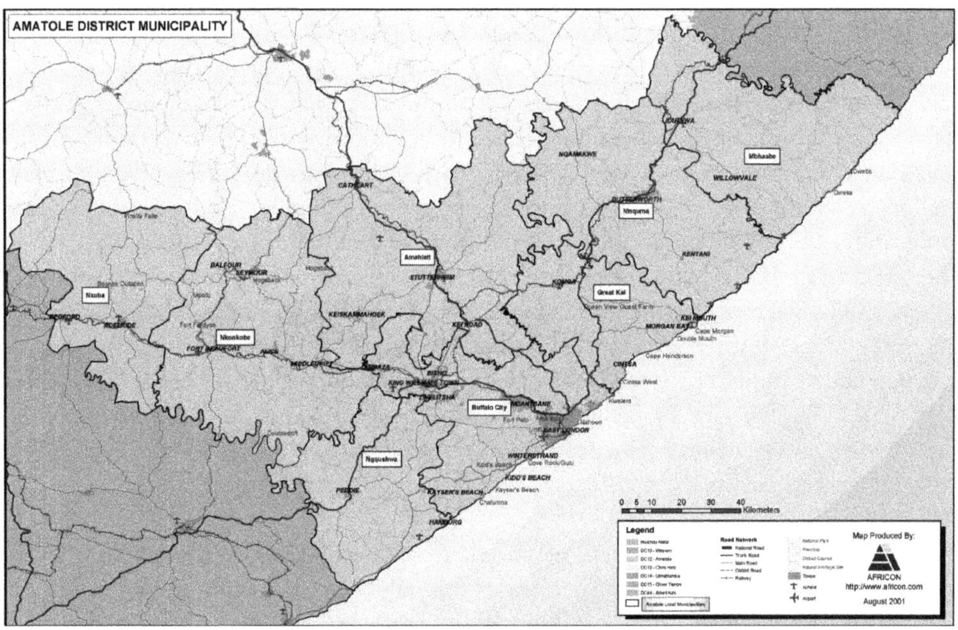

Figure 3.1. Map of the Amatole District showing Stutterheim, Eastern Cape Province, South Africa.

apartheid system. Stutterheim is a district centre surrounded by nine rural communities. Clearly distinguishable rural and peri-urban locations within the district justify its inclusion for purposes of assessing income distribution and inequality, among other issues. Two communities were enumerated to represent a rural setting (Ndakana) and a peri-urban setting (Mlungisi).

3.4 Physical settings

The key factors influencing agriculture in the study area are the soils and climate. Stutterheim experiences summer rainfall which is highly variable from year to year. The district is divided into two fairly distinct zones, a drier Eastern zone and a wetter Western zone. Mean annual rainfall is generally less than 600 mm per year and drops to less than 400 mm near the Kei River. The climate can be considered as semi-arid, being dry and warm throughout most of the year. Stutterheim experiences slightly extreme conditions with temperatures of 5-35 °C. Extensive areas are characterized by poor, shallow soils which are not conducive to intensive, arable farming. The dominant vegetation communities are subtropical thicket, a little afro-montane forest, and grasslands. Stutterheim as a whole is largely a stock-farming district with a large number of livestock (cattle, goats and sheep). A local extension officer estimates that some 4,700 ha is under cultivation, of which 1000 ha is under irrigation.

3.5 Socio-economic activities

The town and its surrounding rural communities boast a wide range of economic activities and services. Agriculture is mostly small scale crop farming and open grazed livestock. The key economic activities include home-based business like selling mobile phone air time, tuck shops, tarvens, informal drinking places, etc., farming (livestock, vegetables, crops and citrus) and forestry and associated industries such as saw milling. Forestry and saw milling are the dominant economic activities employing the highest proportion of the local population. The bulk of the population is employed in the forestry, and most are dependent on social/old age grants (pension and child grants). Aside from local employment in forestry and sawmills, Stutterheim town is the major source of employment. It hosts a number of service providing institutions. Migrant workers also constitute a substantial proportion of the population though they seem to be decreasing as reported by Beinart and Wotshela (1995). Besides local employment in Stutterheim, some of the people are employed in the surrounding cities such as Port Elizabeth, East London and King Williams Town. These migrant workers are the key providers of remittances to Stutterheim.

Farming is largely for subsistence rather than commercial sale, although some black commercial farmers are present. The district contains the best grazing lands suitable for livestock production in the province. Households produce their food and generate income from farming activities. Both livestock and crop farming are practised in the area but mostly at a lower scale. Beinart and Wotshela (1995) noted that the rate of cultivation in Stutterheim has dropped during the past few years probably because very few landowners or tenants are able to command sufficient land, capital and labour to produce significant amounts of output.

The study employed household and community data collected from Stutterheim in March 2008. Primary data were generated by means of structured questionnaires administered on a total of 79 households, as well as interview schedules and checklists for purposes of collecting community-level data. Secondary data were gathered by means of extensive document analysis and discussions with knowledgeable persons, including extension workers serving the two study communities.

3.6 Results and discussions

In this and subsequent sections, the results of the study are presented and discussed. Following this introduction, the socio-economic characteristics of the survey households are presented while the diverse sources of rural incomes are examined in the sub-section that follows. The dynamics of poverty reduction and how these are affected by the choice of activity are then discussed on the basis of those findings. The final part of the section takes a look at the determinants of poverty.

3.6.1 Socio-economic characteristics of survey households

The characteristics of the sample households were analyzed by means of descriptive statistics as shown in Table 3.1. Characteristics included gender, age, marital status, main source of income and education of household head interviewed, in addition to the size of the household's field. From the evidence, the average household size was more or less the same in both study communities although the Mlungisi households seemed marginally larger. On average, household heads in the rural community were older than those in the peri-urban location, reflecting the tendency for individuals to move back to their villages as they grew older. There were correspondingly more widowed persons in the rural area, where household heads were relatively older, less educated, and cultivated relatively larger fields than those in the peri-urban location of Mlungisi (Table 3.1).

3.6.2 Sources of rural income

Households in the two communities derive income from several activities, both farm and non-farm activities (Table 3.2). Most households pursued at least one of the following activities: crop farming, livestock rearing, own business, wage employment, and augmented income by means of pensions, grants and remittances. The average total income for the 12 months recorded for each community was computed. According to the data, households

Table 3.1. Characteristics of the sample households in Ndakama and Mlungisi (Field data, 2008).

		Ndakana	Mlungisi
Number of households		33	46
Gender:	Male	15	30
	Female	18	16
Household size		6.5	7
Age of head of household		57.8	51
Marital status:	Single (%)	6.1	17.4
	Married (%)	20	29
	Widow (%)	30	15.2
Education:	Not educated	13	9
	Primary level	8	21
	Secondary level	12	15
	Tertiary level	0	1
Farm households (%)		57.6	43.5
Land size (ha)		1.4	0.9

Table 3.2. Distribution of households by average income and activity type in Mlungisi and Ndakama (Field data, 2008).

	Livelihood strategy	Frequency	Percentage (%)	Average income (R)
Ndakana	Own agricultural	19	57.6	1,941.84
	Own business	6	18	9,733.35
	Wage employment	10	30.3	24,120.00
	Remittances	15	45.5	7,240.00
	Pensions and grants	21	63.6	13,371.43
Mlungisi	Own agricultural	20	43.5	1,396.01
	Own business	6	17.4	10,399.99
	Wage employment	15	32.6	12,640.00
	Remittances	10	21.7	5,520.00
	Pensions and grants	35	76.1	9,836.57

in Ndakana earned a total of R21,996.86 per annum, on average, while households in Mlungisi received R14,769.57 per annum, on average.

Table 3.2 provides a breakdown of the various income sources and their relative importance. More than 30% of the respondents from Ndakana had at least a member engaged in wage employment. In the case of Mlungisi, participation in wage employment was approximately 3% higher than Ndakana among the survey households. A total of 39 respondents, representing more than 50% of all respondents, engaged in agricultural activities as a means of livelihood in both communities. In Ndakana, the more rural community where field sizes are relatively larger, a larger proportion of the respondents, about 58%, engaged in agricultural activities, and earned an average income of R1,942, while in the peri-urban area of Mlungisi, about 44% of the respondents from that sub-sample engaged in agricultural activities, earning slightly lower average agricultural income of R1,396. Interestingly, for both communities, despite a large proportion of households engaging in agriculture, agricultural incomes turned out to be extremely low relative to other income sources.

The results show that households derived income from other sources such as operation of own businesses, pensions and grants, and remittances from family members living outside the community. The dominant own businesses observed in the communities were owning of tuck-shops, vending of cellular phone re-charge vouchers (or 'airtime'), selling of vegetables, running of taverns and shebeens (an unlicensed drinking place for alcoholic beverages) and food processing. The data from the two communities indicate that about 18% of the respondents in Ndakana operate their own business enterprises and 17% in Mlungisi, and in each case, this activity contributed about 17% of household income (see Table 3.2). As Table 3.2 revealed, pensions and grants emerged as important sources of income in both communities, accounting for the bulk of incomes received by 64% in Ndakana and 76% in Mlungisi. Remittances were clearly more important in the rural community than the peri-urban community (Table 3.2).

3.6.3 Dynamics of poverty reduction

The estimates of poverty incidence in this paper employ Foster, Greer and Thorbecke (FGT) method (1984) mathematically expressed as:

$$P_\alpha = 1/n\ \Sigma y_i{<}z\ [z\text{-}y_i / z]^\alpha$$

where y_i is the real per capita income of a household i, is the total number of household members, z is the poverty line, and $\alpha = 0$, P_0 is the headcount ratio or the proportion of household population whose annual per capita income falls below the poverty line. If $\alpha = 1$, P_1= is the poverty gap ratio representing an index of the depth of poverty, which is the sum of all individual poverty gaps expressed as a fraction of the poverty line divided by the total number of households. If α is 2, P_2 is the squared poverty gap ratio representing an index of the severity of poverty, which is the total number of households. P_2 gives greater weight to the income shortfall of the poorest of the poor. Poverty line employed in this analysis is the official poverty line based on Statistics SA's Income and Expenditure Survey (IES) data and CPI estimates. The poverty line per person is R388 per month in 2008 constant rand. Table 3.3 below shows the poverty situation in our study area.

The poverty situation is better in Mlungisi than Ndakana, with a head count ratio of 50% compared to 63% in Ndakana. The poverty situation in Ndakana is higher than the estimated figure for rural areas of South Africa: 59.3%. Both the poverty gap ratio and severity of poverty are high in Ndakana than Mlungisi. Mlungisi is located next to the town and wage employment has emerged as the main sources of income for a greater population in the same community. Labour employment opportunities in the non-farm sector are common near town. The main asset owned by most of the households in the study area is labour. Therefore the level of poverty is seemingly controlled by participation in wage employment which is mainly accessible to households located next to the town. Overall, it seems that poverty is more prominent in communities with agriculture as their main activity. Thus the notion that agriculture is the sole driver of poverty in rural South Africa should be changed. The desperate poverty situation in South Africa can be reduced by promoting both farm and non-farm activities.

Table 3.3. Poverty situation in Mlungisi and Ndakana.

Community	Mlungisi	Ndakana
Incidence of poverty	50%	63%
Poverty gap ratio	23%	31%
Severity of poverty	13%	17%

3.6.4 Income by activity

In order to identify the determinants of income, we regress each of the four sources of income (own business, own agriculture, wage income and remittances) on the full range of dependent variables. The results of the model are shown in Table 3.4. The findings point to the key role played by physical capital, social capital and education in the determination of almost all the sources of income in the study area. Expectedly, physical capital proved not to be important in the level of remittances. In all cases, except in the case of the effect of physical capital on the remittances equation, all the coefficients were positive. With the exception of wage income, an increase in social capital has statistically significant positive impact on all the other types of income. These results are not exceptional. Hebinck and Lent (2007) and Katungi *et al.* (2007) recognise the importance of social capital in rural South

Table 3.4. OLS estimates of determinants of income by activity type.

Variables	Own business		Own agriculture		Wage labour		Remittances	
	Coefficient	t-ratio	Coefficient	t-ratio	Coefficient	t-ratio	Coefficient	t-ratio
Age	-10.664	-0.230	-25.135	-1.793*	-264.660	-1.728*	1,450.793	1.965*
Physical capital	1,298.481	1.693*	467.412	-1.659*	17,579.438	0.477	-102.842	-0.249
Social capital	2,090.591	1.883*	326.983	1.737*	3,717.960	0.000	825.318	1.889*
Education	522.634	1.756*	-261.028	-1.322	1,434.535	1.678*	1,366.43	1.902*
Gender	662.589	0.555	-308.461	-0.944	-3,244.915	-0.815	-34.639	1.522
Labour	-193.724	0.579	128.117	1.302	-489.977	0.443	2,371.173	-0.122
Dependency ratio	515.096	0.179	403.582	-0.495	-22,893.95	-2.20**	144.225	1.002
Land size	-202.780	0.460	156.009	1.299	-100.646	0.062	950.083	-0.292
Geographic location	972.958	0.854	174.234	1.675	-4,355.692	1.160	7,719.835	2.733
Constant	1,936.203		1,975.611		24,555.700		-7,719.835	
R-square	0.10		0.23		0.36		0.37	

* Indicates significance at the 10% level; Number of observations = 79.

Africa. Similar conclusions in these results indicate the need to create and promote social capital both as a way of improving household income and promoting rural livelihoods.

Location also plays an important role in determining household income by activity group. Sanchez (2005) found that households in communities located away from urban areas have own agriculture and remittances as their highest sources of income relative to income from wage employment and own business. De-Janvry and Sadoulet (2001) found that geographical location affects household income and specific sources of income. The results of this study confirm the assertion of Sanchez (2005). Positive coefficients were found for the community variable in agriculture, remittances, own business and wage income equation for Ndakana village which is the more rural settlement. The community variable only reflects a negative coefficient for income in the wage employment equation. This probably reflects some level of regional comparative advantage in activities across communities based on their location. Households located next to towns are likely to receive relatively higher income from employment activities whilst those in rural settlements benefit most from own agriculture, own business and remittances.

3.6.5 Determinants of poverty

The Probit model was used examine the determinants of poverty among the survey household the estimation of the total household income equation was carried out. The model examined a total of nine variables namely, age, education, gender, labour availability, possession of physical assets, social capital, dependency ratio, land size, and geographical location. Of these, the results suggest that education, gender, labour availability, possession of physical assets, and social capital affect household poverty status significantly. Table 3.5 provides the poverty determination equation from the reduced form Probit model for poverty status in the study area.

Education significantly affects household's poverty status. Better-paid jobs and most income generating activities require formal schooling, usually to completion of primary school or beyond (Davis, 2003). Positive relationship between education and household's income level and consequently poverty status is often reported. This is results tally well with findings by Bryceson (2002) in both Malawi and Ethiopia, Sanchez (2005) in Bolivia, and Matsumoto *et al.* (2006) in Kenya. They showed that the level of education indicated in years of formal education, enhances household members' capability to exploit better income generating opportunities thus transition out of poverty.

A gender perspective adds significant insight into rural poverty and livelihood issues (Davis, 2005). Gender has emerged as an important factor influencing household poverty status in

Table 3.5. OLS estimates of determinants of poverty in Ndakana and Mlungisi villages.

Variables	Coefficient	t-ratio
Age	24.114	1.806
Education	123.050	0.200*
Gender	-67.187	-1.683*
Labour	557.37	2.072*
Physical assets	1,047.185	2.826*
Social capital	3,549.103	1.725*
Dependency ratio	1,243.220	0.394
Land size	1,123.087	0.394
Geographical location	435.22	2.254
R-square	0.38	
F-value	4.24	

* Indicates significance at the 10% level; Number of observations = 79.

the study area with the female heads of households most likely to be poor. Regarding gender and marital status, Matsumoto *et al.* (2006) in their studies in Ethiopia, Kenya and Uganda, found that larger proportions of poor are female households. Compared to women, male individuals are more willing to explore opportunities, and are more mobile than female heads of households, thus exhibiting greater propensity for migrating for wage employment.

Physical assets are the key requirements for most income generating activities in most rural areas. Economic theory of investment supports the view that household endowment in assets places them in a relatively better position to exploit income from various sources (Parkin, 2008). Consistent with Holden *et al.* (2006), we found that physical assets significantly affect household poverty status. Wealth-differentiated barriers to entry into non-farm activities, as noted by Holden *et al.* (2004), have been noted in Burkina Faso, Côte d'Ivoire, Ethiopia, Kenya, Rwanda, South Africa, and Tanzania.

The results with respect to social capital and physical capital are consistent with findings elsewhere (Schwarze, 2004), and reflect the crucial role of a strong asset base in wealth creation. The results with respect to labour contrast sharply with results obtained elsewhere by Nargis and Hossain (2006) in Bangladesh, and Matsumoto *et al.* (2006) in Ethiopia, Kenya and Uganda. This is probably due to the minimal use of family labour in income generating activities in rural South Africa (Hebinck and Lent, 2007). Although household size might be large in some instances, few household members contribute to household's labour supply for economic purposes, suggesting an inherent underutilization of labour in the system. In terms of model adequacy, despite a low R^2, the model as a whole seems well fitted on the basis of the significant F-statistic.

3.7 Conclusion and policy implications

The foregoing results have demonstrated that the gross local income in the survey communities was derived from both farm and non-farm activities. Besides agriculture, wage employment, own business, remittances, and pensions/grants were revealed as important sources of rural income. The clear indication that agriculture is not the sole income source is quite interesting and substantiates recent doubts about the sector's declining role in rural livelihoods. It also emerged that the possession of education, social capital, physical capital, and the availability of labour positively influences income. On the other hand, age and gender of the head of household were shown to negatively influence income.

The paper therefore argues for a change of direction in the sectoral focus of efforts to improve rural livelihoods and reduce poverty. Rural income growth and reduction of poverty requires fundamental reforms that stress the importance of human, physical capital and social capital for building a strong base for rural income growth and poverty reduction in South Africa. It is clear, therefore that both farm and non-farm sectors should be equally emphasised, with attention paid to programmes that enhances rural households' control of productive assets (land, physical assets, and all the factors of production), improve the quality of human capital (provision of education to secondary level, knowledge on technical aspects) and, promote a highly productive rural non-farm sector.

Policies aimed at improving rural households successful participation in activities of their choice must be supported by provision of adequate capacity and an adequate institutional environment for the rural households. They must seek to promote the mobilization not only of capital, human and institutional resources, but also capacities, relationships and knowledge needed to initiate, develop and conduct new types of projects in the secondary and tertiary sectors such as agro-processing, provision of services, recreation, and environmental services.

References

Adams, R.H. Jr. (1999). Nonfarm income, inequality and land in rural Egypt. Working paper. World Bank Washington, DC, USA. Available at: http://www.worldbank.org/html/dec/Publications/Workpapers/wps2000series/wps2178/wps2178.pdf.

Barrett, C.B., T. Reardon,and P. Webb (2001). Nonfarm income diversification and household livelihood strategies in rural Africa: Concepts, dynamics, and policy implications. Cornell University, Ithaca, USA.

Beinart, W. and L. Wotshela (1995). Border-Ciskei District study on land reform; land holding, agriculture, land availability and land reform negotiations in Isidenge (Stutterheim) District-Eastern Cape Province. University of Fort Hare, Alice, RSA.

Bryceson, D.F. (2002). Multiplex livelihoods in rural Africa: recasting the terms and conditions of gainful employment. Journal of Modern African Studies 40, 1-28.

Byerlee, D., C. De Haan, S.K. Kane, E. Pehu, C. Ragasa and A. Winters-Nelson (2005). agricultural growth for the poor: an agenda for development. The World Bank, Washington, DC, USA.

Davis, R. Jr. (2003). The rural non-farm economy, livelihoods and their diversification: Issues and options. Natural Resources Institute. Report No: 2753.

Davis, R. Jr. and D. Pearce, (2001). The non-agricultural rural sector in Central and Eastern Europe. Natural Resources Institute Report No. 2630. Available at: http://www.nri.org/rnfe/pub/papers/2630.pdf.

De Janry, A. and E. Sadoulet (2001). Income strategies among rural households in Mexico: the role of off-farm activities. World Development 29, 467-480.

Fan, S., L. Zhang and X. Zhang (2002). Growth, inequality, and poverty in rural China. Role of public investments. International Food policy Research Institute. Washington, DC, USA.

Fraser, G., N. Monde, and W. Van Averbeke (2003). Food security on South Africa: a case study of rural livelihoods in the Eastern Cape Province. In: L. Nieuwoudt and J. Groenewald (eds.), The challenge of change: agriculture, land and the South African economy. University of Natal Press: Pietermaritzburg, RSA, pp. 171-183.

Foster, J.E., J. Greer and E. Thorbecke (1984). A class of decomposable poverty measures. Econometrica 52, 761-766.

Hebinck, P. and P.C. Lent (2007). Livelihoods and landscapes: The people of Guquka and Koloni and their resources. Brill: Leiden, the Netherlands.

Holden, S., B. Shiferawi, and J. Pender (2004). Non-farm income, household welfare, and sustainable land management in a less-favoured area in the Ethiopian highlands. Discussion paper. Department of Economics and Resource Management, Agricultural University of Norway.

Holden, S.T., C. Barrett and F. Hagos (2006). Food-for-work for poverty reduction and promotion of sustainable land use: can it work? Environment and Development Economics 11, 15-38.

Katungi, E., C. Machethe, and M. Smale (2007). Determinants of social capital formation in rural Uganda: Implications for group-based agricultural extension approaches. African Journal of Agricultural and Resource Economics 1, 167-190.

Matsumoto, T., Y. Kijima, and T. Yamano (2006). The role of local nonfarm activities and migration in reducing poverty: evidence from Ethiopia, Kenya and Uganda. Agricultural Economics 35, 449-458.

Mehta, G.S. (2002). Non-farm economy and rural development. Giri Institute of Development Studies, Lucknow. Available at: http://developmentfirst.org/india/planningcommission/special_study_reports/nonfarm_eco_ruraldev.pdf.

Nargis, N and M. Hossain (2006). Income dynamics and pathways out of poverty in Bangladesh, 1988-2004. Agricultural Economics 35, 425-435.

Parkin, M. (2008). Economics. Addison-Wesley, London, UK.

Perret, S. (2002). Livelihood strategies in rural Transkei (Eastern Cape Province): how does wool production fit in? University of Pretoria, Pretoria.

Sanchez, V. (2005). The determinants of rural non-farm employment and incomes in Bolivia. Masters of Agricultural Economics Thesis, Michigan State University, MI, USA.

Schwarze, S. (2004). Determinants of income generation activities of rural households in Central Sulawesi, Indonesia. Georg-August University Göttingen, Institute of Rural Development, Germany.

Statistics of South Africa (StatsSA) (2001). Household Survey Statistical Release P0317. Available at: http://www.statssa.gov.za/RELEASE/Households/98/P0317.htm.

Yúnez-Naude, A. and E. Taylor (2001). The determinants of nonfarm activities and incomes of rural households in Mexico, with emphasis on education. World Development 29, 561-572.

World Bank, (2006). World Development report 2007: Development and the next generation. Word Bank, Washington, DC, USA.

4. Rural income dynamics in post-apartheid South Africa: implications for reduction of poverty and income inequality

Ajuruchukwu Obi and Simbarashe Ndhleve

Abstract

Development interventions aimed at reducing poverty and addressing inequality are popular in post-apartheid South Africa. This chapter analyzes rural income distribution using data from a household survey conducted in two rural communities of South Africa. Data were collected on various socio-economic aspects and sources of income using a semi-structured questionnaire. The estimation of inequality is carried out using both Lorenz curves and the Gini decomposition technique. After disintegrating total household income according to source, empirical results indicate highly skewed income distribution that mirrors observed patterns elsewhere in South Africa and the rest of the region. Own agriculture income and pensions/grants exhibit much higher inequality reducing effect followed by wage income. Own business and remittances appeared to increase inequality in both communities. There are important lessons for policy makers and practitioners concerned about poverty reduction.

4.1 Introduction

Progress in eliminating poverty and inequality remain elusive in South Africa. Both urban and rural areas of South Africa have continued to attract considerable policy and academic interest in the post-apartheid era as programmes established to redress past wrongs fail to produce the desired results. Recent years have seen a number of researches that have revealed sometimes shockingly high levels of poverty, high levels of inequality and in most cases poor standards of living in various areas of South Africa. World Bank (2006) reports that, in spite of the strong inequality addressing policies, South Africa still remains one of the highest in the world in terms of income inequality. Quite appalling figures on inequality are always reported for South Africa. The level of inequality is confirmed by the Gini coefficient. A Gini coefficient of one indicates perfect income inequality, while a Gini coefficient of 0 indicates perfect equality (Provide, 2006). Provide (2006) estimated high income inequality figures ranging from 0.65 to 0.68 for all the provinces of South Africa. Furthermore, national figures show that income inequality increased by 5% from 0.64 to 0.67 in the period 1997-2007, post-apartheid era.

High levels of income inequality produces unfavourable environment for economic growth and development (Provide, 2006). Households' income-generating activities in several rural areas reveal an amalgam of activities, including farm and non-farm activities. Analyses of rural livelihoods in Africa, Asia, and Latin America, show that rural households derive a significant proportion of their livelihoods from a mix of activities (Adams, 1999; Barrett *et al.*, 2001; Escobal, 2001; Fraser *et al.*, 2003). Attempts have been made by researchers to classify rural populations' sources of income. In South Africa, Perret *et al.* (2005) and Fraser *et al.* (2003) came up with five sources of income, namely own business, own agriculture, wage income, remittances and pensions/grants in the former homeland areas of Limpopo, North West Provinces and Eastern Cape. Income generating activities in these Provinces manifest themselves in the form of either farm or non-farm activities. Maldistributions of income from different sources among households affect overall income distribution at community level. This according to Leibbrandt *et al.* (2000) explains why household income varies internationally, nationally and even among neighbouring households. Despite efforts to address unequal distribution of income in both rural and urban South Africa, there is no agreement on which income-generating activities should be promoted and/or how development agencies should allocate available limited resources to reduce inequality. The government of South Africa has established numerous programmes in the post-apartheid era to address inequality e.g. Black Economic Empowerment (BEE), but the picture of rural impoverishment and unequal distribution of income persists. With the information available, it is possible to conclude that only a tiny minority have seen phenomenal increases in personal wealth from these programmes.

In view of the bleak economic picture of growing poverty and inequality in South Africa, this paper presents an inventory of rural income levels and levels inequality in two communities of South Africa. It intends to examine the nature and dimensions of income inequality through disintegrating it by both source and community of origin. Specifically, this objective will be achieved by:

a. identifying and cataloguing the various sources of income in the study area;
b. determine the level of income inequality in the study area;
c. determine the contribution of each income source to overall income inequality.

The paper is set out as follows, the next section discusses the study site followed by data and model structure (Gini decomposition technique); thereafter we look at the distribution of income in the study area and then we apply the above-mentioned methodology to calculate income distribution by source. As conclusion, the paper provide comments about underlying trends ascertainable from these estimates and what these imply for the income distribution of both communities.

4.2 Study area

The study area, Stutterheim, is situated in the Eastern Cape province of South Africa, located along the N6 road to East London (see Figure 3.1). Approximately sixty kilometres from King Williams Town, the area was formerly part of the Ciskei Bantustan under the apartheid system. Stutterheim is a district centre surrounded by nine rural communities. Clearly distinguishable rural and peri-urban locations within the district justify its inclusion for purposes of assessing income distribution and inequality, among other issues. Two communities were enumerated to represent a rural setting (Ndakana) and a peri-urban setting (Mlungisi).

4.2.1 Physical settings

The key factors influencing agriculture in the study area are the soils and climate. Stutterheim experiences summer rainfall which is highly variable from year to year. The district is divided into two fairly distinct zones, a drier Eastern zone and a wetter Western zone. Mean annual rainfall is generally less than 600 mm per year and drops to less than 400 mm near the Kei River. The climate can be considered as semi-arid, being dry and warm throughout most of the year. Stutterheim experiences slightly extreme conditions with temperatures of 5-35 °C. Extensive areas are characterized by poor, shallow soils which are not conducive to intensive, arable farming. The dominant vegetation communities are subtropical thicket, a little afro-montane forest, and grasslands.

Stutterheim as a whole is largely a stock-farming district with a large number of livestock (cattle, goats and sheep). A local extension officer estimates that some 4,700 ha is under cultivation, of which 1000 ha is under irrigation.

4.2.2 Socio-economic activities

The town and its surrounding rural communities boast a wide range of economic activities and services. Agriculture is mostly small scale crop farming and open grazed livestock. The key economic activities include home-based business like selling mobile phone air time, tuck shops, taverns, informal drinking places, etc., and farming (livestock, vegetables, crops and citrus), forestry and associated industries such as saw milling. Forestry and saw milling are the dominant economic activities employing the highest proportion of the local population. The bulk of the population is employed in the forestry, and most are dependent on social old age grants (pension and child grants). Aside from local employment in forestry and sawmills, Stutterheim town is the major source of employment. It hosts a number of service providing institutions. Migrant workers also constitute a substantial proportion of the population though they seem to be decreasing as reported by Beinart and Wotshela (1995). Besides local employment in Stutterheim, some of the people are employed in the surrounding cities such as Port Elizabeth, East London and King Williams Town. These migrant workers are the key providers of remittances to Stutterrheim.

Farming is largely for subsistence rather than commercial sale, although some black commercial farmers are present. The district contains the best grazing lands suitable for livestock production in the province. Households produce their food and generate income from farming activities. Both livestock and crop farming are practised in the area but mostly at a lower scale. Beinart and Wotshela (1995) noted that the rate of cultivation in Stutterheim has dropped during the past few years probably because very few landowners or tenants are able to command sufficient land, capital and labour to produce significant amounts of output.

The study employed household and community data collected from Stutterheim in March 2008. Primary data were generated by means of structured questionnaires administered on a total of 79 households, as well as interview schedules and checklists for purposes of collecting community-level data. Secondary data were gathered by means of extensive document analysis and discussions with knowledgeable persons, including extension workers serving the two study communities.

4.2.3 Data and model structure

The study employed a multi-stage sampling technique with stratified and random procedures. Purposive sampling was used to choose one peri-urban community and one rural community. Two communities namely Mlungisi and Ndakana were randomly selected for the study. Finally, households were selected randomly for enumeration. Data was collected on both socio-economic aspects and different sources of income using a semi-structured questionnaire.

Household income in this study refers to the actual income earned from different sources in 2007, including monetary income or income in kind. The 12 months were chosen because the period is not too long for household members to recall their income with a good degree of precision. Income components for this study are broadly categorised as farm and non-farm income. Classification of sources of income in this study follows almost similar classification as that adopted by Fraser *et al.* (2003) in Guquka and Koloni villages. These villages are in the same district with the present study area. The following classification of activities was used for income.

1. Own agriculture income (income earned from agriculture-crop, poultry and livestock).
2. Own business income (income from activities such as industry, transportation, construction, and services).
3. Remittances (money or goods received from migrant household member by the household).
4. Wage income (income earned from formal or informal wage employment, including salaries, allowances, bonuses, and other kinds of remuneration).
5. Inheritance and grants – incomes without quid pro quo, such as pensions, transfers, grants/subsidies, rents, and financial income.

The analyses employed here to measure the effect of differentiated sources of income on inequality was suggested for similar studies by Adams (1999), Piesse *et al.* (1999) and Leibbrandt *et al.* (2000). Decomposability of income inequality allows inequality to be partitioned either over subpopulations or sources (Adams, 1999). In this technique total inequality is divided into a weighted sum of inequality by various income sources (for example, non-farm and agricultural income) and it encompasses source decomposition of the Gini coefficient. The Gini coefficient is frequently used for analysis of distribution of income because it can be decomposed by income source, illustrating the effects of alternative income sources on total income equality. In their study, Van de Berg and Kumbi (2006) used a similar approach to obtain estimates of the contribution of selected sources of income on inequality in Oromia, Ethiopia. Their analysis follows the common expression for the Gini coefficient (G) for the distribution of total income within the group. See the Appendix at section 2 for the steps involved in carrying out estimations for the baseline-related study presented in this volume.

4.3 Results and discussions

This section presents the results of the analysis with respect to rural income distribution and the application of Gini decomposition technique in the limited context of one rural and one peri-urban settlement. The policy implications of these results are then examined.

4.3.1 Socio-economic characteristics of survey households

The characteristics of the sample households were analyzed by means of descriptive statistics (previously shown in Table 3.1). Characteristics included gender, age, marital status, main source of income and education of household head interviewed, in addition to the size of the household's field. From the evidence, the average household size was more or less the same in both study communities although the Mlungisi households seemed marginally larger. On average, household heads in the rural community were older than those in the peri-urban location, reflecting the tendency for individuals to move back to their villages as they grew older. There were correspondingly more widowed persons in the rural area, where household heads were relatively older, less educated, and cultivated relatively larger fields than those in the peri-urban location of Mlungisi (see Table 3.1).

4.3.2 Sources of rural income

Households in the two communities derive income from several activities, both farm and non-farm activities (see Table 3.2). Most households pursued at least one of the following activities: crop farming, livestock rearing, own business, wage employment, and augmented income by means of pensions, grants and remittances. The average total income for the 12 months recorded for each community was computed. According to the data, households in Ndakana earned a total of R21,996.86 per annum, on average, while households in Mlungisi received R14,769.57 per annum, on average.

More than 30% of the respondents from Ndakana had at least a member engaged in wage employment (see Table 3.2). In the case of Mlungisi, participation in wage employment was approximately 3% higher than Ndakana among the survey households. A total of 39 respondents, representing more than 50% of all respondents, engaged in agricultural activities as a means of livelihood in both communities. In Ndakana, the more rural community where field sizes are relatively larger, a larger proportion of the respondents, about 58%, engaged in agricultural activities, and earned an average income of R1,942, while in the peri-urban area of Mlungisi, about 44% of the respondents from that sub-sample engaged in agricultural activities, earning slightly lower average agricultural income of R1,396. Interestingly, for both communities, despite a large proportion of households engaging in agriculture, agricultural incomes turned out to be extremely low relative to other income sources.

The results show that households derived income from other sources such as operation of own businesses, pensions and grants, and remittances from family members living outside the community. The dominant own businesses observed in the communities were owning of tuck-shops, vending of cellular phone re-charge vouchers (or 'airtime'), selling of vegetables, running of taverns and shebeens (an unlicensed drinking place for alcoholic beverages) and food processing. The data from the two communities indicate that about 18% of the respondents in Ndakana operate their own business enterprises and 17% in Mlungisi, and in each case, this activity contributed about 17% of household income (see Table 3.2). As Table 3.2 revealed, pensions and grants emerged as important sources of income in both communities, accounting for the bulk of incomes received by 64% in Ndakana and 76% in Mlungisi. Remittances were clearly more important in the rural community than the peri-urban community.

4.3.3 Income distribution

This study also examined the distribution of rural households' sources of income. Income distribution implies the level of dispersion of the gross community income from a given source, assessing whether it is equally distributed or unequally distributed (Schwarze, 2004). Provide (2006) found that unequal pattern in the distribution of income received by rural households' leads to high levels of income inequality and this produces unfavourable environment for economic growth and development.

Various income distribution or inequality measures exist in the literature. One approach to measuring inequality employs the Lorenz curves. It plots cumulative total income from each source against cumulative share of households, where households are ranked according to share of total income. A 45-degree line is used as a reference, and this is known as the line of perfect equality (Schwarze, 2004; Provide, 2006). The further a Lorenz curve is below the 45-degree line, the greater the inequality. Provide (2006) reported that the Eastern Cape Province's Lorenz curve runs along the same 'path' as the South African Lorenz curve though marginally above in the initial stages, it crosses at the 90th percentile. In this study, modified Lorenz curves are used to illustrate income distribution by source in the two communities. Figure 4.1 shows the distribution of five sources of income (own agriculture income, own business income, wage labour, remittances and pensions and grants in both communities combined. Modified Lorenz curves for all the noted sources of income are presented in Figure 4.1.

The inequality that exists in the two communities, and particularly in the distribution of income from each source, is reflected in the Lorenz curves presented in Figure 4.1. Combining incomes from both communities, virtually all the income sources are unequally distributed. Relative to other sources of income, income from pensions and grants is more equally distributed, followed by remittances. Productive sources of income; own business, remuneration and own agriculture incomes are more unequally distributed. Diversity in households' participation and income is responsible for the level of inequality displayed in

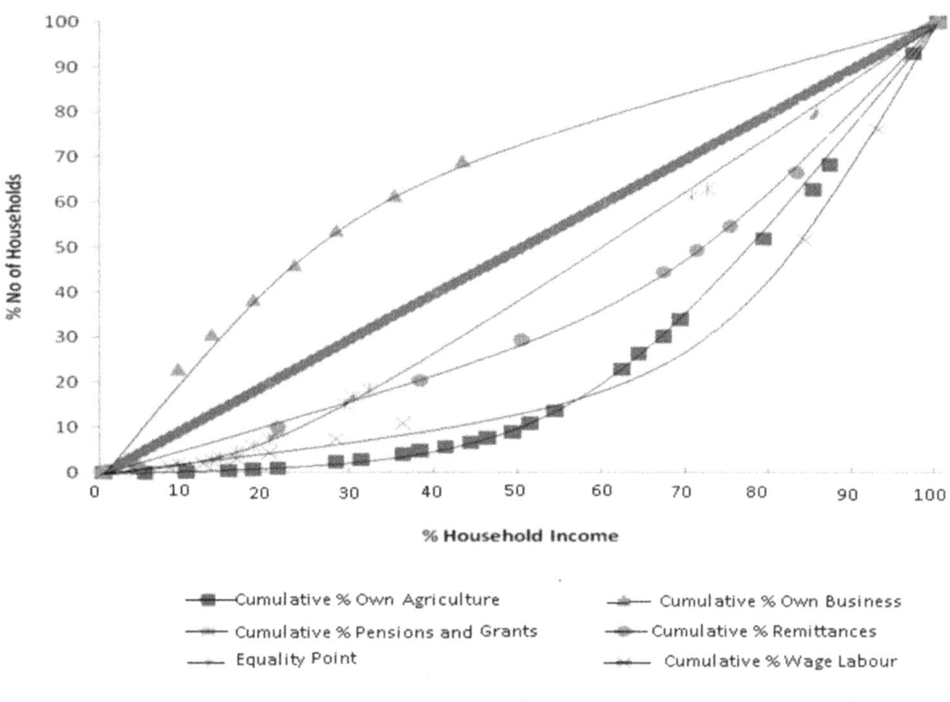

Figure 4.1. Income distribution by source of income (modified Lorenz curve) (Field data, 2008).

Figure 4.1. Different sources of income exist among households from both communities and these mutually exist with differentiated levels of income. The combined effect of these two levels of differentiation is responsible for high inequality in the distribution of income from selected sources. Besides the distribution of income shown in Figure 4.1, Table 4.1 presents the calculated Gini coefficients for all the sources of income both within communities and across communities.

The calcualted Gini coefficient for total household income of the two communities combined is 0.40. This is a bit higher than that of 0.34 for Mlungisi. Higher figure have been reported for South Africa, Provide (2006) reported that South Africa's Gini coefficient was estimated at about 0.69 in 2004. The Gini coefficients for all the five sources of income were calculated for both communities. All the communities show evidence for unequality as shown by higher Gini coefficients for all the sources of income. In Ndakana, wage income is unequally distributed (0.62) followed by own agriculture (0.43), own business (0.36), and lastly remittances (0.28) and pensions/grants (0.28) being the two most equally distributed sources of income. Pensions and grants are paid by the government at a fixed rate and all recipients receive the same amount once their eligibility is confirmed. It is therefore not surprising that these are the most equally distributed incomes. Lower figures were found for the Gini coefficients for wage employment and remittances in Mlungisi. The probable reason being several households in Mlungisi were employed at the time of the survey. This has some equalising effect on the distribution of this source of income. Inequality for own agriculture in Mlungisi (0.55) was a bit higher than that for Ndakana (0.43). This can be explained by the number of household participation in own agriculture in both communities. A larger proportion (more than 50%) of households in Ndakana received income from agriculture and the higher figure for Mlungisi was probably due to lower participation in agriculture which results in most households receiving no income from own agriculture.

Table 4.1. Gini coefficients by source of income for Ndakana and Mlungisi (Field data, 2008).

Sources of income	Gini coefficients		
	Ndakana	Mlungisi	Both communities
Own business	0.36	0.32	0.36
Own agriculture	0.43	0.55	0.50
Wage employment	0.62	0.31	0.50
Remittances	0.28	0.29	0.29
Pensions and grants	0.28	0.18	0.24
Total household income	0.40	0.34	0.40

4.3.4 Decomposition of income by source

Using decomposition techniques, the Gini coefficient was used to assess the contribution of the different sources of income to total household incomes. In this way, it is possible to determine how a change in a particular source of income affects the overall income distribution within a given community. Table 4.2 presents a decomposition of total income calculated on the basis of Gini decomposition technique. For each community the rows present statistics associated, respectively with own agriculture, own business, wage labour, remittances and pensions/grants. Those statistics are include the corresponding share in total in total income, the Gini coefficient, the Gini correlation with total income, the pseudo-Gini and relative contribution of each income source to the Gini of total income. The last column of the table presents the effects on total income inequality of marginal changes in each income source.

Source elasticity of overall inequality in Table 4.2 shows the effects of a 1 percent increase in a particular income source on overall income inequality. Increases in both pensions/grants and own agriculture incomes are likely going to reduce income inequality in Ndakana and Mlungisi. A 1% increase in income received from pensions and grants reduces the overall income inequality in Ndakana by 30% and the same increase in income in Mlungisi

Table 4.2. Gini decomposition by source of income for Ndakana and Mlungisi (Field data, 2008).

Income source	Income share (S_k)	Gini (G_k)	Correlation with the distribution of total income (R_k)	Pseudo-Gini (G_kR_k)	Share of total income inequality (P_k)	Source elasticity of total inequality (g_k)
Ndakana						
Own agriculture	0.19	0.36	0.06	0.02	0.0950	-0.1805
Own business	0.06	0.43	0.07	0.03	0.0045	-0.0555
Wage labour	0.27	0.62	0.31	0.19	0.1283	-0.1418
Remittances	0.15	0.28	0.92	0.26	0.0975	-0.0525
Pensions/grants	0.33	0.28	0.14	0.04	0.0330	-0.2970
Mlungisi						
Own agriculture	0.20	0.32	-0.27	0.09	0.0529	-0.3900
Own business	0.05	0.55	0.43	0.24	0.0353	-0.0147
Wage labour	0.37	0.31	0.67	0.21	0.2285	-0.1415
Remittances	0.10	0.29	0.56	0.16	0.0470	-0.0529
Pensions/grants	0.28	0.18	0.56	0.10	0.0823	-0.1976

G = 0.41 (Ndakana) and 0.34 (Mlungisi).

causes a reduction in inequality by 20%. We find that agriculture has an equalising effect in both communities of 18% in Ndakana and 39% in Mlungisi. Own business has the least inequality reducing effect in both communities. This is mainly because it is in the hands of a few households capable of investing. Own agriculture and pensions/grants are the major inequality reducing sources of income in both communities followed by wage labour. These two sources of income are the major determinants of inequality in the two communities followed by the latter; any efforts to change income from these sources highly affect income distribution. Analyses of the share to total income and contribution to total inequality shows a positive relationship between income share from a particular source and its contribution to total inequality. The higher the share of income received from a particular source in a community the higher it contributes to total inequality. This is particularly true for income from pensions/grants for Ndakana and income from own agriculture for Mlungisi. High corresponding figures were found for both share of income and elasticity for these two sources of income (Table 4.2). This probably implies that efforts should be made to encourage household participation in own agriculture as a way of improving agriculture's share in total household income thus addressing inequality.

Besides contributing a high share to total income in both communities, wage income has the highest figure for share to total income inequality in both communities as well (Ndakana 13%, Mlungisi 23%). Leibbrandt *et al.* (2000) reported similar results on wage labour for South Africa. Income received from remittances also brought positive change towards addressing income inequality in both communities. A 1% increase in income from remittances accounts for 5% increase in total income inequality and this was the lowest for all the sources of income of both communities. The results above implies that efforts to improve income from those sources with both high share in total income and high elasticity (wage income and own agriculture) will probably positively affect both inequality and welfare of rural households in the study area.

4.4 Conclusion and policy implications

Increasing income inequality and poverty continue to be the most challenging economic problem facing most developing countries. In this paper, we used household survey data collected from two communities, one rural and one peri-urban, to analyze the sources of income inequality in rural South Africa. The results show that households have diverse sources of income of both farm and non-farm origin. The analysis shows that the overall Gini coefficient of income inequality is 0.41 for Ndakana and 0.34 for Mlungisi. All the sources of income in the study area are unequally distributed. Own agriculture, pensions/grants followed by wage labour were found to have an income equalizing effect relative to all the other sources. This implies that public administration that fosters access to income from these sources of income by the low income sect of the society has the potential to address inequality. Policies aimed at improving agricultural income either through arranging markets for agricultural produce and/or extension services can address inequality. Equality can be promoted by removing entry barriers to activities with high labour demand and regulations regarding working hours and minimal wage rate should be strictly enforced for the welfare of those involved in wage labour. Educational programmes, access to credit and provision of infrastructure can help.

References

Adams, R.H. Jr. (1999). Nonfarm income, inequality and land in rural Egypt. Working paper. World Bank Washington, DC, USA. Available at: http://www.worldbank.org/html/dec/Publications/Workpapers/wps2000series/wps2178/wps2178.pdf.

Barrett, C.B., T. Reardon, and P. Webb (2001). nonfarm income diversification and household livelihood strategies in rural Africa: concepts, dynamics, and policy implications. Cornell University, Ithaca, USA.

Beinart, W. and L. Wotshela (1995). Border-Ciskei District study on land reform; land holding, agriculture, land availability and land reform negotiations in Isidenge (Stutterheim) District-Eastern Cape Province. University of Fort Hare, Alice, RSA.

Escobal, J. (2001). The determinants of nonfarm income diversification in rural Peru. World Development 29, 497-508.

Fraser, G., N. Monde, and W. Van Averbeke (2003). Food security in South Africa: a case study of rural livelihoods in the Eastern Cape Province. In: L. Nieuwoudt and J. Groenewald (eds.), The challenge of change: agriculture, land and the South African economy. University of Natal Press: Pietermaritzburg, RSA, pp. 171-183.

Leibbrandt, L., C. Woodlard and I. Woodlard (2000). The contribution on income components to income inequality in the rural former homelands of South Africa: a decomposable Gini Analysis. Journal of African Economies 9, 79-99.

Perret, S., W. Anseeuw, and N. Mathebula (2005). Poverty and livelihoods in rural South Africa. Investigating diversity and dynamics of livelihoods. Case studies in Limpopo. Unpublished Project report num.05/01, Kellogg's Foundation / University of Pretoria, 65 p.

Piesse, J., J. Simister and C. Thirtle (1999). Modernisation, multiple income sources and equity: a gini decomposition for the communal lands in Zimbabwe. University of Reading, UK.

Provide (2006). A profile of the Eastern Cape province: demographics, poverty, inequality and unemployment. Provide Background Paper 2005: 1(2). Elsenburg, RSA. Available at: www.elsenburg.com/provide.

Schwarze, S. (2004). Determinants of income generation activities of rural households in Central Sulawesi, Indonesia. Institute of Rural Development, Georg-August University Göttingen, Germany.

Van De Berg, M. and G.E. Kumbi (2006). Poverty and the rural non farm income in Oromia, Ethiopia. Agricultural Economics 35, 469-475.

World Bank, 2006. World Development report 2007: Development and the next generation. Word Bank, Washington, DC, USA.

Part III
Socio-economic and commodity marketing factors among small farmers in South Africa

5. Influence of institutional factors on smallholder farmers' marketing channel choices

Bridget Jari, Gavin Fraser and Ajuruchukwu Obi

Abstract

Smallholder farmers are faced with a range of institutional challenges in produce marketing. In turn, these challenges influence their selection of marketing channels. In this chapter, a multinomial logistic regression model was used to test the significance of institutional factors in marketing. The results are based on a case study for smallholder farmers in the Kat River Valley, in South Africa. It is shown that the statistically significant variables are access to market information, expertise on grades and standards, availability of contractual agreements, existence of extensive social capital, group participation and reliance on traditions.

5.1 Introduction

Markets often do not serve the interest of smallholder farmers because these farmers face difficulties in accessing markets. Factors such as poor infrastructure, lack of market transport, dearth of market information, insufficient expertise on, and use of grades and standards, inability to conclude contractual agreements and poor organisational support have led to inefficient use of markets, hence, results in commercialisation bottlenecks. Furthermore, smallholder farmers lack vertical linkages in the marketing channels, which result in their exclusion from the use of formal markets (Fenwick and Lyne, 1999; Delgado, 1999; Makhura, 2001; Wynne and Lyne, 2003). Smallholder farmers have weak financial and social capital and limited access to legal recourse, implying that it is difficult to change these negative market factors individually (FAO, 2004; Fenwick and Lyne, 1999). As a result, they are trapped and continue to operate within the given market constraints and they do not receive rewarding incomes from their agricultural activities.

This chapter tests the significance of institutional factors in choosing marketing channels. The chapter starts by explaining the multinomial logistic regression model which was used for analysis, which is followed by the definition of the variables. The results are then presented and conclusions drawn.

5.2 The project area

The Kat River Valley is situated northeast of Grahamstown, in the foothills of the Winterberg and the Amatole Mountains (Magni, 1999). It forms part of the Nkonkobe Local Municipality, which falls, under Amatole District Municipality. Before the change of government in 1994, the upper part of the Kat River Valley was part of the Ciskei homeland – one of several black racial reserves created during the apartheid era. The Kat River Valley is approximately 80 kilometres in length and 1,700 km^2 in area (Motteux, 2001). Its catchment includes the areas of Seymour, Balfour, Fort Beaufort and other smaller rural communities.

According to McMaster (2002), the Kat River Valley can be divided into three different sub regions (Figure 5.1), namely: the Upper Kat (UK), Middle Kat (MK) and the Lower Kat (LK).

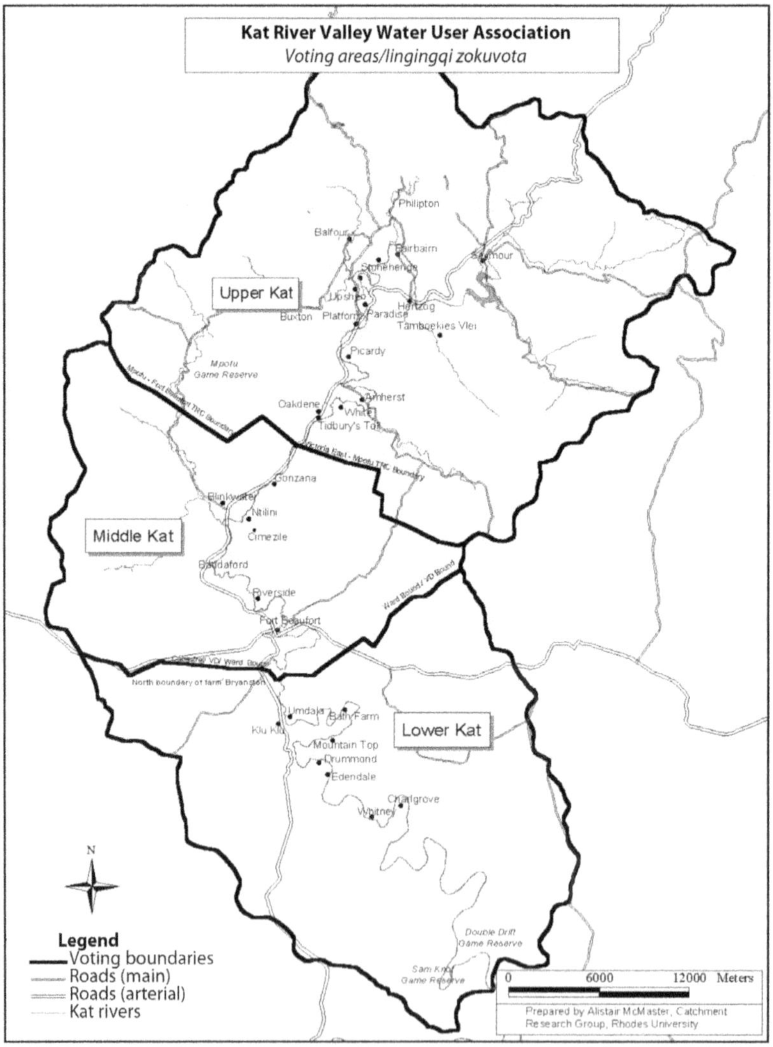

Figure 5.1. Kat River Valley map (McMaster, 2002).

5.3 Topography and climate

The Kat River is a tributary of the Great Fish River, which lies on the eastern part of South Africa. Its altitude increases from approximately 600 metres to 1,600 metres at the top of the escarpment (Shackleton and Shackleton, 2006). The Kat River Valley's topography is heterogeneous because it has a 'basin' topography with significant areas of terrace and foot slope bottom lands, surrounded by steep mountain slopes (Motteux, 2001). The topography limits development in the area because the steep slopes and escarpments restrict tillage and road development. The availability of dongas in the area makes some of the land unusable.

According to Magni (1999), Kat River Valley's climate can be described as mild. The rainfall is unevenly distributed within the area. It ranges between 400 mm and 1,200 mm, where the least rainfall is received at the confluence with the Great Fish River and the highest, in the mountainous northern region of the catchment (Magni, 1999). Although the rainfall is relatively high in the mountainous region, much of the area in the catchment can be regarded as sub-humid to semi-arid. Kat River Valley receives both summer and winter rainfall. Approximately 75% of the mean annual precipitation is received between October/November and February/March, where the highest rainfall figures are recorded in March. The temperatures range from moderately hot summers to cool moderate winters (Motteux, 2001).

5.4 Socio-economic factors

Population in the Kat River Valley is composed of different races, and the racial composition is the result of history and apartheid. The Upper and Middle Kat belong mainly to black *Xhosa* speakers and coloured Afrikaans speakers, whereas the Lower Kat belong mainly to white English speakers. Of the total population in the area, approximately 94.28% are black *Xhosa* people, 4.12% coloured and 0.76% white (Motteux, 2001). The towns of Seymour and Fort Beaufort are principally inhabited by black *Xhosa* speakers. Fort Beaufort is the major town and rural service centre for the Kat River Valley. Farmers, especially smallholder farmers, get most of their farming inputs from and sell their produce in Fort Beaufort.

The Upper Kat River Valley resembles the characteristics of typical former homeland areas of South Africa. Human population densities are high, employment levels are low and there is a high dependence level in the Upper Kat. Poor inhabitants rely on gifts, state pensions and migrant labour remittances for household survival (Nel, 1998). However, some emerging and smallholder farmers are farming for the market, and the Kat River Valley can be regarded as one of the areas where smallholder farmers are farming successfully. Proportions of males to females, amongst the emerging and smallholder farmers, differ with farming practices. For instance, in citrus farming, there are higher proportions of males whereas there are higher proportions of females in vegetable farming (Magni, 1999).

5.4.1 Agricultural potential

Kat River Valley is characterised by a variety of land uses, ranging from export-oriented citrus farming, commercially oriented rangeland stock farming to small-scale vegetable and crop production and stock farming (McMaster, 2002). Commercial farmers are mainly located in the Middle Kat and Lower Kat, whereas smallholders and emerging farmers mostly practice agriculture in the Upper catchment. According to Nel (1998), during apartheid, citrus farming was only practiced by white commercial farmers. As part of black empowerment drive, some black farmers were helped operate citrus farms by Ulimocor, after apartheid. In addition to citrus farming, Farolfi and Rowntree (2005) noted that vegetable gardening is an important occupation amongst emerging and smallholder farmers. Most of these vegetables are grown on fertile plots lying adjacent to rivers and streams. Whereas some farmers practice sprinkler irrigation, irrigation by hand is also practiced by farmers who lack irrigation infrastructure.

The soil, on which most cultivation occurs in the valley, is alluvium, which is suitable for agriculture. According to Smit (2003), the fertile valley land can be utilized only through irrigation, using water from the Kat River. Furthermore, Smit (2003) explained that even though the soil is suitable for agriculture, phosphorous and potassium deficiencies have been identified in the alluvial soil profiles of the Kat River basin. These deficiencies will only become effective threats if the pH level rises, because the soil will be at risk of losing necessary iron, manganese and boron needed for successful plant growth (Magni, 1999). Regardless of the potential threats, the potential for cultivation in the catchment is strengthened because the alluvium soil type within the Kat River is relatively uniform between the upper and lower areas of the river.

There is a corresponding change in vegetation from Eastern Thorn Bushveld dominated by *Acacia karroo* in the valley, to more succulent thicket in the south. The *Acacia* bush is the predominant vegetation type at the valley bottom owing to lower rainfalls. This vegetation type is capable of supporting livestock, explaining animal farming in the area. *Acacia* bush is also suitable for game farming, even though the area has not yet gained tourism importance from the four game reserves in the area (Motteux, 2001). Apart from different farming types, lack of sufficient rainfall remains a limiting factor to agricultural development in the catchment. In addition to rainfall problems, environmental problems in the Kat River Valley include over-fertilization, litter, water-pollution, erosion and river siltation, reduced tree and grass cover and increasing sediment output (Smit, 2003).

Emerging and smallholder farmers in the Kat River Valley have a potential to increase their production size if production difficulties are lessened. One of the main production difficulties that the emerging and smallholder farmers are facing is lack of title deeds on the land they farm. Unavailability of title deeds makes it difficult to access credit and, in turn, it becomes difficult to invest on the farming land (Mbilase, pers. comm., 2007). In addition, the emerging farmers need human development as they lack skills and knowledge, especially on improved technology use.

5.4.2 Markets accessibility

In the Kat River Valley, marketing appears to be difficult amongst smallholder farmers, even though successful farming has been pursued. Communication appears to be slow, where the majority of the smallholder farmers depend on the word of mouth for market information (Nel *et al.*, 1997). Transport problems remain a challenge to the farmers, resulting in a reduction in rural-urban linkages and an increase in rural-rural linkages, where transportation is unnecessary. However, periodic urban markets arise during payout days, where the farmers take their produce to the payout points. In some cases, the shop owners who want the produce have to drive and collect it from the farms. Not so many shop owners go to the farms for the produce because the road is poorly developed and sometimes impassable (Farolfi and Rowntree, 2005). It does not make economic sense to collect produce from the farms, taking into consideration that the farmers produce small volumes.

The majority of smallholder farmers in the Kat River Valley do not belong to any legal organisations, which makes it difficult to get marketing assistance. It is even difficult to gain market access because of unreliable supplies and inconsistent produce quantity and quality from the farmers (Farolfi and Rowntree, 2005). The poor quality of the produce does not make smallholder farmers' produce an attractive competitor in the formal markets. Other farmers, particularly the citrus emerging farmers who sell their produce abroad, make use of middlemen. The main problem facing such farmers is that they do not have knowledge of such markets and sometimes face problems of fruit rotting before they are sold (Mbilase, pers. comm., 2007).

5.4.3 Infrastructure

The Kat Dam is the main bulk water infrastructure in the Kat River Valley. It supplies water for domestic use to the towns of Seymour and Fort Beaufort, as well as for the irrigation purposes, with citrus being the predominant crop (Magni, 1999). The dam was commissioned in 1970 with the purpose of irrigating alluvial soils on the banks of the Kat River Valley. According to Motteux (2001), the Kat Dam is a concrete multiple arch dam with a dam wall, which is approximately 55.6 m high.

Nel *et al.* (1997) highlighted that road development is a challenge in the Kat River Valley because it is influenced by parallel escarpments. Apart from the road to the market, farmers use poor roads from production areas (orchards, garden and fields) to the loading zones. The roads are slippery during rainy seasons, making it even more difficult to move produce from production areas to market places. The emerging and smallholder farmers in the Kat River Valley lack proper marketing infrastructure (Magni, 1999). For instance, vegetable farmers opt to sell from their homes because they do not have marketing sheds and proper storage facilities.

5.5 The methodology

The multinomial logistic regression model was used to test the institutional and technical factors that influence households from using greater depth marketing methods, which have the potential of increasing their incomes. According to Matungul *et al.* (2002), the greater the depth in marketing methods used by households, the greater the expected income.

Multinomial logistic regression can be used to predict a dependent variable, based on continuous and/or categorical independent variables, where the dependent variable takes more than two forms (Hill *et al.*, 2001). Furthermore, it is used to determine the percent of variance in the dependent variable explained by the independent variables and to rank the relative importance of independent variables. Logistic regression does not assume linear relationship between the dependent variable and independent variables, but requires that the independent variables be linearly related to the logit of the dependent variable (Gujarati, 1992). Pundo and Fraser (2006) explained that the model allows for the interpretation of the logit weights for the variables in the same way as in linear regression.

The model has been chosen because it allows one to analyse data where participants are faced with more than two choices. In this study, smallholder farmers are faced with three choices, which are; formal market participation, informal market participation and non-market participation. Firstly, the farming households are assumed to decide whether to market their products or not. When they choose to market, they then decide on the marketing channel to be used (either formal markets or informal markets). These decisions are made based on the option, which maximizes their utility, subject to institutional and technical constraints.

As such, the utility maximizing function can be given as:

$$\text{Max } U = U\,(C_k, R_{fk}, R_{ik}; H_u) \tag{1}$$

Where:
Max U denotes the maximum utility that can be attained from agricultural production;
C_k represents the consumption of produced goods by the household;
R_{fk} represents revenue gained from formal market participation;
R_{ik} represents revenue gained from informal market participation;
H_u represents a set of institutional factors shifting the utility function.

From the utility maximizing function, it can be seen that households make decisions to produce, consume and market, subject to institutional factors. It follows that if the costs that are associated with using a particular channel are greater than the benefits, households will be discouraged from using it, shifting to another option that maximizes their utility. For instance, if there are institutional challenges specific to formal markets that increase marketing cost above the revenue, households will be discouraged from using formal markets. They then, analyse the costs associated with informal markets. If the institutional factors that are unique to informal markets increase marketing costs above returns, then households will decide not to sell their produce. In this case, consumption of own production is considered as the option that provides the maximum utility. In the utility function, the amount of good *k* that is consumed or sold does not have to exceed the amount that is produced.

O' Sullivan *et al.* (2006) pointed out that it is difficult to measure utility directly; therefore, it is assumed that households make participation choices depending on the option that maximizes their utility. Thus, decisions to participate in either formal or informal markets or even not participating signify the direction, which maximizes utility. With the given assumption, multinomial regression was used to relate the decisions to participate in formal markets, informal markets or not participating and the factors that influence these choices.

A typical logistic regression model, which was be used is of the form:

$$\text{Logit}\,(P_i) = ln\,(P_i / 1 - P_i) = \alpha + \beta_1 X_1 + \ldots + \beta_n X_n + U_t \qquad (2)$$

Where:
$ln\,(P_i / 1 - P_i)$ = logit for market participation choices;
P_i = not participating in markets;
$1-P_i$ = participating in markets;
β = coefficient;
X represents covariates;
U_t = error term.

In the model, market participation choice, with three possibilities, *viz.* formal markets, informal markets and not participating in markets, has been set as the dependent variable. The variable of non-market participation was accepted as the baseline group; therefore, it took the value of zero. Informal market participation took the value of one and formal market participation was equated to two. The multinomial model used in the paper was used to determine the odds of both formal market participation versus non-participation and informal market participation versus non-participation. It follows that P_i represents the probability of not participating in produce marketing and $(1 - P_i)$ represents the probability of either informal market participation or formal market participation. The probability that the farmer prefers one market compared to the other is restricted to lie between zero and one ($0 \leq P_i \leq 1$). It should be noted that logit (P_i) ranges from negative infinity to positive infinity (Gujarati, 1992).

5.6 Specification of the model

The independent variables which were used in the multinomial logistic model are defined in Table 5.1, showing their expected signs.

When the variables are fitted into the model, the model is presented as:

$$ln\,(P_i / 1 - P_i) = B_0 + B_1MKTINFO + B_2GRDS + B_3EXT + B_4ORGMEM + B_5FMNGTYP + B_6RDINFR + B_7TRANS + B_8ADDVAL + B_9MKTINFR + B_{10}STOR + B_{11}CONTRCT + B_{12}SOCIALK + B_{13}PART + B_{14}TRAD + Et$$

The variable reflecting access to market information (MKTINFO) was measured by the household's ability to get market information in time and the ability to interpret it correctly. In order to capture this variable closely, households were interviewed on the communication networks that are accessible to them. Ability to communicate in either English or Afrikaans was used to measure the accuracy of information interpretation by

Table 5.1. Definition and description of variables used in the model.

Variable	Variable description	Anticipated sign
Access to market information (MKTINFO)	have access = 1, 0 otherwise	+
Expertise on grades and standards (GRDS)	have expertise = 1, 0 otherwise	+
Access to extension contact (EXT)	have Access = 1, 0 otherwise	+
Organizational support services (ORGMEM)	belong to an organization = 1, 0 otherwise	+
Type of farming (FMNGTYP)	arable farming = 1, 0 otherwise	+/-
Road infrastructure (RDINFR)	good = 1, 0 otherwise	+
Market transport (TRANS)	own transport = 1, 0 otherwise	+
Add value (ADDVAL)	add value = 1, 0 otherwise	+
Market infrastructure (MKTINFR)	good = 1, 0 otherwise	+
Storage facilities (STOR)	good = 1, 0 otherwise	+
Contractual agreements (CONTRCT)	have contract = 1, 0 otherwise	+
Social capital (SOCIALK)	extensive social capital = 1, 0 otherwise	+
Groups or individual participation (PART)	participate in a group = 1, individual participation = 0	+
Guided by tradition and beliefs (TRAD)	guided = 1, 0 otherwise	-

households. Access to information has been set as a dummy variable, where a household with access to information takes the value of one and a household that has no access to information takes a value of zero. Access to information was expected to influence market participation positively; implying that households with access to information are more likely to participate in marketing, making more use of formal markets. Another variable that is closely linked to information availability is access to extension services (EXT) such as access to farming advice and knowledge through extension officers. This variable was also allocated dummy values where households with access to extension services took the value of one and zero if otherwise.

According to Reardon and Barrett (2000), smallholder farmers have difficulties in meeting market grades and standards, leading to exclusion of such farmers from participating in mainstream agriculture. In this study, expertise on grades and standards (GRDS) was recorded in order to investigate whether it influences marketing participation choices or not. Households were asked if they were aware of market grades and standards, and whether they had problems meeting such standards. The households with knowledge on grades and standards, and had no problems meeting them were set to have expertise on grades and standards. Such households took the value of one and those households with no expertise on grades and standards were equated to zero. Households with expertise on grades and standards are expected to use formal markets more than those without, thus an expected positive relationship.

As cited by Kherallah and Kirsten (2001), collective action is important in agricultural marketing because it contributes towards reduced transaction costs and it strengthens farmers' bargaining and lobbying power. Collective action is measured by two main variables in this study, which are organizational support services (ORGMEM), and groups or individual participation (PART). Respondents were asked whether they belonged to an organization or not and whether they sold output in groups or individually, the responses were allocated dummy values. Both the variables are anticipated to impact positively on market participation choice among the smallholder farmers.

The availability and condition of both road and market infrastructures are thought to have an influence on marketing efficiency. Where the infrastructure is unavailable or poor, farmers are discouraged from using it, thereby limiting market participation. Therefore, the availability of good road and market infrastructures are expected to exert a positive influence on market participation. Road infrastructure (RDINFR) is measured by the adequacy of the road networks that are accessible to households and their conditions. Market infrastructure (MKTINFR) is measured by the availability of infrastructure, such as marketing stalls and their condition. Dummy values are used to define the variables, where in both cases, one indicates good condition and zero indicates either unavailability or poor condition.

Ownership of market transport (TRANS), specifically vehicles, was used to measure the availability of produce transportation facilities by households. Moreover, the availability of transportation facilities helps reduce long market distance constraint, offering greater depth in marketing choices. In cases where households owned a vehicle, the variable took the value of one, and zero if the household did not own any form of vehicle. This variable is expected to have a positive influence on the market participation choices.

Social capital (SOCIALK) in this study refers to personal social networks that encourage market participation. It is through these networks that trust is developed, which, in turn, encourages cooperation and regular exchanges. Therefore, social networks reduce transaction costs, leading to diversified market participation choices. Again, information and production resources can be transmitted through these networks. In order to capture this variable, respondents were asked about their connection with their customers and whether they had any regular customers. The availability of an extensive social capital structure is expected to impact positively on the dependent variable.

The availability of contractual agreements (CONTRCT) ensure the availability of a guaranteed market for the farmers, thus promoting market participation and including the smallholder farmers in mainstream agriculture. In other words, the existence of a guaranteed market reduces the costs that are associated with searching for potential buyers, thereby encouraging participation in formal markets. This variable is expected to have a positive relationship with the dependent variable

The types of farming (FMNGTYP) have been divided into either arable farming or livestock farming, where the former takes a value of one and the latter takes the value of zero. This variable was included into the model in order to capture the differences in the nature of produce from different farming types. Thus, in some types of farming, formal market penetration may be easier than in the other types. For instance, Matungul *et al.* (2002) pointed out that formal livestock markets are readily accessible to both commercial and small-scale livestock farmers in South Africa, owing to public investment in sales yards. The variable can take either a positive or a negative value.

The ability to add value to agricultural produce is captured by the variable (ADDVAL). Dummy values are used to define the variable, where those households who add value to their produce, take the value of one and those who do not, equal to zero. It is hypothesized that the ability to add value exerts a positive impact on market participation. This positive relationship is because households with the ability to add value can sell their produce in an improved state, which can be more appealing to customers. The variable storage facilities (STOR), is closely related to value adding. Good storage facilities reduce loss of produce and urgency of produce selling, and maintain the physical state of produce. Thus, households with good storage facilities are more likely to participate in formal markets, hence a positive relationship.

Traditions and beliefs are part of informal institutions that can influence marketing choices. In the model, the dummy variable, guided by traditions and beliefs (TRAD) is used to determine effects of traditions and beliefs on marketing. Households were asked on the extent to which tradition plays a role in their agricultural activities. They were also asked if they were open to new farming and marketing methods offered by non-family members. The variable was allocated dummy values where households with strong traditional guidance took the value of one and zero if otherwise. The variable is expected to take a negative value because household depending on traditions and beliefs are normally not liable to change (Dorward *et al.*, 2005). Such households would rather stick to what they know especially if the marketing environment is changing rapidly. Thus, such households are less likely to participate in the formal markets.

5.7 Empirical results

This section presents the results of the multinomial logistic regression model and discusses the results of the significant variables that determine market participation choices in the Kat River Valley. The variables that were discussed in the previous section were considered for the model and tested for their significance. The multinomial logistic results of informal and formal market choices as compared to non-market participation are presented in Table 5.2. The table shows the estimated coefficients (β values), standard error, significance values and odd ratio of independent variables in the model.

According to Gujarati (1992), the coefficient values measure the expected change in the logit for a unit change in each independent variable, all other independent variables being equal. The sign of the coefficient shows the direction of influence of the variable on the logit. It follows that a positive value indicates an increase in the likelihood that a household will change to the alternative option from the baseline group. On the other hand, a negative value shows that it is less likely that a household will consider the alternative (Gujarati, 1992; Pundo and Fraser, 2006). Therefore, in this study, a positive value implies an increase in the likelihood of changing from not participating in marketing to either informal or formal market participation choice. In this study, the variables were tested at the 5% significance level.

5.7.1 Significant variables in the model

As indicated in Table 5.2, some predictor variables influence market participation choices significantly. Of the 14 independent variables used in the model, five and six variables in informal and formal market choices respectively, are statistically significant at the 5% significance level. In all but one of the cases, the signs of the estimated coefficients are consistent with the *a priori* expectations.

Table 5.2. Multinomial logistic results for informal and formal market choices as compared to non-marketing choice.

Variable	Informal market choice				Formal market choice			
	Coefficient	Std. error	Significance	Odds ratio	Coefficient	Std. error	Significance	Odds ratio
MKTINFO	2.686	1.050	0.011*	14.673	4.217	1.385	0.006*	67.83
GRADES	-1.623	0.905	0.073	0.197	3.830	0.848	0.016*	46.06
EXT	0.606	0.868	0.485	1.833	1.166	2.942	0.282	3.209
ORGMEM	0.788	0.786	0.316	2.199	1.324	1.384	0.330	3.758
FMNGTYP	-0.248	0.754	0.742	0.780	-1.164	1.854	0.530	0.312
RDINFR	0.862	0.841	0.305	2.368	2.992	2.171	0.168	19.92
SOCIAILK	0.222	0.948	0.050*	1.248	1.180	2.863	0.031*	3.254
ADDVAL	1.352	1.079	0.210	3.865	0.392	0.218	0.860	1.479
MKTINFR	2.557	1.030	0.013*	12.897	-0.687	0.026	0.735	0.503
STOR	0.584	0.777	0.453	1.793	0.259	1.873	0.890	1.296
CONTRCT	0.844	0.755	0.263	2.326	2.803	1.912	0.047*	16.49
TRANS	0.843	0.774	0.276	2.323	0.449	1.644	0.785	1.567
PART	1.899	0.854	0.026*	6.679	1.997	1.418	0.039*	7.367
TRAD	2.477	1.441	0.031*	11.905	-2.144	2.296	0.007*	0.117
INTERCEPT	-4.934	2.860	0.048	-	-17.069	4.466	0.044	-

Goodness-of-Fit	Chi-Square	df	Sig.
Pearson	111.372	168	0.150
Deviance	84.301	168	0.988

* Statistically significant at $P \leq 0.05$.

Access to market information has a positive sign for both formal and informal market choices, which is consistent with the *a priori* expectations. The significance values of 0.011 for the informal market choice and 0.006 for the formal market choice imply that there is enough evidence to support that an increase in the availability of market information results in an increase in both informal and formal market participation. The larger values in odds ratios show that households are likely to increase participation in both informal and formal markets with the availability of market information.

Expertise on grades and standards is significant for the formal market choice with a significance value of 0.016. A positive sign on its coefficient indicates that an improvement in expertise on grades and standards results in an increase in the formal market participation choice by households. An unexpected sign was noted between expertise on grades and standards, and the informal market choice. A negative relationship between the two variables and an odds ratio value less than one explain that it is less likely for households to change from non-market participation to informal market choice with an increase in expertise on grades and standards. A possible explanation that can be given for this relationship is that expertise on grades and standards involves capital commitment (Reardon and Barrett, 2000). Therefore, households who acquire it will be willing to market their produce in the formal markets, which are more rewarding than informal markets. Unfortunately, the negative relationship is not significant at the 5% level but is at the 10% level.

A positive and significant (0.047) relationship was found between formal market participation and the availability of contractual agreements. The relationship implies that households tend to increase in formal market participation with the availability of contractual agreements. This relationship is most likely due to the influence of the citrus farmers. The value of the odds ratio (16.49) supports the higher probability of the variable influence on the formal market choice.

The variable existence of extensive social capital is significant for both informal (0.050) and formal (0.031) market choices. The positive relationship in both formal and informal market participation choices explains that an increase in social capital results in households shifting from non-participation to formal and informal market participation. The odds ratios for both formal and informal market choices suggest a higher probability of shifting to formal and informal marketing with an increase in social capital. Therefore, it can be concluded that social networks are important in produce marketing, regardless of the choice of market being used.

It was expected that the availability of good market infrastructure could have a positive influence on alternative market participation choices, away from not participating in marketing. However, the *a priori* expectations hold true for the informal market choice only. There is sufficient evidence (significance value of 0.013) to support that the availability of good market infrastructure is likely to encourage households to market their produce through informal channels. Unlike formal channels where market infrastructure is not important for farmers, as they supply their produce in bulk once harvested to the higher level of the marketing channel (Takavarasha and Jayne, 2004).

Group participation in marketing was expected to have a positive influence on the dependent variable. The results shown in Table 5.2, for this variable are consistent with the *a priori* expectations. For both formal and informal market choices, there is enough evidence to support that when households market their produce in groups, there is a higher chance of participating in either formal or informal markets. Thus, group participation encourages market penetration among smallholder farmers who find it difficult individually to gain market access.

A positive and significant (0.031) relationship was found between informal marketing and guidance from tradition and beliefs. These results are in contradiction with the *a priori* expectations, where guidance from traditions and beliefs was expected to influence the dependent variable negatively. The positive relationship between the variables may be possibly explained by traditional wisdom and skills passed on in families and creation of marketing links through traditions and beliefs. For instance, some households may prefer to sell their produce (especially in cattle marketing) to people they are familiar with. On the other hand, there is a negative and significant (0.007) relationship between formal marketing and guidance from traditions and beliefs. The explanation to this relationship may be that the marketing environment is ever changing (Kherallah and Kirsten, 2001); therefore, if farmers are to be part of the formal markets, they have to be receptive to changes.

The goodness-of-fit test for a logistic regression model measures the suitability of the model to a given data set. An adequate fit corresponds to a finding of non-significance for the tests (Hill *et al.*, 2001). The results for the goodness-of-fit test shown in Table 5.2 indicate that the model fits the data well. Thus, the results for both Pearson and Deviance chi-squared methods show that the multinomial logistic regression model is well suited to predict the influence of independent variable on the dependent variable.

5.7.2 Summary of variables determining market participation choice

The previous section discussed the influence of the significant predictor variables on the dependent variable. It can be concluded that the variables that have a higher probability of shifting households from non-market participation to informal marketing are access to market information, availability of good market infrastructure, existence of extensive social capital, group participation and guidance from tradition. All of the five variables positively influence informal marketing, implying that households are likely to shift from non-marketing to informal market participation with an increase in any one of the variables.

The variables that are likely to shift households from non-market participation to formal market participation include group participation, access to market information, expertise on grades and standards, availability of contractual agreements and existence of extensive social capital. Thus, an increase in each of the variables results in a higher probability of households changing from not participating in produce marketing to formal market participation. In addition, households willing to participate in the formal markets have to be open to new marketing ideas rather than reliance on traditions.

Based on the results presented in this section, it is shown that marketing channel choices among smallholder farmers are influenced by institutional factors. Where the institutional factors are poorly developed, farmers have difficulties in marketing their produce through formal and informal markets. It should be acknowledged that there is only enough evidence to support the influence of the significant variables, but that does not make the insignificant variables irrelevant. A possible explanation is that without the significant variables, it makes it difficult for households to be involved in different marketing choices. For the insignificant variables, whereas they are important, households can still withstand produce marketing without changing such variables. For instance, without market information, it is difficult to market produce even if the household owns transport. On the other hand, households can still market if they have market information but without their own transport.

5.8 Summary

This chapter provided empirical evidence of factors influencing market participation choices among smallholder farmers in the Kat River Valley. The institutional factors influencing market participation choices were defined and tested using a multinomial logistic regression model. The statistically significant variables, at the 5% level are access to market information, expertise on grades and standards, availability of contractual agreements, existence of extensive social capital, availability of good market infrastructure, group participation and reliance on traditions.

Based on the results of this study, several suggestions can be made on how smallholder farmers can be actively involved in produce marketing. Generally, the findings suggest that an adjustment in each one of the significant variables can significantly influence the probability of market participation. That is, technological growth and institutional developments that affect such variables can help farmers improve participation and encourage formal market participation. It is important to identify the ideal technological changes and institutional development pathways that best fit the smallholder farmers. In coming up with different ways of incorporating smallholder farmers in mainstream agriculture, it has to be accepted that smallholder farmers cannot individually compete against commercial farmers in markets. In addition, it is difficult for them to get contractual agreements individually, owing to a small marketable output. Therefore, smallholder farmers have to be encouraged to work in groups.

References

Delgado, C. (1999). Sources of growth in smallholder agriculture in sub-Saharan Africa: the role of vertical integration of smallholders with processors and marketers of high-value added items. Agrekon 38 (Special Issue), 165-189.

Dorward, A., J. Kydd, J. Morrison and C. Poulton (2005). Institutions, markets and economic development: Linking development policy to theory and praxis. Development and Change 36, 1-25.

FAO (2004). Changing patterns of agricultural trade. The evolution of trade in primary and processed agricultural products. Available at: http://www.fao.org/docrep/007/y5419e/y5419e05.htm.

Farolfi, S. and K. Rowntree (2005). Negotiation and game theory for water management in the Kat Basin, South Africa: Development of a cooperative game theory model and a comparison with a role-playing game results derived of a negotiation approach. Available at: http://www.ceepa.co.za/docs/Desole.pdf.

Fenwick, L.J. and M.C. Lyne (1999). The relative importance of liquidity and other constraints inhibiting the growth of small-scale farming in KwaZulu-Natal. Development Southern Africa 16, 141-155.

Gujarati, D. (1992). Essentials of econometrics. MacGraw–Hill: New York, USA.

Hill, R.C., W.E. Griffiths and G.G. Judge (2001). Econometrics (2nd Edition). John Wiley and Sons, New York, USA.

Kherallah, M. and J. Kirsten (2001). The new institutional economics: applications for agricultural policy research in developing countries. Markets and Structural Studies Division, Discussion paper No. 41, International Food Policy Research Institute, Washington, DC, USA.

Magni, P. (1999). Physical description of the Kat River valley. Geography Department, Rhodes University, Grahamstown, RSA.

Makhura, M.T. (2001). Overcoming transaction costs barriers to market participation of smallholder farmers in the Northern Province of South Africa. Unpublished PhD thesis, University of Pretoria, Pretoria, RSA.

Matungul, P.M., G.F. Ortmann and M.C. Lyne (2002). Marketing methods and income generation amongst small-scale farmers in two communal areas of KwaZulu-Natal, South Africa. Paper presented at the 13th International Farm Management Congress, Wageningen, the Netherlands.

McMaster, A. (2002). GIS in participatory catchment management: a case study in the Kat River Valley, Eastern Cape, South Africa, Unpublished MSc thesis, Rhodes University, Grahamstown, RSA.

Motteux, N. (2001). The development and co-ordination of catchment fora through the empowerment of rural communities. Catchment Research Group. WRC Report K5/1014. Water Research Commission, Pretoria, RSA.

Nel, E.L. (1998). An evaluation of community driven economic development, land tenure and sustainable environmental development in the Kat River Valley. Report to the Programme for Human Needs, Resources and the Environment, HSRC Publishers, Pretoria, RSA.

Nel, E., T. Hill and T. Binns (1997). Development from below in the 'new' South Africa: the case of Hertzog, Eastern Cape. The Geographical Journal 163, 57-64.

O'Sullivan, A., D. Sheffrin, and S. Perez (2006). Economics: principles, applications and tools (5th Edition). Prentice Hall: New York, USA.

Pundo, M.O. and G.C.G. Fraser (2006). Multinomial logit analysis of household cooking fuel choice in rural Kenya: The case of Kisumu district. Agrekon 45, 24-37.

Reardon, T. and C. Barrett (2000). Agro-industrialization, globalization, and international development: An overview of issues, patterns, and determinants. Journal of Agricultural Economics 23, 195-205.

Smit, T. (2003). Catchment management forums in the Eastern Cape Province of South Africa. Available at: http://campus.ru.ac.za/download.php?actionarg=986.

Takavarasha, T. and T.S. Jayne (2004). Toward improved maize marketing and trade policies to promote household food security in Southern Africa. Available at: http://www.aec.msu.edu/maizemarket/JayneTakavarashaSummary.pdf.

Wynne A.T. and M.C. Lyne (2003). Rural economic growth linkages and small scale poultry production: a survey of poultry producers in KwaZulu-Natal. Paper presented at the 41st Annual Conference of the Agricultural Economic Association of South Africa (AEASA), Pretoria, RSA.

Part IV
Marketing constraints and opportunities in specific commodity lines

6. Marketing challenges and opportunities faced by the Nguni cattle project beneficiaries in the Eastern Cape Province of South Africa

Lovemore Musemwa and Abyssinia Mushunje

Abstract

Market off-take rate is low in the communal cattle production system of Africa, with off-take rates of between 5 and 10% compared to 25% in the commercial sector. Pre-tested structured questionnaires were administered to 183 smallholder farmers to determine factors that influence market off-take rate in three randomly selected municipalities in the Eastern Cape Province of South Africa. The results show that off-take rates in the sampled municipalities averaged 12%. Education, market distance, body condition and herd size were significantly associated with municipality ($P<0.05$). Chris Hani had the highest number of the interviewees with at least primary education. Farmers in Amatole municipality sold their cattle in far away markets. The body condition of cattle in Chris Hani was the best. Lack of information access reduced cattle sales. Presence of markets that farmers like in or near their communities facilitated cattle sales. Smallholder farmers sold more cattle as transport became more available. The probability of smallholder farmers selling their cattle decreased as the body condition of cattle increases. As the household head changed from being male to being female, the probability of selling cattle increased. It is, therefore, evident that an integrated approach is likely to underpin an efficient livestock marketing system. This requires a joint effort by the government, municipalities, smallholder farmers, producer organizations and private sector role players. Group marketing, decentralization of cattle information centres and the involvement of communal farmers' in the dissemination of information plays a critical role in improving market off-take rate of smallholder farmers.

6.1 Introduction

In terms of market access, Jooste and Van Rooyen (1996) concluded that the transition of the small scale sector towards commercial production will ultimately be determined by its access to markets. According to Stroebel (2004), several constraints affect the efficient marketing of livestock in the Eastern Cape Province of South Africa, chief of which are related to market availability and high transactions costs. Currently there is poor response of consistent cattle supplies to existing market outlets by small-scale farmers on communal lands (USAID, 2003). This raises serious concerns as to why these farmers do not sell their cattle.

This chapter empirically tests marketing constraints and household characteristics that are hypothesized to have the largest potential to cause small-scale cattle farmers not to sell cattle. It further tests association of municipalities with market availability, market distance, transport availability, transport affordability, information access, current market price and condition of cattle, herd size, method of payment, household size, gender, age and level of education of head of household (the hypothesized marketing constraints). It is therefore of utmost importance to determine these constraints so as to identify mitigation strategies that are consistent with the broader objectives of poverty alleviation and black economic empowerment. Solving marketing constraints will, in turn, increase market off take thereby increasing household income earnings and standards of living of the beneficiaries of livestock improvement programmes.

6.2 Overview of the Nguni cattle project

The Nguni cattle development project was initiated in 1998 by the University of Fort Hare in collaboration with a number of rural development agencies in the Eastern Cape Province of South Africa. Under the project, farmers in selected communities receive an in-kind grant of bulls and heifers to allow them build up a nucleus herd (Fuller, 2006). Each community receives two bulls and 10 in-calf heifers and the existing bulls in the community are replaced by registered Nguni bulls. After five years, the community gives back to the project two bulls and 10 heifers, which are then passed on to another community (Raats *et al.*, 2004). It works on the 'pay it forward' system, a concept built around the principle of voluntarism and random act of kindness without expectation of direct payback to the original source of the kind act. The cycle continues, with each community paying the dividends of its original gift forward to another one (Fuller, 2006). One of the conditions of the project is that communities should have fenced grazing areas, a rangeland management committee and practice rotational resting at specified stocking rates (Mapiye *et al.,* 2007). The model's long-term goal is to develop a niche market for Nguni beef and skins and to position the communal farmers for the global beef market through organic production and product processing (Raats *et al.,* 2004).

To date, the project has benefited about 45 communities out of the target of 100 (Raats *et al.,* 2004). The University of Fort Hare project discourages use of nondescript cows and enforces the removal or castration of the existing exotic bulls in the communal areas which are replaced by pure registered Nguni bulls. The project adopts a participatory approach which provides a quick, viable and sustainable mechanism for the establishment of nucleus Nguni herds in the communal areas (Mapiye *et al.,* 2007). The University of Fort Hare is also responsible for planning, coordinating and training of livestock managers and extension officers in the province and adjoining provinces such as Kwazulu-Natal, the Free State, and Limpopo. A project development committee made up of interested stakeholders is in charge of the development of infrastructure, training of farmers and the redistribution of animals. The implementation of the model in communal areas is conducted in collaboration with the Department of Agriculture (Raats *et al.,* 2004). In 2007, a collaboration agreement was signed with the National University of Lesotho that provided for that University receiving the Nguni cattle from Fort Hare and participating in the process of popularizing the breed in the neighbouring country of Lesotho which shares a number of agro-ecological characteristics with the Eastern Cape Province (Obi, 2007).

6.3 Qualities, characteristics and traits of Nguni cattle

Nguni cattle are small-framed and are well adapted to the environment because they have evolved in those surroundings for thousands of years and are endemic to the South of Africa. (Muchenje *et al.*, 2008). They are known by their high fertility rates and excellent resistance to ticks and immunity to tick-borne diseases. Disease incidence and mortality are low (Mapiye *et al.*, 2007). The Nguni cattle is characterized by its multi-coloured skin, which can present many different patterns, but their noses are always black-tipped (Figure 6.1).

Due to its diminutive, almost stunted, physical appearance and its popularity as draught animal, farmers consider the Nguni cattle to have little or no value as a source of beef. Under sound management conditions it is becoming increasingly popular as a beef breed (Muchenje

Figure 6.1. Nguni cattle / Herd of Nguni cattle at the University of Fort Hare experimental farm.

et al., 2007). Besides its resistance to tropical diseases and parasites, the Nguni is highly adaptable to poor quality grazing and conditions of excessive heat and humidity. The Nguni also has adaptive traits such as walking ability, which enables it to walk long distances in search of grazing and water. They are excellent foragers and can graze and browse on steep slopes and in thick bush alike. According to Bester *et al.* (2005), the evolutionary development of the Nguni has resulted in a breed with good temperament and mothering ability.

6.4 Contribution of Nguni cattle to communal livelihoods

Communal farmers keep cattle for multiple purposes. Rural households depend on cattle for milk, meat, hides, horns and income (Chimonyo *et al.*, 1999; Dovie *et al.*, 2006; Simela *et al.*, 2006). Cattle provides dung for manure, fuel and floor polish/seal, and draught power for cultivation of crops and transport of goods in communal areas (Shackleton *et al.*, 1999; Bayer *et al.*, 2004). Cattle are an inflation-free form of banking for resource-poor people and can be sold to meet family financial needs such as school fees, medical bills, village taxes and household expenses (Dovie *et al.*, 2006; Simela *et al.*, 2006). They are a source of employment, collateral and insurance against natural calamities. Some farmers keep cattle for prestige and pleasure (Shackleton *et al.*, 1999). More importantly, indigenous cattle are valuable reservoirs of genes for adaptive and economic traits, providing diversified genetic pool, which can help in meeting future challenges resulting from changes in production sources and market requirements (FAO, 2007).

Socio-cultural functions of cattle include their use as payment for bride price (known as *lobola)* and to settle disputes (as fine) in communal areas (Chimonyo *et al.*, 1999). They are reserved for special ceremonies such as marriage feasts, weddings, funerals and circumcision (Bayer *et al.*, 2004). Cattle are given as gifts to visitors and relatives, and as starting capital for youth and for the newly married man about to start a family. They are used to strengthen relationships with in-laws and to maintain family contacts by entrusting them to other family members (Dovie *et al.*, 2006). The quality of care given to the entrusted animal is a measure of the value placed on the relationship, and families are known to have fractured irretrievably as a result of disagreement on the way cattle so entrusted has been looked after. Cattle plays an important role in installation and exorcism of spirits and frequently the object of choice for assorted sacrificial offerings to appease ancestors or avenging spirits (Bayer *et al.*, 2004).

The relative importance of each of the functions mentioned above varies with production system, rangeland type, region and socio-economic factors such as gender, marital status, age, education and religion of the keepers (Chimonyo *et al.*, 1999; Simela *et al.*, 2006). The differences in farmers' objectives and perspectives to communal cattle production hamper the formulation of effective livestock policies aimed at improving the livelihoods of the resource-poor farmers (Bayer *et al.*, 2004) across all regions or countries. Efforts to improve communal cattle production should, therefore, emphasize the understanding of farmers' objectives, perceptions and experiences. From this knowledge, constraints and opportunities faced by resource poor communal cattle farmers can be identified and sustainable developmental strategies formulated (Dovie *et al.*, 2006).

6.5 Variables used in the analysis

In the context of cattle marketing, a marketing constraint can be considered to be a variable that precludes farmers from selling their cattle or negatively affects the profitability of cattle sales and thus acts as a disincentive to expanded trade. This study identified thirteen independent variables measured as continuous and discrete variables. The identified variables include market availability, market distance, transport availability, transport affordability, information access, current market price, and condition of cattle, herd size, method of payment and household characteristics which are size, sex, age and level of education of head of household. Diseases were not included in the model as there was no indication that they were important to all respondents covered in the survey.

6.5.1 Market availability

Lack of marketing facilities imposes a serious constraint on the marketing of livestock (Mahabile *et al.,* 2002). The presence of markets that farmers prefer most in or near their communities will facilitate cattle sales. Where farmers are willing to sell but cannot locate a market where profitable exchange is feasible, it is expected that this will have a dampening effect on the willingness to sell and also on actual sales concluded. Market availability was included in the model as a dummy variable (where, market availability is assigned the number 1 and zero otherwise).

6.5.2 Market distance

Remote location of small scale farmers results in them staying far away from markets. A farmer may be aware of a market but may find the physical distance extremely demoralizing. In Botswana, Makhura (2001), Mahabile *et al.* (2002) and Nkhori (2004) noted that even if farmers are in areas with good road linkages, the distance to the markets tends to influence transaction costs. The further the farmers are from markets, the higher the transaction costs farmers face because of transport costs involved in transporting large stock. In such circumstances, cattle sales are limited with respect to that particular market. If no other market is within easy reach of the farmer, then total cattle sales undertaken will be low. Distance to market was treated as a continuous variable.

6.5.3 Transport availability

Due to poverty in many rural areas, only few community members own a motorized transport vehicle of adequate size, such as trucks or pick-up vans (locally known as *bakkie*) for purposes of transporting cattle to distant markets (Musemwa *et al.*, 2007). This is a serious constraint especially to those farmers that may be ready to pay for transportation of their cattle to the market. Transport availability was treated as an important variable because it was hypothesized that there would be a direct relationship between this variable and the extent of cattle sales. Transport availability was calibrated as a dummy variable (transport available = 1 and zero otherwise).

6.5.4 Transport affordability

Most farmers in rural areas cannot afford to buy even the most basic commodities. This therefore implies that even if transport were available to carry their cattle to the market, the financial means to hire transport may be a binding constraint. The less money a farmer has, the less the cattle sales to the preferred market. Transport availability was treated as a dummy variable (money available = 1 and zero otherwise).

6.5.5 Information access

Insufficient or poor provision of agricultural market information is a key factor constraining agricultural development in developing countries (Bailey *et al.,* 1999). The farmers' market information needs are those that enable her/him make rational, relevant decisions and strengthen much-needed negotiating ability during transactions with buyers in order to prevent or minimize possible exploitation by better informed buyers (Coetzee *et al.,* 2004). Word of mouth is still used as the main source of information. It was, therefore, hypothesized that the more the access the farmer has to information the more cattle he/she will sell. Information access was treated as a dummy variable (access to information = 1 and zero otherwise).

6.5.6 Current market price

Before making a decision about how to market a product and to whom to sell, it is assumed that cattle producers first determine the price that they expect to receive. Market prices tend to affect cattle sales. It is hypothesized that the higher the cattle prices the more the farmers are motivated to sell their cattle. Market prices were treated as continuous variables and were measured in South African Rand.

6.5.7 Condition of cattle

The poor condition of livestock results in lower farm gate prices, especially during dry spells (Coetzee *et al.*, 2005). Livestock auctioneers and speculators often raise concerns that they cannot pay competitive prices for animals that are in poor condition or not ready for the market (Nkhori, 2004). When lower than expected prices are offered by buyers, small-scale livestock farmers are discouraged from selling their cattle. On the part of the buyer, the poor condition of animals creates doubts in the minds of the buyers about the wholesomeness of the animal and may reduce the willingness to purchase the cattle. Condition of cattle was treated as a categorical variable and entered the model as a dummy.

6.5.8 Method of payment

The method of payment determines after how long the farmer will get his/her cash after selling cattle. The delay between when the cattle are sold and when the farmer gets liquid cash is an important cost. According to Nkhori (2004), the shorter the time lag the greater the greater the willingness on the part of the farmers to sell their cattle since, in most cases, small scale farmers sell their cattle to cover up for emergences (Musemwa *et al.*, 2007). It is therefore expected that the better the method of payment, measured by how fast the transaction is concluded and materializes as cash, the greater the frequency of cattle sales undertaken. Method of payment was treated as a categorical variable.

6.5.9 Herd size

Total herd is a continuous variable that reflects household wealth and a necessary condition for market participation. According to Fidzani (1993), large herds generate a higher marketable surplus than small herds. Fidzani (1993) also observed that larger herds mean that the farmer would have a stronger bargaining power when negotiating on price with prospective buyers. It is therefore expected that the larger the herd the more the farmer is willing to sell.

6.5.10 Household characteristics

Four variables related to household characteristics were included in the model since it is hypothesized that the household heads' backgrounds, circumstances, attitudes and disposition may have a constraining effect on cattle sales. These include household size, gender of household head, age of household head and level of education of household head.

6.5.11 Household size

Household size, according to Montshwe (2006), is a useful unit of analysis given the assumptions that within the household resources are pooled, income is shared, and decisions are made jointly by responsible household members. Household requirements are many and one person in most cases cannot handle them alone. Marketing of livestock is in most cases one of the least prioritized activities in a household. It is therefore expected that the smaller the household size the less the cattle sales will be because of fewer household needs that require substantial cash. Household size was coded as continuous variables.

6.5.12 Gender of household head

The sex of the household head is expected to influence cattle sales. Men are expected to sell more than women as they are able to have better access to information through their much wider networks, on average, than women. Women are also involved in more domestic chores than men and are therefore not able to access market information relevant to cattle sales as readily as men. Gender was coded as a dummy variable (male = 1 and zero otherwise).

6.5.13 Age of household head

Musemwa *et al.* (2007) argues that the higher the age of the head of household, the more stable the economy of the farm household, because older people relatively have richer experiences of social and physical environments as well as greater experience of farming activity. However these older people cannot walk long distances hence cannot sell their cattle in far away markets and have limited access to information since most of them are illiterate. Most of the older people view cattle farming not as a business but for prestige and social purposes (Chimonyo, 2000). It is therefore expected that the older the household head the less frequent the cattle sales. Age of the household head was treated as a categorical variable.

6.5.14 Level of education of household herd

Another important factor to consider is the level of education of the heads of households since they are the decision makers in matters concerning the farming activities and other family income earning activities. According to Nkhori (2004), education increases the ability of farmers to use their resources efficiently and the allocative effect of education enhances farmers` ability to obtain, analyse and interpret information. The more the farmer is educated the more she/he sells cattle at the market. Education was treated as a categorical variable.

6.6 Results of association tests

Out of the thirteen variables included in the study, four were significantly associated with municipality ($P<0.05$) namely education, market distance, body condition and herd size. The Pearson Chi Square test results are summarized in Table 6.1, while the ANOVA test results are in Table 6.2. Level of education of head of household varies across municipalities, the likely reasons being variation in training programmes and number of schools available in the three municipalities. For instance in Alfred Nzo there are very few schools, to the

Table 6.1. Relationship between geographical location and market factors based on Pearson's correlation tests.

Municipality	Value	Df	Asymp. Sig (2-sided)
Gender	0.971	2	0.615
Age	9.011	4	0.061
Education	11.360	4	0.023
Information access	5.674	2	0.059
Market price	10.045	6	0.123
Market distance	13.143	4	0.011
Market availability	1.375	2	0.503
Transport availability	5.319	2	0.070
Transport affordability	1.591	2	0.451
Method of payment	0.345	2	0.841
Body condition	33.778	4	0.000
Market opportunities	57.986	8	0.000

Table 6.2. Analysis of variation in household and herd sizes in the survey municipalities.

		Sum of squares	Mean square	F	Sig
Household size	Between groups	37.6	18.8	1.203	0.303
	Within groups	2,813.62	15.631		
	Total	2,851.22			
Herd size	Between groups	2,269.92	1,134.96	11.376	0
	Within groups	17,958.5	99.77		
	Total	20,228.4			

extent that school children walk very long distances to schools. This results in most of the people in the area not going to school.

Distances that farmers travel when selling their cattle is significantly associated with municipality ($P<0.05$). This can be explained by location of active cattle markets in the three municipalities. While in Amatole, active cattle markets (major towns in the municipality) are far from the sampled communities, for Chris Hani and Alfred Nzo District Municipalities the bulk of the surveyed communities are close to major towns. Chris Hani Municipality provides market infrastructure to small scale farmers in most of the communities as opposed to Amatole that provides little or no marketing infrastructure. The result of this is that farmers in Amatole District sell their cattle in more distant markets since buyers are unable to access the farmsteads due to absence of good quality marketing infrastructure. According to Montshwe (2006), distance to market can be reduced by bringing buyers closer to small scale farmers. Condition of cattle significantly varies across municipalities ($P<0.05$). This is due to the fact that rainfall and soil fertility, which play an important role in determining animal condition, differ from one municipality to the other. In Alfred Nzo, the body condition of cattle was poor compared to Amatole probably due to poor condition of grazing lands, low rainfall, snow and poor soils.

There is an association between municipality and herd size ($P<0.05$). Different municipalities have different agro-ecological conditions and this leads to variations in herd sizes since other places are more conducive to cattle farming than others. Mean herd size difference for Alfred Nzo significantly differs from that of Amatole and Chris Hani (Table 6.3). This may be due to extreme climatic conditions and high levels of stock theft experienced in Alfred Nzo.

6.7 Results of the logistic regression

Out of the thirteen variables included in the model, five were significant ($P<0.05$) and these are market availability, transport availability, body condition of cattle, information access and gender of household head as shown in Table 6.4. In this section all the variables are discussed in general and to what extent they conform to *priori* expectations.

6.7.1 Availability of markets

Market availability significantly affects cattle sales positively, *ceteris paribus*. This positive and significant relationship was expected. These results corroborate those of Nkhori (2004) and Montshwe (2006). As market becomes more available and accessible to farmers, the more the market would offer farmers better and more competitive prices. Good price will result in farmer being willing to sell since rational people want good offers and this will in turn increase cattle sales. As expected, the results suggest that those households that

Table 6.3. Distribution of mean herd size differences among survey municipalities.

Dependent variable	Municipality (I)	Municipality (J)	Mean difference (I-J)	Std. error	Significance
Household size	Amatole	Chris Hani	-0.53	0.72	0.464
		Alfred Nzo	-1.2	0.79	0.134
	Chris Hani	Amatole	-0.53	0.72	0.464
		Alfred Nzo	-0.66	0.93	0.477
	Alfred Nzo	Amatole	1.2	0.79	0.134
		Chris Hani	0.66	0.93	0.477
Herd-size	Amatole	Chris Hani	-2	1.83	0.275
		Alfred Nzo	-9.57*	2.01	0
	Chris Hani	Amatole	2	1.83	0.275
		Alfred Nzo	-7.57*	2.36	0.002
	Alfred Nzo	Amatole	9.57*	2.01	0
		Chris Hani	7.57*	2.36	0.002

* The mean difference is significant at 5%.

Table 6.4. Maximum likelihood estimates for the Nguni Cattle marketing model.

Parameter	Estimate	Error	Chi-Square	*P*-value
Intercept	14.5488	4.9590	8.6071	0.0033
Gender	4.7246	2.0549	5.2865	0.0215
Age	-1.3229	0.7723	2.9339	0.0867
Education	1.2223	0.8806	1.9268	0.1651
Household size	0.00248	0.1216	0.0004	0.9837
Herd size	0.00133	0.0606	0.0005	0.9824
Information access	-7.2121	2.2111	10.6390	0.0011
Market price	0.0465	0.5126	0.0082	0.9276
Market distance	-0.3879	0.6506	0.3556	0.5510
Market availability	2.8836	1.3306	4.6965	0.0302
Transport availability	4.2060	1.4369	8.5683	0.0034
Transport affordability	1.8118	1.2262	2.1831	0.1395
Method of payment	0.2698	1.1086	0.0593	0.8077
Body condition	-5.0995	1.6142	9.9805	0.0016

had access to markets are likely to sell more cattle than those that had little or no access to markets. The presence of markets that farmers like in or near their communities will facilitate cattle sales. The constraint of not having a market has also been reported to reduce cattle sales (Nkhori, 2004) as the lack of markets also increases transactional costs.

6.7.2 Access to information

Keeping everything else constant, insufficient access to information reduced cattle sales. Thus, a negative and significant relationship was not expected. This deviation from expectation could be due to the farmers becoming more informed about market prices and trends to enable them make rational, relevant decisions and strengthen their negotiating ability during transactions with buyers and consequently prevent possible exploitation by better informed buyers as noted by Coetzee *et al.* (2004).

Access to information will then result in small scale farmers not completing deals with buyers who, in most cases, would be exploiting farmers. The failure to complete deals, consequently, reduces cattle sales. The study showed that more than 30% of the respondents preferred to sell their cattle through auctions of which most of the buyers at auction are speculators and butcheries who would want to maximize profits by exploiting the farmers and as the farmers become informed about market prices, they would not sell their cattle if buyers offer them prices lower than the prevailing market prices, resulting in the reduction of cattle sales.

6.7.3 Availability of transport

Another positive and significant factor that affected the probability of farmers selling their cattle was the availability of transport. Using the law of demand, the more the transport is available the cheaper the cost of transporting the cattle and the more the farmers can sell. Cattle sales are high if transport costs are low as farmers will be able to sell even in far away markets as transactional costs would be low. The positive probability of farmers selling as transport becomes more available is in line with the theoretical expectation as well as with Nkhori (2004).

6.7.4 Condition of cattle

All things being equal, an increase in the condition of cattle decreases the probability of small scale farmers selling their cattle. The negative relationship was not expected since small scale farmers, in most cases, sell animals in poor condition (Musemwa *et al.,* 2006). The better the condition of the cattle the more the farmer is motivated to retain the animals rather than sell them. The condition of cattle has a significant effect on the probability of selling cattle in rural areas of the Eastern Cape Province.

6.7.5 Gender of household head

There is a positive and significant relationship between the probability of cattle sales and gender of head of household. As the household head change from being male to being female, the probability of selling cattle increases. This result is not in line with the *a priori* expectations of the probability of males selling being more than that of females. Females are normally involved in many household activities and most of them in the small scale farming sector are not employed hence they do not have any other source of income. Upon the death of the spouse, a widow will in most cases be forced to sell cattle in order to meet the household needs and since women are involved in many household activities adding other activities related to cattle production and marketing such as cattle herding and taking cattle to auctions would be difficult. This therefore forces the new household head (the wife: female) to sell cattle. This probably explains the relationship between gender of household head and the probability of selling cattle.

6.8 Variation of market opportunities across municipalities

The Pearson Chi-square value for market opportunities across municipalities was less than 0.05, meaning that opportunities significantly varied across municipalities (Table 6.1). This is in line with *a priori* expectations that municipalities emerged from different administrations, with differential support services to farmers, and thus might lead to differences in market opportunities. According to the Eastern Cape Development Corporation (ECDC) (2003), the Chris Hani District has a broader base in agriculture with limited agro-processing industries than Amatole. This therefore results in variations in market opportunities in these two municipalities.

6.9 Evaluation of model performance

The Hosmer and Lemeshow Goodness of Fit test statistic is 0.794 and it indicates that the logistic model is a fairly good fit. If the Hosmer and Lemeshow Goodness of Fit test statistic is 0.05 or less, we reject the null hypothesis that there is no difference between the observed and predicted values of the dependent variable; if it is greater, we fail to reject the null hypothesis that there is no difference, implying that the model's estimates fit the data at an acceptable level. While the model obviously does not explain much of the variations in the dependent variable, it however does so to a significant degree.

6.10 Conclusion

Education, market distance, body condition and herd size are significantly associated with municipality while age of head of household, information access, market availability, transport availability and body condition of cattle significantly affect the probability of selling cattle. The model's estimates fit the data at an acceptable level. Mean herd size difference for Alfred Nzo significantly differs with that of Amatole and Chris Hani District Municipalities. Nguni cattle market opportunities vary across municipalities.

References

Bailey, D., C.B. Barrett, P.D. Little and F. Chabari (1999). Livestock markets and risk management among East African pastoralists: A review and research agenda. Utah University, USA.

Bayer, W., R. Alcock and P. Gilles (2004). Going backwards? – Moving forward? – Nguni cattle in communal Kwazulu-Natal. 'Rural poverty reduction through research for development and transformation'. A scientific paper presented at a conference held at Agricultural and Horticultural Faculty, Humboldt-Universität, Berlin. October 5-7, 2004. pp. 1-7.

Bester, J., L.E. Matjuda, J.M. Rust and H.J. Fourie (2005). The Nguni: a case study. Community-Based Management of Animal Genetic Resources. Proceedings of the workshop held in Mbabane, Swaziland, 7-11 May 2001.

Chimonyo, M., N.T. Kusina, H. Hamudikuwanda and O. Nyoni (1999). A survey on land use and usage of cattle for draught in a smallholder farming area of Zimbabwe. Journal of Applied Sciences for Southern Africa 5, 111-121.

Coetzee, L., B.D. Montshwe. and A. Jooste (2004). Livestock marketing: constraints, challenges and implications for extension services. Paper presented at the 38th Conference of the South African Society for Agricultural Extension. May 6, 2004.

Coetzee, L., B.D. Montshwe and A. Jooste (2005). The marketing of livestock on communal lands in the Eastern Cape Province: constraints, challenges and implications for the extension services. South African Journal of Agricultural Extension 34, 81-103.

Dovie, D.B.K., C.M. Shackleton and E.T.F. Witkowski (2006). Valuation of communal area livestock benefits, rural livelihoods and related policy issues. Land Use Policy 23, 260-271.

ECDC (2003a). ECDC sector profile: dairy opportunities. Available at: http://www.ecdc.co.za/sectors/sectors.asp?pageid=108.

ECDC (2003b). ECDC sector profile: leather and leather products. Available at: http://www.ecdc.co.za/sectors/sectors.asp?pageid=115.

FAO (2007). Sub-regional report on animal genetic resources: Southern Africa. Annex to the State of the World's Animal Genetic Resources for Food and Agriculture. Rome, Italy. pp. 1-37.

Fidzani, N.H. (1993). Understanding cattle offtake rates in Botswana. PhD Thesis, Boston University, USA.

Fuller, A. (2006). The sacred hide of Nguni; the rise of an ancient breed of cattle is giving South Africa new opportunity. Miracles that are changing the Nation. Industrial Development Corporation (IDC) Newsletter. pp. 3-4.

Jooste, A. and C.J. Van Rooyen (1996). Constraints to small farmer development in the red meat industry. Red Meat Producers Conference, Roodevallei, Pretoria, 27 March 1996.

Mahabile, M., M. Lyne and A. Panin (2002). Factors affecting the productivity of communal and private livestock farmers in Southern Botswana: A descriptive analysis of sample survey results. Agrekon 41, 326-338.

Makhura, M.T. (2001). Overcoming transaction costs barriers to market participation of smallholder farmers in the Northern Province of South Africa. PhD Thesis, University of Pretoria, RSA.

Mapiye, C., M. Chimonyo, V. Muchenje, K. Dzama, M.C. Marufu and J.G. Raats (2007). Potential for value-addition of Nguni cattle products in the communal areas of South Africa: a review. African Journal of Agricultural Research 2, 488-495.

Montshwe, D.B. (2006). Factors affecting participation in mainstream cattle markets by small-scale cattle farmers in South Africa. MSc Thesis, University of Free State, RSA.

Muchenje, V., K. Dzama, J.G. Raats, M. Chimonyo and P.E. Strydom (2007). Growth performance, carcass characteristics and selected meat quality traits of Nguni, Bonsmara and Aberdeen Angus steers raised on veld in The Eastern Cape, South Africa. In: Proceedings of the 8th Research Council of Zimbabwe Symposium in Harare from 31 January 2007 to 2 February 2007, pp. 23-30.

Muchenje, V., K. Dzama, M. Chimonyo, J.G. Raats and P.E. Strydom (2008). Meat quality of Nguni, Bonsmara and Aberdeen Angus steers raised on natural pasture in the Eastern Cape, South Africa. Meat Science 79, 20-28.

Musemwa, L., C. Chagwiza, W. Sikuka, G. Fraser, M. Chimonyo and N. Mzileni (2007). Analysis of cattle marketing channels used by small scale farmers in the Eastern Cape Province, South Africa. Livestock Research for Rural Development 19 (9), Article #131.

Nkhori, P.A. (2004). The impact of transaction costs on the choice of cattle markets in Mahalapye district, Botswana. MSc Dissertation, University of Pretoria, RSA.

Obi, A. (2007). Institutional constraints to economic development in Africa: a comparative analysis of trade patterns for small farmer development in SACU and non-SACU countries of Southern Africa. Inception Report on the FIRCOP Project No. 6, Department of Agricultural Economics, University of Fort Hare, Alice, RSA.

Raats, J.G., A.M. Magadlela, G.C.G. Fraser and A. Hugo (2004). Re-introducing Nguni nucleus herds in 100 communal villages of the Eastern Cape Province. A proposed co-operative project between the University of Fort Hare, Agricultural and Development Research Institute (ARDRI) and the East Cape Department of Agriculture and the Kellogg Foundation, Unpublished report.

Shackleton, C.M., S.E. Shackleton, T.R. Netshiluvhi, F.R. Mathabela and C. Phiri (1999). The direct use value of goods and services attributed to cattle and goats in the Sand River Catchment, Bushbuckridge, CSIR-Environmentek Report No. ENV-P-C 99003, CSIR, Pretoria, RSA.

Simela, L., B.D. Montshwe, A.M. Mahanjana and M.P. Tshuwa (2006). The livestock production environment in the South African smallholder sector. New challenges for the animal science industries. SASAS 41st Congress Abstracts. pp. 66.

Stroebel, A. (2004). Socio-economic complexities of smallholder resource-poor ruminant livestock production systems in Sub-Saharan Africa. PhD Thesis, University of Free State, RSA.

USAID (2003). Agri-link II Project 2003, Monthly progress report #22, South Africa.

7. Welfare and incentive effects of possible changes in the regulatory environment of the maize market in Swaziland

Ajuruchukwu Obi and Nkosazana N. Mashinini

Abstract

Maize production in Swaziland has been falling in recent years, forcing the country to increasingly rely on imports from neighbouring South Africa. Weather patterns that have featured unrelenting droughts in the country, as in the other Southern African countries, have contributed in no small way to this situation. But it is also considered that excessive regulation of the maize market by the statutory body, the National Maize Corporation (NMC) has played a major role. For instance, it has the sole responsibility to anticipate domestic maize demand based on which imports are programmed. In turn, importation of maize can only be done upon receipt of an import license which is not only costly but also involves very cumbersome processes. With the threat of food insecurity growing, suggestions have been made to deregulate the market for maize in order to make the system more efficient possibly by eliminating inefficient production and service units and transferring resources to their best alternative uses. However, to date, no studies have explicitly investigated the likely effects of the current marketing arrangements and how they will differ in a fully deregulated environment. The purpose of this paper is therefore to present the results of a study that examined the welfare effects of regulation in the maize market of Swaziland as a basis for determining the likely impacts of a possible deregulation of the industry, emphasis being placed on the social welfare effects. The study utilized a partial equilibrium model to quantify the distortions in the maize industry. It also drew from the methodological guidance provided by the Comparative Advantage Studies for Southern Africa sponsored by the United States Agency for International Development (USAID). The study established that the maize industry of Swaziland is uncompetitive and that this situation gives rise to the serious distortions in the maize market. It was also observed that the system features high degree of efficiency losses as a result of the misallocation of resources in the economy. Prospects for any improvement look quite bleak. The paper concludes by recommending the full deregulation of the maize market and dismantling of the statutory structures that currently manage the system. Implications for effective food chain management are highlighted.

7.1 Introduction

Maize is the staple food grain in Swaziland, the small, land-locked, former British colony to the North-East of the Republic of South Africa. About 90% of the crop is grown primarily under subsistence, largely traditional, rain-fed cultivation systems on communal land known as the Swazi Nation Land which is held in trust and allocated by traditional authorities (Magagula and Faki, 1999; Mkhabela *et al.*, 2005). With such high dependence on natural moisture sources, maize production is influenced to a large extent by the weather patterns which have been quite erratic in recent times, featuring frequent drought conditions. According to Ministry of Finance (2005), the last 15 years have been particularly difficult years for maize production, with domestic production consistently failing to meet food consumption requirements of the population.

In order to meet the shortfalls in the domestic production of maize, the country has resorted to huge food imports which now dominate the international trade with neighbouring South Africa. It is estimated that at the minimum, the country derives as much as 60% of its domestic maize requirements from imports from South Africa (Conway and Tyler, 1995; Thompson, 2004; Ministry of Finance, 2005). Despite this, however, frequent food shortages still occur and continue to threaten national food security.

At the same time, the government has pursued a food self-sufficiency policy which features the setting of producer prices above world market prices, the subvention of government marketing institutions, and control of maize imports. As Magagula and Faki (1999) and Thompson (2003) have noted, this arrangement presents a real threat to national food security. In fact, more than any other factors, the present maize crises have exposed the high degree of efficiency losses arising from the gross misallocation of productive resources as a result of the regulatory regime being implemented.

The suggestion has therefore been made that full deregulation of the market must be implemented without delay in order to begin to reverse the damage done by excessive protection of the market manifested in the on-going food crisis in the country. At a basic level, the rationale for market deregulation is that it eliminates inefficient production and service units by transferring resources to their best alternative uses. But to date no studies have explicitly investigated the effects of the current arrangements and the potential effects of full deregulation. It is expected that the present study will fill this gap in knowledge and stimulate discussions on the feasible options for developing the smallholder maize sector in particular, and the agricultural sector in general.

7.2 Objectives

The main objective of the present study was to examine the welfare effects of regulation in the maize market of Swaziland, and on the basis of that, determine the likely impact of a possible deregulation of industry, with emphasis on the social welfare effects. Specifically, the study concentrated on the following aspects:

1. review of the Swazi Maize Market in the context of marketing policies for possible market reform;

2. evaluation of policies and other factors affecting maize production, distribution, and consumption;
3. assessment of the scope for better food security and maize market competitiveness;
4. projections on the impact of the total removal of statutory restrictions on the maize market, particularly examining the welfare effects to producers, consumers and government and other issues relating to the enhancement of market access.

7.3 The country socio-economic context and the place of maize

The small nation of Swaziland is one of the oldest monarchies on the African continent. Landlocked between Mozambique and South Africa, it has a total area of 17,364 km^2 out of which water makes up 160 km^2. The population which currently stands at 1,083,289 is growing at an annual rate of 2.02% (Thompson, 2004). Though statistically classified as a middle-income country due to a nominal per capita GDP of US$ 4,900, Swaziland is a highly fragile economy with few employment opportunities and as much as 69% of the population living below a nationally-defined poverty line of SZL 128.60 (US$ 21.43) per month (Ministry of Finance, 2005).

The mainstay of the economy is agriculture which is dominated by small-scale arable farming and raising of livestock such as goats, pigs, chicken, sheep, etc. The major crops grown under rain-fed, low technology conditions are maize, cotton, vegetables and groundnuts. About 65,800 hectares of land is under maize production each year on land held under communal tenure by majority of the farming population. Annual production of maize is about 94,618 tonnes.

7.3.1 Maize production

On average, about 65,800 hectares of land is under maize production each year, producing on average 94,618 tonnes per year with a yield average of 1.5 tonnes per hectare. It is estimated that about 90% of the maize farms in the Swazi Nation Land are smallholder farms (Mkhabela *et al.*, 2005). Figure 7.1 shows the estimated area planted and production and yield averages of maize in 2003/2004 across the six agro-ecological zones of Swaziland, namely the Highveld, the Upper Middleveld, the Lower Middleveld, the Western Lowveld, the Eastern Lowveld, and the Lubombo Plateau. The attributes of the agro-ecological zones are described in Table 7.1.

The estimated area planted, production and yield averages of maize from 1997 to 2003 in the four main agro-ecological zones of the countryare shown in Table 7.2. Estimates of the maize production, consumption, imports, and self-sufficiency in Swaziland over a 15-year period, from 1987 to 2002, are shown in Table 7.3. Thus, from the 1999/2000 farming

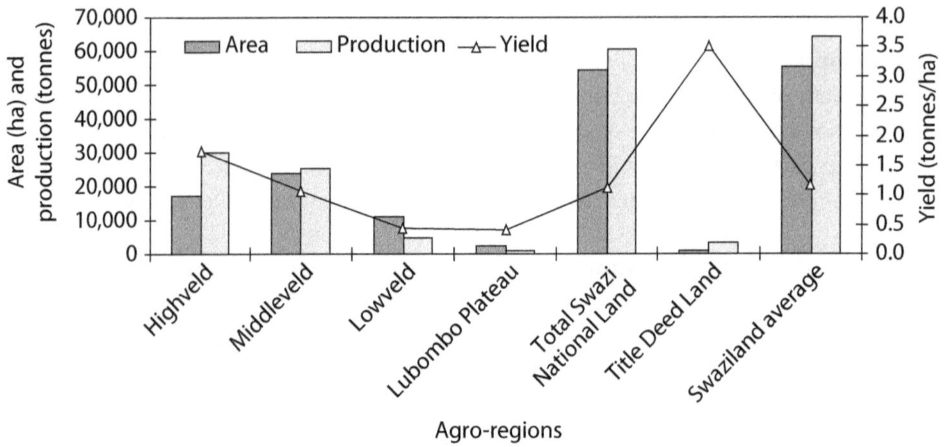

Figure 7.1. Area planted production and yield by agro-ecological zone in 2003/04 (FAO/WFP, 2003, 2004).

Table 7.1. Attributes of Swaziland agro-ecological zones (Magagula and Faki, 1999).

Zone (% of total area)	Altitude range (min-max)	Annual rainfall (80% reliability)	Soils	Farm activities
Highveld (33%)	900-1,400 (600-1,850)	1000-1,200	acidic, low in N, P and Mg; erosion	cattle grazing; small-scale farming, maize is the main crop
Upper Middleveld (14%)	600-800 (400-1000)	800-1000	deep clay loam	main agricultural zone; crops: citrus, pineapple, cotton, maize
Lower Middleveld (14%)	400-600 (250-800)	700-850	sand and sandy loam	crops: groundnuts, beans, vegetables
Western Lowveld (20%)	250-400 (200-500)	450-500	good to fair soils	crops: sugar cane, cotton
Eastern Lowveld (11%)	200-300 (200-500)	400-450	vertisols	crops: groundnuts, sorghum
Lubombo Plateau (8%)	250-600 (100-700)	550-700	escarpment, limited arable land (20%)	main activities: ranching, maize, cotton, minor crops

season, maize production has been declining quite steeply and the country's ability to meet domestic demand from own production has weakened even further.

According to the data, both area planted and total output have been on the decline. For instance, whereas area planted to maize grew from 55,970 ha in 1998/99 to 67,687 ha in 2002/2003, with a 5-year average of 63,587 ha, the area planted in 2003/2004 planting season had declined to 54,470 ha, representing about 86% of the 5-year average. The

Table 7.2. Estimated total maize planted area (hectares) in 2003/04 compared with the 1998/99-2002/03 averages by agro-ecological zone (Central Statistical Office).

Agro-ecological zone	1998/99 (hectares)	1999/00 (hectares)	2000/01 (hectares)	2001/02 (hectares)	2002/03 (hectares)	Five year average (hectares)	2003/04 (hectares)	Percentage average (%)
Highveld	20,025	20,338	20,672	24,358	16,700	20,419	17,236	84
Middleveld	21,241	27,003	19,434	24,254	22,940	22,994	23,642	103
Lowveld	12,096	18,886	14,771	15,831	22,142	16,745	11,064	66
Lubombo Plateau	2,608	2,306	2,974	3,355	5,900	3,429	2,528	74
Swaziland	55,970	68,533	57,851	67,898	67,687	63,587	54,470	86

Table 7.3. Maize production, imports and self-sufficiency in Swaziland (1987/88-2003/04 marketing year) (NMC annual report, 2002).

Year	Consumption (tonnes)	Production (tonnes)	Imports (tonnes)	Self-sufficiency (%)
1987/88	143,845	110,345	33,500	77
1988/89	141,058	122,858	18,200	87
1989/90	104,064	88,964	15,100	85
1990/91	101,539	89,639	11,900	88
1991/92	122,920	59,320	63,600	48
1992/93	123,671	92,971	30,700	75
1993/94	94,215	83,815	10,400	89
1994/95	113,357	83,567	29,700	74
1995/96	158,390	149,190	10,200	94
1996/97	113,114	119,028	29,700	88
1997/98	111,900	107,185	23,215	96
1998/99	125,500	112,456	30,764	90
1999/00	139,000	72,000	24,794	52
2000/01	155,700	72,500	34,330	47
2001/02	156,700	72,600	48,000	46.5

reductions in the area planted seemed to have been most dramatic for the Lowveld and the Lubombo Plateau (Table 7.2). With regard to production, the picture is equally disturbing, with maize output in 2003/2004 being a mere 69% of the average production over the previous 5 years. As shown in Table 7.4, total production of maize in 1998/99 was 107,340 tonnes whereas in 2003/2004 it had fallen to 60,608 tonnes.

The past few years have been rather difficult for the maize industry. According to an FAO mission, the past three years have been marked by unfavourable weather patterns, featuring below average rainfall which has drastically affected maize production in Swaziland, (FAO, 2004). The upshot seems to have been the withdrawal of many farmers from maize production in the face of persistent crop failures and losses in agricultural incomes. In no small way, this phenomenon has exacerbated the poverty situation already complicated by the high rates of unemployment and underemployment in the country.

As Table 7.4 shows, the foregoing situation has posed serious threats for food security in a country where a food self-sufficiency policy is clearly failing. The estimated domestic cereal supply of 78,100 tonnes for the 2003/2004 marketing year falls short of the total national consumption requirement that stood at 205,800 tonnes, resulting in an import requirement of 127,700 tonnes. Commercial imports are estimated at 103,400 tonnes and food aid at 24,300 tonnes. Against this requirement, current government allocations, World Food Programme (WFP) stocks and anticipated relief stock amounted to 15,300 tonnes, still leaving a deficit of 9,000 tonnes to be covered by additional food aid contributions (MOAC-NEWU, 2004).

The production problem is complicated by the high skewedness of the income distribution pattern which puts 43% of the national product in the pockets of a mere 10% of the

Table 7.4. Estimated maize production (tonnes) in comparison to the 1997/1998-2002/2003 by agro-ecological zone (MOAC; NEWU, 2003).

Agro-ecological zone	1997/98 (tonnes)	1998/99 (tonnes)	1999/00 (tonnes)	2000/01 (tonnes)	2001/02 (tonnes)	2002/03 (tonnes)	5 years average (tonnes)	Percentage average (%)
Highveld	44,741	45,486	38,721	33,493	31,404	36,063	38,769	93
Middleveld	50,661	39,939	43,514	28,995	23,921	22,091	37,406	59
Lowveld	24,562	17,358	27,627	16,860	7,223	7,185	18,726	38
Lubombo Plateau	5,240	4,557	2,917	3,187	2,436	3,762	3,667	103
Swaziland	125,204	107,340	112,779	82,535	64,984	62,101	98,568	70

population (FAO, 2004), and the high incidence of HIV/AIDS. With the resulting grossly low purchasing powers of the majority of the population, high cereal prices occasioned by the frequent shortages and scarcities of the basic grain pose a serious risk to national consumer welfare (FAO, 2002 and 2004). As a result, at least since 2002, the country has featured on the priority lists of all major food aid organizations dealing with the current food and humanitarian crisis in Southern Africa. In February 2005, unrelenting drought conditions reportedly destroyed 70% of the maize crop, putting an estimated 260,000 persons at risk of starvation (IRIN, 2005). In May 2009, the National Early Warning Unit of the Swaziland Ministry of Agriculture released estimates that put maize production for the year at about 70,000 tonnes. While these figures represent an 11% increase over the 2008 performance, they are still less than domestic requirements or levels attained in the late 1990s (Table 7.4) and leave a yawning gap of about 127,000 tonnes to be filled through imports over the coming year (FAO, 2009).

7.3.2 Maize policy framework

The major role player in the Swaziland maize industry is the National Maize Corporation (NMC). The NMC is the brainchild of the Ministry of Agriculture and Cooperatives (MOAC) and was established in 1985 as a government parastatal to guarantee a market for maize farmers at competitive prices and to provide high quality, reasonably priced maize meal to the people of Swaziland. The objectives changed in 1995 when NMC contracted out its milling functions to the Swaziland Milling Company (SMC), retaining only its commodity trading role, i.e. the purchase, storage and marketing of maize. The expectation was that this marketing arrangement would assure market access to local farmers for their maize produce (NMC, 1997). Government has entrusted NMC to carryout the following key responsibilities:

- To guarantee a year round competitive market for Swazi maize farmers. This is government strategy to keep more farmers in maize production to maintain the food self-sufficiency objective.
- To guarantee year-round supplies of maize at reasonable costs to the nation.
- To reduce marketing barriers and cost to Swazi producers by improving the logistics of marketing services by running silos efficiently, registering producers, providing drying and shelling services and providing price information.
- To increase the efficiency of the maize market in Swaziland by promoting the availability of white maize to consumers at reasonable costs in all regions of the country.

Without a doubt, NMC's chief purpose is to keep a tight control on the maize industry and to serve as the instrument for achieving the objectives of the government of developing the industry. According to Thompson (2003), NMC stimulates and promotes commercial maize production by improving the domestic marketing system and infrastructure such as the decentralisation of the purchasing function and rehabilitation of the regional silos, which has reduced growers' marketing and transport costs. In addition, NMC has set aside SZL 2 million, which is managed by the Enterprise Trust Fund to assist maize growers. Towards the end of 2001, 150 farmers had benefited from this investment and more are still to benefit. Much as farmer support is necessitated by the imperatives of the domestic farming environment, the fact remains that scarce resources are being diverted from elsewhere in the economy towards supporting a loss-making industry. How this policy has been articulated, at least in respect of maize marketing, is discussed in the next section.

Despite a relatively low share of the GDP, at about 16% in 2003, agriculture is considered a leading sector that plays an important role in rural employment creation and subsistence production. Expectedly, government has put policy emphasis on enhancing sustained and equitable development of the sector as part of a programme of self-sufficiency in food production. Maize policy in Swaziland focuses on interventions to control the marketing, milling, importation of maize and the implementation of a guaranteed minimum price for producers as elaborated below:

7.3.3 Import control policy

The government of Swaziland operates a tight import control regime for maize. The importation of maize meal is completely prohibited. Only the National Maize Corporation (NMC) is permitted by statute to import maize into the country and domestic millers purchase from the organization. Domestic millers also purchase maize from the few rural surplus producers. The complex chain linking maize and maize meal supplies to the urban and rural consumers is illustrated in Figure 7.2. When NMC's stock falls below what is required to meet the demand of local millers, the National Agricultural Marketing Board (NAMBOARD) issues permits to the millers to import maize according to guidelines regarding quantity, quality, grade and class which it strictly supervises. The NAMBOARD also prescribes and collects permit fees and imposes such levies as may be appropriate on the value of scheduled products. The Ministry of Agriculture and Cooperative (MOAC) plays the role of facilitator and arbitrator in the process, especially where disagreements arise between the NAMBOARD and the permit applicants.

7.3.4 Pricing policy

Until 1994 the main pricing policy instrument used by the government to promote maize production was the pan-territorial and pan-seasonal floor prices gazetted each year. The objective of this policy was to ensure that domestic prices do not fall below import parity prices for farmers. Farmers benefited from selling to NMC, which acted as the buyer of last resort that pays the guaranteed price for delivery to its silos (NMC, 1997). Since 1994, NMC has introduced a system of setting prices that are varied according to the levels of domestic supply and demand. Even though NMC continues to set producer and miller

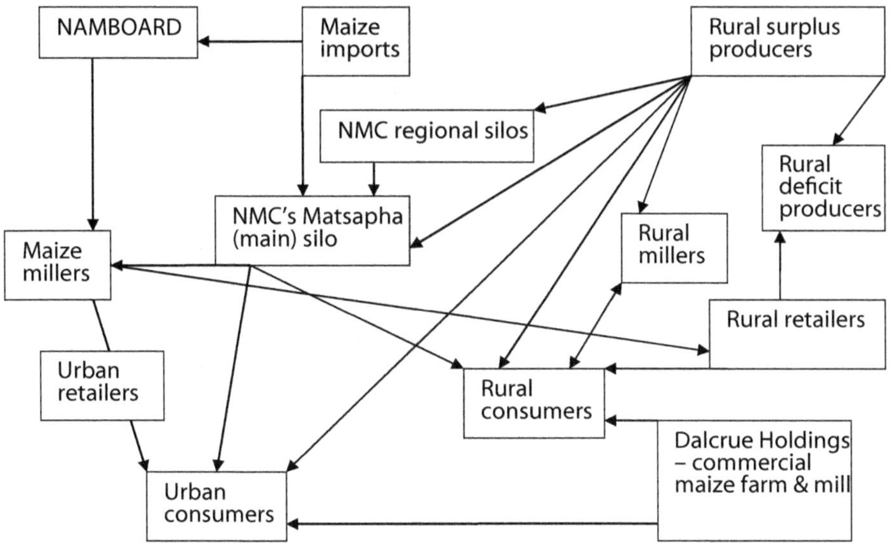

Figure 7.2. The maize marketing chain in Swaziland.

prices, these are based on the SAFEX (South African Futures Exchange) trading prices. The rationale for using SAFEX prices in determining domestic market prices in Swaziland is that the import parity prices used in the country are a\lso derived from SAFEX prices. The implication is that the NMC-imposed SAFEX-based prices places domestic prices at levels higher than import parity prices in order to protect domestic producers against foreign competition.

7.3.5 Marketing reforms

Prior to 1985, commercial maize milling in Swaziland was a monopoly of the private milling company, Swaziland Milling Company (SMC) owned by the SWAKI group. In 1985, because of cheap imports from South Africa, SWAKI threatened to terminate its agreement and close the mill, and this resulted to the establishment of NMC as a parastatal that was responsible for buying and milling local maize plus co-ordinating maize imports as required by the country. However, in 1987 NMC leased the milling monopoly to SMC and the government through NMC bared the risk of purchasing at or above the gazetted prices for locally produced maize which was higher than SA prices, and SMC continued it milling duties as before (Magagula and Faki, 1999).

The government began to reform the maize industry from 1995 under the Maize Marketing Improvement Project (MMIP) which was launched to address concerns over the lack of competition in the maize market and the resulting high prices of maize and maize meal paid by consumers. The first stage of the MMIP was implemented by an institutional reform of NMC which entailed the revocation of its milling responsibilities through the SMC and the restructuring of the organization to a neutral maize purchasing, handling and storage company. The MOAC also transferred the ownership and responsibility for the operations and maintenance of government silo to the NMC.

After the implementation of the MMIP reform, the government pricing policy changed because from the 1994/95 marketing year NMC was no longer willing to purchase at prices above the set floor price, thus changing the floor price to now represent the actual buying price offered by NMC. To date, NMC prices are set from the trading prices and the prices are no longer pan-seasonal as before, though they are still pan-territorial in nature (NMC, 1997).

7.3.6 Government support services affecting the industry

Ministry of Agriculture and Cooperatives, through the Rural Development Administrations (RDAs) provides heavily subsidized tractor services to farmers all over the country. These tractors are used for field preparation and planting. The Ministry of Agriculture and Cooperatives has discontinued the subsidization of maize production inputs like seeds and fertiliser. Since the withdrawal of this subsidy, maize production in the country has declined (NMC, 1997). Today, there is Inhlanyelo Fund, a non-governmental, microfinance institution that provides small low-interest loans on inputs to impoverished Swazis to grow maize and other crops for business. Farmers are using this scheme to finance their production. One of its key strengths is that it is working through existing administrative and community leadership structures which provide the channels through which it reaches individual beneficiaries at the grassroots level. Also, MOAC's Agricultural Research Division (ARD) undertakes research and testing/screening of new crop varieties to ascertain their suitability to local conditions. The ARD also carries out trials related to pest and disease control across the country and makes the results available to farmers through the extension services.

7.3.7 Theoretical framework and methodology

Governments have normally intervened in cross border trade through the use of tariffs, quotas and non-tariff barriers. Reductions in these trade restrictions have been common practice in recent years, and it is mostly achieved through international negotiations (Freenstra, 1995). Today's most celebrated economic theory in worldwide trade negotiations is that free trade is more beneficial than trade protection. It is now widely accepted that trade barriers distort optimal allocation of production resources and ultimately result in reduced output. This is the static argument. The dynamic argument in favour of dismantling trade restrictions is that economic freedom stimulates competition and ensures an environment beneficial for economic growth (Bale and Greeenshields, 1978). Despite widespread freer trade advocacy and the numerous global efforts to lighten the economic costs of trade protection, a number of governments still intervene in trade.

In many instances, those in favour of protection plead the food-price dilemma, vis-à-vis the conflicting goals of maintaining food prices that are profitable for producers and affordable to consumers. In Southern Africa as elsewhere on the continent, this classic food price dilemma historically has been addressed through controlled marketing systems, in which food prices are artificially raised for producers and lowered for consumers through subsidies. However, this practice has become fiscally unsustainable in many countries. As a result, food market reforms are currently being implemented in over 20 African countries (Jayne and Argwings-Kodhek, 1997; Jayne and Jones, 1997; Donovan, 1996; World Bank, 1994).

However, the process of food market liberalisation has often proceeded without the full confidence of key policy makers. Serious concerns have been raised throughout Africa over the effects of market liberalisation and the elimination of food subsidies on low-income consumers' access to food. Consumer behaviour is a key determinant of the impact of market liberalisation on household food security and, as a result, controls on food prices and trade have frequently been revived during drought years when fears of food shortages and high prices are heightened (Jayne and Argwings-Kodhek, 1997; Lewa and Hubbard, 1995).

Whilst much research has been devoted to understanding how farmers, traders and consumers have responded to food market reform, little is known about the responses of the Swaziland maize industry to such market reforms. This study therefore opts to shade light on the welfare effects of the way the maize market industry is currently organized in Swaziland.

The study on which the present paper is based used the partial equilibrium model to quantify the distortions in the maize industry of Swaziland since the goal is to gain practical understanding of one-sector policies rather than an economy-wide analysis for which a general equilibrium framework would have been appropriate. Similar approaches have been adopted in the literature as clearly reported by Bale and Greenshields (1978), Bale and Lutz (1981), Wright and Nieuwoudt (1993), Van Schalkwyk *et al.* (1997). The methodological manual developed for the regional Comparative Advantage Studies for Southern Africa under the auspices of the United States Agency for International Development (USAID), prescribed similar approaches (Hassan *et al.*, 1999) adapted from the seminal works of Monke and Pearson (1989) and Tsakok (1990). In a recent book, Aksoy and Beghin (2005) suggest that partial equilibrium analysis at country and commodity levels provide sufficient guidance for credible policy formulation.

Data for the present analysis were obtained from National Maize Corporation (NMC), the National Early Warning Unit for food security and the Food and Agricultural Organisation (FAO) statistical database. The welfare calculations were made for a six-year marketing period 1998/1999 to 2003/2004. The domestic price P_d represents the actual price that producers receive, while the border price P_b represents the import prices.

By means of a set of nine variables, the social impact of the current regulatory regime in the maize market is assessed, looking specifically at efficiency losses in production and consumption, changes in producer and consumer surpluses, and revenue over a six-year period, 1998-2004. The variables and their derivations are set out in Table 7.5 while the analytical procedures to determine the welfare effects are shown in Table 7.6.

In this structure, supply and demand elasticities and price data are used to calculate the financial implications of a change in commodity price, the welfare transfers between producers and consumers and the net gains and losses in economic efficiency. In line

Table 7.5. Variables for assessment of social impact of maize market regulation in Swaziland.

Variable	Definition	Formulae
P_d	Domestic price for maize	
P_b	Border price for maize	
e_s	Price elasticity of supply	
n_d	Price elasticity of demand	
NPC	Nominal Protection Coefficient	P_d/P_b
t	Implicit tariff	NPC-1
t′	Nominal tariff rate	tP_b/P_d
V′	Value of domestic production at domestic price	P_d×dom. Prod.
W′	Value of domestic consumption at domestic price	P_d×total supply

Table 7.6. Specification of welfare effects of maize market regulation in Swaziland.

Welfare effects	Definition	Formulae
NEL_p	Deadweight loss in production	$0.5 \times e_s \times t^2 \times V'$
NEL_c	Deadweight in consumption	$0.5 \times n_d \times t'^2 \times W'$
WG_p	Change in producer surplus	$t'V' - NEL_p$
WG_c	Change in consumer surplus	$-(t'W' + NEL_c)$
^GR	Change in revenue	$t'(W'-V')$
Loss/capita	Loss per capita	WGc/total.pop

with Tsakok's approach, the quantification of price changes relating to policy changes is considered to be complementary to measuring coefficients of protection and comparative advantage which all contribute to effective policy analysis and design.

The treatment of prices needs to be highlighted at this point. The prices received by maize producers represent the domestic prices employed in this analysis. These are also the prices prevailing at the mill gate. Border price is the price prevailing at the point of entry (or exit) for an internationally tradable commodity and represents its shadow price or opportunity cost. For imported commodities, adjustments are made to accommodate the expenditures on transportation, processing, and other post-harvest/marketing activities, in order to equilibrate border prices with prices received by local farmers for the commodities. Internal margins arising from milling must also be added if border prices are adjusted to mill-gate prices. Due to shortage of sufficient time-series data, it was not possible to estimate the price elasticities econometrically. However, these variables are expected to be quite low for the following reasons:

1. The price elasticity of supply is expected to be close to zero because maize dominates the cropping system and few food crops are available as close substitutes. Again, most of the households grow maize for own-consumption so that price-effects on production are minimal.
2. Maize production and consumption in Swaziland are governed by strong cultural factors which means that price variations play a minimal role, especially in the dominant subsistence sector.
3. The price elasticity of demand is also expected to be close to zero because maize is a dietary staple with no close substitutes in the food system of Swaziland.

As a result, price elasticities of demand and supply were obtained from previous research conducted to measure price distortions in the maize marketing system of South Africa (Wright and Niewoudt, 1993). Results for South Africa are used because they are low enough, being derived from studies that included communal areas where conditions and dietary patterns are no different from what obtains in Swaziland. The price elasticity estimates are also the nearest to a normal competitive marketing environment to allow for the analysis of the welfare effects of alternative marketing arrangements for Swaziland. The remaining parameters/variables were derived using the formulae as shown in Table 7.5.

7.3.8 Welfare effects of maize pricing and marketing policies in Swaziland

The results of the analysis with respect to welfare effects of the maize pricing and marketing policies are presented in Table 7.7 for the six marketing seasons, 1998-2004 covered by the study. Table 7.7 shows that both domestic and border prices have been increasing steadily since the beginning of the reference period although they fell slightly at the beginning of 2003 after attaining a peak in 2002. It should be noted that the recent humanitarian and

Table 7.7. Welfare analysis of the maize market system in Swaziland.[1]

Variable	Label	Formulae	Price/Value					
			1998/99	1999/00	2000/01	2001/02	2002/03	2003/04
Parameters:								
P_d	Domestic price for maize		700.00	850.00	750.00	1,276.00	2137	1,200.00
P_b	Border price for maize		663.31	803.13	665.45	1,225.00	1,623.46	1,030.63
e_s	Price elasticity of supply		0.151	0.151	0.151	0.151	0.151	0.151
n_d	Price elasticity of demand		-0.513	-0.513	-0.513	-0.513	-0.513	-0.513
NPC	Nominal Protection Coefficient	P_d/P_b	1.055	1.058	1.127	1.902	1.202	1.164
T	Implicit tariff	NPC-1	0.055	0.058	0.127	0.902	0.202	0.164
t′	Nominal tariff rate	tP_b/P_d	0.052	0.055	0.113	0.474	0.168	0.141
V′	Value of domestic production at domestic price	P_d×dom. Prod.	75,138,000	91,239,000	80,505,000	250,102,200	209,457,909	128,808,000
W′	Value of domestic consumption at domestic price	P_d×total supply	87,850,000	106,675,000	94,125,000	292,415,000	244,894,425	150,600,000
Analysis:								
NEL_p	Deadweight loss in production	$0.5 \times e_s \times t^2 \times V'$	15,586.13	20,944.21	77,239.51	4,246,927.58	446,503.94	193,730.85
NEL_c	Deadweight loss in consumption	$0.5 \times n_d \times t'^2 \times W'$	61,910.00	83,192.94	306,804.73	16,869,313.45	1,773,568.01	769,522.54
WG_p	Change in producer surplus	$t'V' - NEL_p$	3,922,865.06	5,009,991.29	8,997,991.79	114,363,455.30	34,749,029.60	17,986,413.94
WG_c	Change in consumer surplus	$-(t'W'+NEL_c)$	-4,666,676.38	-5,965,272.36	-10,917,402.16	-155,546,443	-42,923,553.66	-22,025,421.28
^GR	Change in mill revenue	$t'(W'-V')$	666,315.20	851,143.92	1,535,366.13	20,066,746.35	5,954,452.11	3,075,753.95
Loss/capita	Loss per capita	WG_c/total.pop	-4.64	-5.81	-10.46	-6.07	-64.02	-20.45

[1] Authors' calculations; price in SZL.

food crisis in Southern Africa actually came to a head in 2002 when almost all the countries in the region faced severe food shortages. This weather-related disaster actually exacerbated a crisis that was already looming as a result of inappropriate policy. For instance, the results show that the import tariff imposed in 1998/99 resulted in an observed producer efficiency loss amounting to some SZL 15,586.13 (Swaziland Lilangani; SZL 1 = 1 South African Rand). By 2001/2002, the efficiency loss (also defined as 'deadweight loss in production') amounted to about SZL 4.2 million before it began to fall over the next two years of the observation period. Without a doubt, this efficiency loss experienced by producers arose from the tariff structure occasioned by government intervention.

The results also show that consumers experienced similar hardships during the period under reference. For instance, as shown in Table 7.7, the deadweight loss in consumption rose steadily from 1998 to 2001/2002, and far exceeded the deadweight losses in production. Even though they have been lower since 2002, they are still substantial and far in excess of the losses encountered by producers. It is safe to say therefore that consumers have been hurt more than producers by the current marketing policy framework for maize. It is in fact possible that the consumer efficiency losses would have been higher in the absence of the subsidy to NMC.

In 1998/99 marketing season domestic maize producers gained an estimated amount of SZL 3,922,865.06 for selling their produce to NMC at higher than the equilibrium prices. In the same period, consumers lost SZL 4,666,676.38. In 2003/04 producers gained SZL 17,986,413.94 while consumers lost an estimated amount of SZL 22,025,421.28 for consuming at higher prices. On the other hand, government has benefited from the receipt of import tariff which yielded a total of SZL 666,315.20 in 1998/99 and SZL 3,075,753.95 in 2003/04.

7.3.9 Incentive effects of the maize marketing policy

Given the foregoing, it was necessary to undertake a systematic assessment of the maize marketing system to determine the incentive effects of the current policy regime for the industry. For this purpose, a Policy Analysis Matrix (PAM) was constructed. The enterprise budget for the production of a tonne of maize in Swaziland is presented in Table 7.8 while the Policy Analysis Matrix framework and the effects of the divergences are shown in Table 7.9. The results are based on production and marketing data derived for a tonne of maize. In line with the PAM methodology, the analysis examined the private profitability, social profitability, and policy divergences associated with the production and marketing of maize within the policy framework under investigation.

Table 7.8. Enterprise budget in SZL for maize production in Swaziland.[1]

		Private values (SZL)	Social values (SZL)
Revenue		1,100	850
Costs			
Tradable inputs			
• Seed		70	58
• Fertilizer: 2.3.2 (38) + Zn		405	331
• LAN 28%		46	38
• Insecticide:	– Decis	29	24
	– Malathion	12	10
	– Stalk-borer controller	90	74
Total tradable inputs		652	535
Non-tradable inputs			
• Labour:	– Weeding	96	133
	– Fertilizer application	60	83
	– Pest control	36	50
	– Reaping and handling	120	166
• Farm machinery:	– Dryer fees	18	45
	– Tractor hire	150	950
	– Sheller hire	300	600
Total non-tradable inputs		780	2,028
Gross margin		-332	-1,713

[1] Authors' calculations.

Table 7.9. Summary estimates of price effects of maize marketing policies in Swaziland (2004 prices).[1]

	Revenue	Tradable inputs	Domestic factors	Profits
Private prices	1,100	652	780	-332
Social prices	850	535	2,028	-1,713
Divergences	250	117	-1,247	1,380

[1] Authors' calculations; values in SZL.

7.3.10 Profitability of the maize industry

The results indicate that the maize enterprise was not financially viable when examined from the standpoint of the individual farmer who grew maize in the Swazi Nation Land. As Table 7.8 shows, a high negative profit was realized during the study period. This situation probably arises from the high domestic costs of production which mean that gross margins remain low despite higher domestic prices for maize. The market prices for labour are those based on the national minimum wage legislation. One reason for this is that costs are higher than revenues, leading to production losses. Since the private profitability parameter is also a measure of competitiveness, it can be concluded that the maize industry of Swaziland is not competitive.

The above finding is quite serious because, despite a general down-scaling of subsidies in recent times, farmers still receive considerable support from the Government of Swaziland. For instance, there are still some subsidies on the use of farm machineries as is obvious from the disaggregation of the production costs in Table 7.8. As the enterprise budget shows (Table 7.8), farm machineries like dryers, tractors and shellers which are hired from the Ministry of Agriculture, come at an average subsidy of about 60%. Another indication of the extent of subsidization of local costs of maize production is given by the large difference between the private domestic costs, at SZL 780, and the social domestic costs, estimated at SZL 2,028 per tonne of maize. The high social costs of maize production also mean that the industry's value adding to foreign exchange earnings is almost negligible. This element was assessed by calculating the domestic resource costs (DRC) which gave a value of 6.44 (Table 7.10). As is well known, DRC>1 implies that the industry is using up more resources by producing locally. According to Monke and Pearson (1989), the DRC serves as a proxy measure for social profits. Its high positive value therefore implies that the industry is socially unprofitable.

Table 7.10. PAM ratios for the Swaziland maize industry.[1]

Private cost ratio	1.74
Domestic resource cost ratio	6.44
Nominal protection coefficient on tradable outputs	1.29
Nominal protection coefficient on tradable inputs	1.22
Effective protection coefficient	1.42
Profitability coefficient	0.19
Subsidy ratio to producers	1.62

[1] Authors' calculations.

Again, the yield figure used for computing the level of maize output per tonne and the revenue is at the top end of the possibilities based on optimal adoption of improved varieties. In addition, government tries to protect farmers from import competition by raising domestic prices above import parity price. In spite of these favourable government programmes which have created conducive production and marketing environments for maize farmers, losses are still occurring in the maize industry. This means that if all subsidies and other forms of support are eliminated, losses will be quite high. How serious this will be is evidenced by the high negative social profitability shown in Table 7.9. We thus have a situation where the maize industry is both privately and socially unprofitable. This means that the maize industry is uncompetitive and the country has no comparative advantage in the production of maize, at least within a regional context.

7.3.11 Distorting policies and market failures

In the analysis of the Swaziland Maize Industry, four distinct divergences have been examined, namely the output transfer, the tradable input transfer, domestic input transfer, and the net transfer price. These transfers measure the effects of distorting policies rather than market failures. Along with these measures, a set of ratios were calculated using the same output, price and value figures employed in the enterprise budget and the associated PAM.

Output transfer looks at the size of the actual revenue realized from maize production in relation to what is potentially realizable. If there are no distortions, and in the absence of market failures, the potential can be realized. The actual revenue is derived by expressing the actual physical production in current market value terms, while the potential revenue is derived by shadow pricing the current physical output. This measure therefore makes a judgement on the effectiveness of the system in meeting the production objective of maximum output at the minimum social costs.

Table 7.9 shows that a positive output transfer of SZL 250 was realized in this study. This means that if there were no government interventions in the sector, revenues would fall by SZL 250. Thus, the distorting policy of government setting domestic prices higher than international prices favours domestic producers while hurting the economy at large. The extent to which the policy favours private prices is indicated by the Nominal Protection Coefficient ratio (Table 7.10).

An NPCO value of 1.29 was obtained for the Swaziland maize industry, implying that the market prices in Swaziland are 29% higher than their true scarcity value. This means that the government of Swaziland has policies in operation that raise the domestic market price for maize. This in fact is true. Earlier, it has been revealed that maize prices in Swaziland are slightly elevated to encourage domestic maize production in line with the national objective of self-sufficiency in maize production.

Input transfers measure the extent to which the valuation of the inputs affects the economy. Different calculations are involved depending on whether we are dealing with tradable factors or domestic, non-tradable factors. In the case of tradable input transfer, the interest is to determine the extent to which the domestic prices differ from import parity prices as a reflection of the existence or otherwise of policies that either restrict or promote importation. The main policy instrument in this case is the import tax. A positive tradable input transfer denotes an implicit tax while a negative input transfer denotes an implicit subsidy. From Table 7.9, it is clear that imported inputs are being taxed, with the result that their prices within the country are higher than their border prices. Examination of the Nominal Protection Coefficient on Tradable Inputs (NPCI) in Table 7.10 confirms this observation and shows that local prices are actually higher than international prices by about 22%. On the other hand, Table 7.9 shows that the domestic factors are highly subsidized by the government as evidenced by the high negative divergence of SZL 1,247. This indication is borne out by the high Subsidy Ratio to Producers (SRP) of 1.62 (see Table 7.10).

The net effect of all the transfers is shown in Table 7.9 as a high positive divergence of SZL 1,380. This indicates that the maize industry is still highly subsidized despite recent attempts at market reforms. This result confirms the indication provided by the SRP in respect of the level of subsidy to maize producers. Another way of looking at the total effects is by measuring the profitability coefficient (PC). In this study a value of 0.19 is obtained for the PC, suggesting that there is an 81% net transfer from the domestic economy to support the particular industry under investigation. A subject of much policy discussion in Swaziland remains the huge amount spent by such a poor country in supporting a parastatal like the NMC which presides over a loss-making maize industry.

7.3.12 Sensitivity analysis

This study incorporated a series of sensitivity analyses to address the inherent weakness of the PAM methodology of assuming fixed input-output coefficients which are clearly unrealistic. The methodology used is in line with procedures recommended by Tsakok (1990) and is consisted with Monke and Pearson's (1989) suggestion that whenever it is possible to obtain reliable information for predicting responses of inputs and outputs to social prices, it should be introduced into the calculations to improve the predictive power of the PAM. As is well-known, the economic variables such as prices, yields and even the policy environment, are in a constant state of flux. Ideally, a General Equilibrium model captures this dynamism in nature to a large extent. Unfortunately, the data requirements of such a model are beyond what can be reasonably and cost-effectively obtained on the issues of interest for this research and also in the time available. For this reason, the sensitivity analyses which, while still employing static procedures, varies a number of the variables, is considered an effective compromise.

On the basis of what is known about their policy relevance and the insights they provide into the competitiveness and comparative advantage questions, the study examined the effect of changes on prices and yield on the same ratios already examined in a the purely static framework. Tables 7.11 and 7.12 present the sensitivity analyses under alternative scenarios of the enterprise budget based on prices and yield levels potentially realizable within the economy. Table 7.13 presents a comparison of all the outcomes, namely the social and private gross margins and the PAM ratios which were examined earlier under the static assumption.

As the results show, the key PAM parameters, as well as the production performance index of gross margin, are highly sensitive to the changes in prices and yield levels. The analysis assumed an annual price increase in line with current trends in the economy and outlook for the future. Official data shows that the rate of inflation is about 5.4% per annum and closely

Table 7.11. Sensitivity analysis under alternative scenarios of enterprise budget (2004-2021).

Year	2004/05	2005/06	2006/07	2007/08	2008/09	2009/10	2010/11	2011/12	2012/13
Revenue	850	978	1,124	1,349	1,619	1,942	2,331	2,797	3,357
Tradable input									
Seed	58	61	64	67	70	74	78	82	86
Fertiliser: 2.3.2(38)+zn	331	348	365	383	402	422	444	466	489
LAN 28%	38	40	42	44	46	48	51	53	56
Insecticide: Decis	24	25	26	28	29	31	32	34	35
Malathion	10	11	11	12	12	13	13	14	15
Stalk-borer controller	74	78	82	86	90	94	99	104	109
Total tradable inputs	535	562	590	619	650	683	717	753	790
Domestic inputs									
Labour: Weeding	133	137	141	145	150	154	159	164	169
Fertiliser application	83	86	88	91	94	96	99	102	105
Pest control	50	51	53	55	56	58	60	61	63
Reaping and handling	166	171	176	182	187	193	199	205	211
Dryer fees	45	46	48	49	51	52	54	55	57
Tractor hire	950	979	1,008	1,038	1,069	1,101	1,134	1,168	1,203
Sheller hire	600	618	637	656	675	696	716	738	760
Total domestic inputs	2,027	2,088	2,151	2,215	2,282	2,350	2,421	2,493	2,568
Gross margins	-1,712	-1,672	-1,617	-1,486	-1,313	-1,091	-807	-449	-2
DRC	6.44	5.02	4.03	3.04	2.36	1.87	1.50	1.22	1.00

Values in SZL.

follows that of South Africa (World Bank, 2005). But over the past two years, domestic prices have been rising by 10% annually (World Bank, 2005). According to the IMF (2003) and the World Bank, the high oil prices and the frequent salary increases put considerable pressure on domestic prices. There was a salary increase for civil servants in 2003 whose implementation process was completed in 2004 (Ministry of Finance, 2004).

The domestic prices relevant for the sensitivity analyses are domestic costs of maize production. The most important of these costs are those related to labour and the operation of machinery for the cultivation of more land expected to be brought into production when broad-based trade liberalization is implemented. In order not to give maize undue advantage in terms of cost structure, the sensitivity analysis has been based on an annual increase in the prices of these inputs at the rate of 10% in the first three years and later increasing to 15% over the remaining 17 years of the forecast period. Product prices and

2013/14	2014/15	2015/16	2016/17	2017/18	2018/19	2019/20	2020/21	2021/22	2022/23	2023/24	2024/25
4,028	4,834	5,800	6,960	8,352	10,023	12,027	14,433	17,319	20,783	24,940	29,928
90	94	99	104	109	115	121	127	133	140	147	154
513	539	566	594	624	655	688	723	759	797	836	878
59	62	65	68	72	75	79	83	87	91	96	101
37	39	41	43	45	48	50	52	55	58	61	64
16	16	17	18	19	20	21	22	23	24	25	27
115	121	127	133	140	147	154	162	170	178	187	196
830	871	915	961	1,009	1,059	1,112	1,168	1,226	1,288	1,352	1,420
174	179	184	190	195	201	207	213	220	226	233	240
108	112	115	119	122	126	130	133	137	142	146	150
65	67	69	71	73	75	78	80	82	85	87	90
217	223	230	237	244	252	259	267	275	283	292	300
59	60	62	64	66	68	70	72	74	77	79	81
1,240	1,277	1,315	1,354	1,395	1,437	1,480	1,524	1,570	1,617	1,666	1,716
783	806	831	855	881	908	935	963	992	1,021	1,052	1,084
2,645	2,725	2,806	2,891	2,977	3,067	3,159	3,253	3,351	3,451	3,555	3,662
553	1,237	2,079	3,109	4,366	5,897	7,757	10,012	12,742	16,044	20,033	24,847
0.83	0.69	0.57	0.48	0.41	0.34	0.29	0.25	0.21	0.18	0.15	0.13

expected revenues are as shown in Table 7.11 in social terms and Table 7.12 in market prices. The base year is taken as 2004/2005 and market prices for the year are as were used in the original calculations.

As indicated earlier, changes in the level of maize yield are assumed over the 20-year period in response to expanded farmer support programmes from the Ministry of Agriculture which will now focus its professional attention on issues such as research, information and

Table 7.12. Sensitivity analysis under alternative enterprise budget scenarios and PAM (2004-2021).

AR	2004/05	2005/06	2006/07	2007/08	2008/09	2009/10	2010/11	2011/12	2012/13
Revenue	1,100	1,265	1,455	1,600	1,760	1,936	2,130	2,343	2,577
Tradable input									
Seed	70	74	77	81	85	89	94	98	103
Fertiliser: 2.3.2(38)+zn	405	425	447	469	492	517	543	570	598
LAN 28%	46	48	51	53	56	59	62	65	68
Insecticide: Decis	29	30	32	34	35	37	39	41	43
Malathion	12	13	13	14	15	15	16	17	18
Stalk-borer controller	90	95	99	104	109	115	121	127	133
Total tradable inputs	652	685	719	755	793	832	874	917	963
Domestic inputs									
Labour: Weeding	96	99	102	105	121	139	160	183	211
Fertiliser application	60	62	64	66	75	87	100	115	132
Pest control	36	37	38	39	45	52	60	69	79
Reaping and handling	120	124	127	131	151	173	199	229	264
Dryer fees	18	19	19	20	23	26	30	34	40
Tractor hire	150	155	159	164	188	217	249	287	330
Sheller hire	300	309	318	328	377	434	499	573	659
Total domestic inputs	780	803	828	852	980	1,127	1,296	1,491	1,714
Gross margins	-332	-223	-92	-7	-12	-23	-40	-65	-100
PCR	1.74	1.38	1.12	1.01	1.01	1.02	1.03	1.05	1.06
NPCO	1.29	1.29	1.29	1.19	1.09	1.00	0.91	0.84	0.77
NPCI	1.22	1.22	1.22	1.22	1.22	1.22	1.22	1.22	1.22
PC	0.19	0.13	0.06	0.00	0.01	0.02	0.05	0.15	49.15
EPC	1.42	1.40	1.38	1.16	1.00	0.88	0.78	0.70	0.63
SRP	1.62	1.48	1.36	1.10	0.80	0.55	0.33	0.14	-0.03

Values in SZL.

extension instead of marketing and distribution functions as has been case to date. Table 7.11 shows that revenues will rise gradually over the next 8 years, with the possibility of the system breaking even by 2012/2013. This will be reflected in a steadily rising gross margin. Although the gross margins will still be negative over this period, the results suggest that the level of losses will continue to fall until 2013. Subsequently, the gross margins will start to rise up to the end of the forecast period in 2021.

2013/14	2014/15	2015/16	2016/17	2017/18	2018/19	2019/20	2020/21	2021/22	2022/23	2023/24	2024/25
2,835	3,118	3,430	3,773	4,151	4,566	5,022	5,524	6,077	6,685	7,353	8,088
109	114	120	126	132	139	146	153	160	168	177	186
628	660	693	727	764	802	842	884	928	975	1,023	1,075
71	75	79	83	87	91	96	100	105	111	116	122
45	47	50	52	55	57	60	63	66	70	73	77
19	20	21	22	23	24	25	26	28	29	30	32
140	147	154	162	170	178	187	196	206	217	227	239
1,011	1,062	1,115	1,171	1,229	1,291	1,355	1,423	1,494	1,569	1,648	1,730
243	279	321	369	424	488	561	645	742	854	982	1,129
152	174	201	231	265	305	351	403	464	533	614	706
91	105	120	138	159	183	210	242	278	320	368	423
303	349	401	461	530	610	702	807	928	1,067	1,227	1,411
45	52	60	69	80	92	105	121	139	160	184	212
379	436	501	577	663	763	877	1,008	1,160	1,334	1,534	1,764
758	872	1,003	1,153	1,326	1,525	1,754	2,017	2,320	2,667	3,068	3,528
1,971	2,267	2,607	2,998	3,448	3,965	4,560	5,244	6,031	6,935	7,976	9,172
-148	-211	-292	-396	-527	-691	-893	-1,143	-1,448	-1,820	-2,270	-2,814
1.08	1.10	1.13	1.15	1.18	1.21	1.24	1.28	1.32	1.36	1.40	1.44
0.70	0.65	0.59	0.54	0.50	0.46	0.42	0.38	0.35	0.32	0.29	0.27
1.22	1.22	1.22	1.22	1.22	1.22	1.22	1.22	1.22	1.22	1.22	1.22
-0.27	-0.17	-0.14	-0.13	-0.12	-0.12	-0.12	-0.11	-0.11	-0.11	-0.11	-0.11
0.57	0.52	0.47	0.43	0.40	0.37	0.34	0.31	0.28	0.26	0.24	0.22
-0.17	-0.30	-0.41	-0.50	-0.59	-0.66	-0.72	-0.77	-0.82	-0.86	-0.89	-0.92

Table 7.13. Comparison of private and social gross margins and PAM ratios (2004-2021).

Year	2004/05	2005/06	2006/07	2007/08	2008/09	2009/10	2010/11	2011/12	2012/13
Private gross margins	-332	-223	-92	-7	-12	-23	-40	-65	-100
Social gross margins	-1,712	-1,672	-1,617	-1,486	-1,313	-1,091	-807	-449	-2
PCR	1.74	1.38	1.12	1.01	1.01	1.02	1.03	1.05	1.06
DRC	6.44	5.02	4.03	3.04	2.36	1.87	1.50	1.22	1.00
NPCO	1.29	1.29	1.29	1.19	1.09	1.00	0.91	0.84	0.77
NPCI	1.22	1.22	1.22	1.22	1.22	1.22	1.22	1.22	1.22
PC	0.19	0.13	0.06	0.00	0.01	0.02	0.05	0.15	49.15
EPC	1.42	1.40	1.38	1.16	1.00	0.88	0.78	0.70	0.63
SRP	1.62	1.48	1.36	1.10	0.80	0.55	0.33	0.14	-0.03

Values in SZL.

In line with the positive developments in the maize production performance, it is shown in Table 7.11 that the domestic resource cost ratio (DRC) will continue to decline, reflecting improvements in the international competitiveness of the sector as the changes in maize pricing and yield continue over the period. A DRC value of 1.0 in 2013 equally suggests that the sector is likely to break even in that year if the positive developments in pricing and yield continue. Tables 7.12 and 7.13 present the results in relation to the other PAM ratios such as PCR, NPCO, NPCI, PC, EPC, and SRP. In all cases, there is evidence that a well-planned and implemented liberalization programme would produce beneficial effects on the sector and the economy as a whole. These trends are graphed in Figures 7.3 and 7.4.

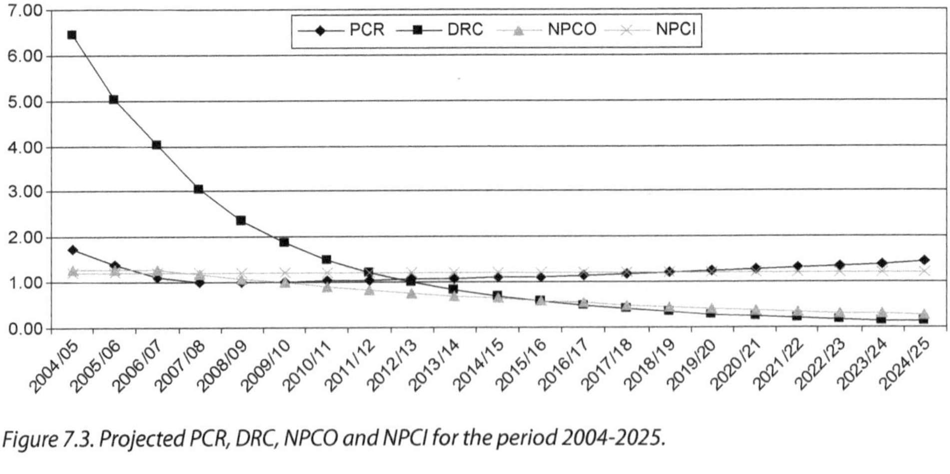

Figure 7.3. Projected PCR, DRC, NPCO and NPCI for the period 2004-2025.

2013/14	2014/15	2015/16	2016/17	2017/18	2018/19	2019/20	2020/21	2021/22	2022/23	2023/24	2024/25
-148	-211	-292	-396	-527	-691	-893	-1,143	-1,448	-1,820	-2,270	-2,814
553	1,237	2,079	3,109	4,366	5,897	7,757	10,012	12,742	16,044	20,033	24,847
1.08	1.10	1.13	1.15	1.18	1.21	1.24	1.28	1.32	1.36	1.40	1.44
0.83	0.69	0.57	0.48	0.41	0.34	0.29	0.25	0.21	0.18	0.15	0.13
0.70	0.65	0.59	0.54	0.50	0.46	0.42	0.38	0.35	0.32	0.29	0.27
1.22	1.22	1.22	1.22	1.22	1.22	1.22	1.22	1.22	1.22	1.22	1.22
-0.27	-0.17	-0.14	-0.13	-0.12	-0.12	-0.12	-0.11	-0.11	-0.11	-0.11	-0.11
0.57	0.52	0.47	0.43	0.40	0.37	0.34	0.31	0.28	0.26	0.24	0.22
-0.17	-0.30	-0.41	-0.50	-0.59	-0.66	-0.72	-0.77	-0.82	-0.86	-0.89	-0.92

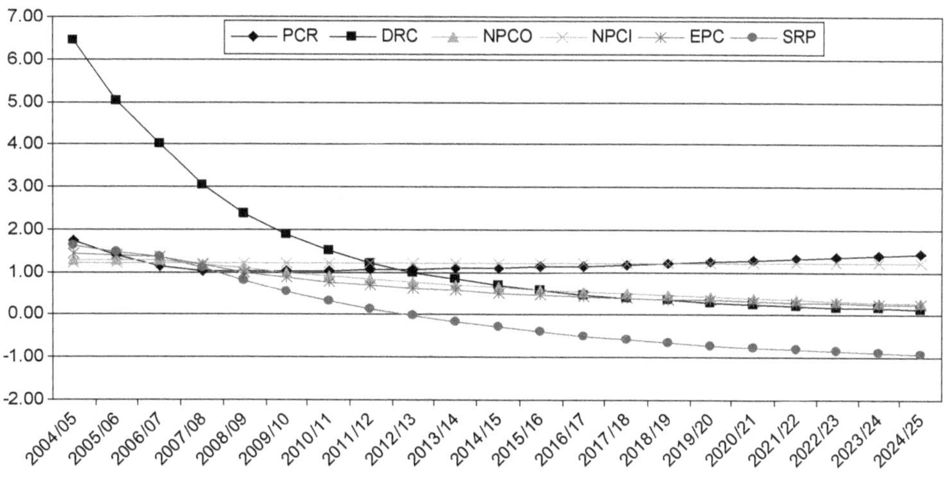

Figure 7.4. Projected PCR, DRC, NPCO, NPCI, EPC, and SRP for the period 2004-2025.

7.4 Summary, conclusions and way forward for Swaziland's maize industry

The welfare effects are defined as the amounts of money that the producers win, consumers lose, government collects as revenue and the whole society loses due to the increase in prices that reduces the demand – deadweight loss (Varblane *et al.*, 2002). In this chapter, the results of an assessment of welfare effects of maize marketing policy in Swaziland were presented for six marketing seasons (1998/99 to 2003/04). In terms of the incentive effects, the key finding is that the maize industry is highly financially unprofitable and that Swaziland has no comparative advantage in the production of maize. The indication provided by the use of the Domestic Resource Cost (DRC) measure is that domestic resources employed in the maize industry are not being used efficiently. This situation seems to arise from the high degree of subsidy that still characterizes the industry despite recent claims that the maize marketing system is liberalized. This is clearly disturbing in the light of the on-going calls from the policy and academic communities for the maize marketing industry to be fully deregulated.

A sensitivity analysis based on expectations of favourable changes in maize pricing and yield shows that both the maize industry and the wider economy will benefit substantially from trade liberalization. The sensitivity analysis was based on a forecast period of 20 years during which the gross margins steadily improved, making the sector to break even by the 8th year and eventually showing reasonable profit up to the end of the forecast period. The evidence based on the conservative assumptions is that the maize industry will become increasingly competitive as the government removes all existing restrictions on the industry and the Ministry of Agriculture concentrates on the expansion of investment in research, information and extension to support the maize producers, especially the small-scale producers on the Swazi Nations Land.

The principal findings can be summarized as follows:

1. The current market policy structure favours producers while taxing consumers. Winners of this market arrangement are the small number of surplus producers who enjoy the benefit of higher product prices than would have prevailed in a less regulated market. Again, the distorted market inhibits the production of more competitive crops. The monopolistic power of the market contributes significantly to the thin rural markets, lack of surplus production by most maize producers, and the absence of a well-developed small-scale processing and trade sector.
2. The surplus gain to production is by far greater than the loss to consumption as a result of the high consumer prices. It would seem that the subsidy given to the National Maize corporation (NMC) has moderated the consumption losses somewhat although they remain high.
3. The government benefits from the current market policy in form of receipts of tax revenue and channels this fund into subsidizing parastatals like NMC, thus sustaining the current inefficiencies.

The agricultural sector of Swaziland is facing a serious crisis with potentially more adverse implications for the country's food security. It is now clear that the current policy of food self-sufficiency is not working and is at the root of the crisis in the sector. Strident calls are now being made for a more deregulated regime for the maize marketing system. The gains from deregulation of the maize sector are numerous and include progress towards ensuring increased food security for the population. It is also widely believed that a deregulated market will drive private sector-led trade, marketing and processing. A possible effect of the full deregulation of the industry is the simplification of the chain that delivers maize to the rural and urban consumer. Such a chain is illustrated in Figure 7.5 which shows more direct routes between domestic and foreign supplies to the consumers, with implications for lower transaction costs and ultimately lower prices paid by the consumers.

In the light of the foregoing, it is recommended that the Government of Swaziland takes immediate steps to deregulate the maize market sector in the following manner:

1. Phase out all the regulatory bodies that restrict the free flow of maize and maize meal with no government interference in maize trade.
2. The National Maize Corporation (NMC) should be privatized; government should no longer be involved in the handling and marketing of maize in the country.
3. Government, through the ministry of agriculture's department of extension service, should concentrate on the provision of education and extension services to farmers with the aim of identifying and promoting investment in the alternative crops for which Swaziland has competitiveness advantage.

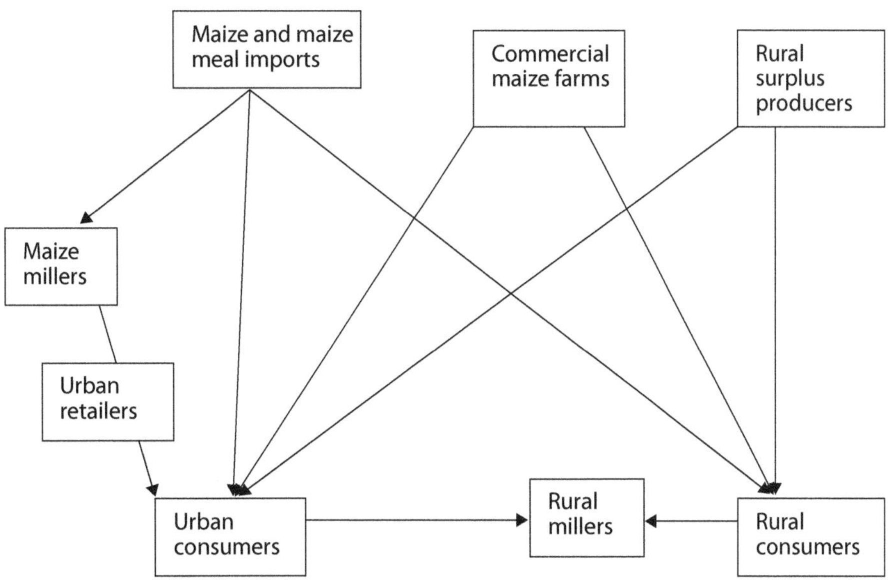

Figure 7.5. Maize marketing chain under deregulated regime in Swaziland.

Acknowledgement

The contributions of Professor Herman van Schalkwyk in the co-supervision of the Masters dissertation from which this chapter is based are gratefully acknowledged by the authors.

References

Aksoy, M.A. and J.C. Beghin. 2005. Global agricultural trade and developing countries. The World Bank, Washington, DC, USA.

Bale, M.D. and B.L. Greenshieds, (1978). Japanese agricultural distortions and their welfare value. American Journal of Agricultural Economics 59, 59-64.

Bale, M.D. and E. Lutz, (1981). Price distortions in agriculture and their effects: an international comparison. American Journal of agricultural Economics 63, 8-15.

Conway, J.A. and P.S. Tyler (1995). Practical implications of grain market liberalisation in Southern Africa. Publication and Publicity services, Natural Resource Institute, Chatham, UK.

FAO (2009). GIEWS country brief on Swaziland (Global information and early warning system). Food and Agriculture Organisation of the United Nations, Rome, Italy.

FAO/WFP (2002). 29th May special report: FAO/WPF crop and food supply assessment mission to Swaziland. FAO/WFP Periodic Publications, Rome, Italy.

FAO/WFP (2003). 9th May special report: FAO/WPF crop and food supply assessment mission to Swaziland: FAO/WFP Periodic Publications, Rome, Italy.

FAO/WFP (2004). 31st July special report: FAO/WPF crop and food supply assessment mission to Swaziland. FAO/WFP Periodic Publications, Rome, Italy.

Freenstra, R.C. (1995). Estimating the effects of trade policy. NBER working papers series 5051. National Bureau of Economic Research, Inc, Cambridge, MA, USA.

IMF (2003). Swaziland: preliminary conclusions of the 2003 article IV consultation mission. Available at: http://www.imf.org/external/np/ms/2003/100803.htm.

IRIN (2005). Swaziland: humanitarian crisis worsening. Integrated Regional Information Networks. available at: http://www.notes.relief.int/rwb.nsf.

Jayne, T.S. and G. Argwings-Kodhek (1997). Consumer response to maize market liberalization in urban Kenya. Food Policy 22, 447-458.

Jayne, T.S. and S. Jones (1997). Food marketing and pricing policy in Eastern and Southern Africa: a survey. World Development 4, 1505-1527.

Magagula, G.T. and H.M. Faki (1999). Comparative economic advantage of alternative agricultural production options in Swaziland. United States of Agency for International Development.

Mkhabela, M.S., M.S. Mkhabela and N.N. Mashinini (2005). Early maize yield forecasting in the four agro-ecological regions of Swaziland using NDVI data derived from NOAA's AVHRR. Agricultural and Forest Meteorology 129, 1-9.

Ministry of Finance (2004). Budget Speech 2004. The Swaziland Ministry of Finance, Publications, Mbabane, Swaziland.

Ministry of Finance (2005). Budget Speech 2005. The Swaziland Ministry of Finance, Publications, Mbabane, Swaziland.

MOAC (2004). Ministry of agriculture and cooperatives publications. Available at: www.gov.sz.

MOAC-NEWU (2003). National early warning unit for food security. April 2003 Bulletin. MOAC, NEWU Periodic Publications, Mbabane, Swaziland

Monke, E.A. and S.R. Pearson (1989). The policy analysis matrix for agricultural development. Cornell University Press: Ithaca, USA.

NMC Publications (1997). Review of role, function and ownership structure of the Swaziland National Maize Corporation. Vol. II background paper. Oxford policy management paper. Oxford, UK.

Thompson, C.T. (2003). Swaziland business plan yearbook. Ministry of Commerce and Industry, Swaziland government publications, Mbabane; Swaziland.

Thompson, C.T. (2004). Swaziland business plan yearbook. Ministry of Commerce and Industry, Swaziland government publications, Mbabane; Swaziland.

Tsakok, I. (1990). Agricultural price policy: A practitioner's guide to partial-equilibrium analysis. Cornel University Press: Ithaca, USA.

Van Schalkyk, H.D., O.T. Van Zyl, P.W. Botha and B. Bayley (1997). Deregulation of Lesotho's maize industry. Agrekon 36, 626-636.

Varblane, U., K. Toming and R. Selliov (2002). The impact of EU-accession on the Estonia trade with food products – partial equilibrium approach. Tartu University Press: Tartu, Estonia.

World Bank (2005). Swaziland at a glance. The World Bank publications. Available at: http://devdata.worldbank.org/AAG/swz_aag.pdf.

Wright, P.D. and W.L. Nieuwoudt (1993). Price distortions in the South African maize economy: A comparative analysis. Agrekon 32, 51-59.

8. Obstacles to the profitable production and marketing of horticulture products in Lesotho: an offset-constrained probit modelling of farmers' perception

Ajuruchukwu Obi and Litsoanelo Mphahama

Abstract

While Lesotho's agriculture stopped being a source of government revenue more than a century ago, it still fulfilled important household subsistence needs. But even that limited role is threatened as farm sizes and participation rates contract even further in the face of a multitude of factors. At the same time, Lesotho's population, especially those residing in the lone modern metropolis, Maseru, continues to be more sophisticated, with consumption patterns that are comparable to those in other modern economies. Why is the country's agricultural sector not rising to the occasion to meet the needs of the growing urban population? The preamble and foreword to the Vision 2020 document tried to address this question but they remain inconclusive. What are the reasons for the poor performance of horticultural products in Lesotho despite growing demand for the products worldwide and in the country? Are there important non-price factors that we should take into account in deciding on appropriate policies for revitalizing the farm sector in a country where few alternative opportunities for employment exist? These were some of the questions this chapter set out to address by examining production and marketing decisions and results in four of Lesotho's ten districts during 2009. There are indications that property rights, the agricultural extension service and the condition of the physical infrastructure may be crucial elements in finding answers to these problems. The results particularly point to the difficulties in accessing markets and land which remain important institutional constraints to horticulture production and marketing in Lesotho. Recommendations are made for these issues to be incorporated into the Vision 2020 process, among other processes underway to address the growing national food insecurity and enhance livelihoods in general.

8.1 Introduction and problem context

Lesotho is a small, predominantly mountainous enclave country, entirely landlocked by South Africa. It is one of the poorest countries in the world with an economy dependent mainly on livestock-based agriculture and remittances from the export of labour services to South Africa. In the late 1990s, water from the massive Lesotho Highlands Water Project (LHWP), emerged as an important source of national budget revenue as the country began to export the commodity to South Africa. The country's mountain grasslands on the eastern boundary with South Africa are very rich in biodiversity.

Lesotho's agriculture has been in decline for several years. It is hardly possible to recall that Lesotho ever produced an agricultural surplus, but the fact is that the country was a net exporter of maize until 1865 (Gill, 1993). In the foreword to the National Vision 2020 document launched in 2005, the Prime Minister of Lesotho, Professor Pakalitha Mosisili, recalled the time following the discovery of diamonds in Kimberly (Northern Cape Province of South Africa), in 1866, when Lesotho earned the reputation of being the *Granary of Southern Africa*, as the major supplier of wheat to meet the unprecedented demand for grains for a population that had come into sudden wealth (Government of Lesotho, 2005). But this seems to have been short-lived as hostilities from the Boers intensified, culminating in the destruction of the grain farms of Lesotho, eventually chasing the Basotho out of the fertile farm areas to the west of the Caledon River (SA History).

Since then, the sector's contribution to GDP has been declining, from over 30% in 1970 to about 20% in the mid-1990s (Moyo *et al.*, 1993), and just about 16% currently (CIA, 2008). Estimates of the sector's contribution are available for several years (FAO/WFP, 2003, 2004) and reflect the persistent downward trend. Correspondingly, agricultural production has been on the decline, resulting in a complete reversal of the country's status, to a net importer of agricultural produce. By the mid-1980s, the country was only able to produce 14% of domestic food consumption needs (Government of Lesotho, 1996). Studies conducted within the last decade suggest that agricultural production has fallen further, with the country importing up to 95% of its domestic food requirements (Van Schalkwyk and Jooste, 2002; Van Schalkwyk and Mhalanga, 2002). The situation is much worse today as the food and humanitarian crises in the region deepens.

Despite this gloomy picture, the sector is the most important employer of labour in the domestic economy. According to Moyo *et al.* (1993), the sector employed about 86% of the population over much of the 1980s. The figure has fallen considerably since then and in 1999 stood at 57% of the domestic labour force (EIU, 2002). Official statistics provided by the Ministry of Agriculture (GoL, 1996) indicate that despite its poor overall performance, as much as 23% of the rural population depend entirely on agriculture for food and income, while a further 32% regard agriculture as a secondary source of livelihood (GoL, 1996).

Government and the development community have responded to the decline of agricultural production in a number of different ways. In the 1970s and 1980s massive investment of resources were made in pursuit of self-sufficiency which reflected the dominant view of food security as a demand-side problem (Sen, 1981 and World Bank 1986). But repeated failures of the type and scale aptly captured by Ferguson (1994) in his book, *The anti-politics machine,* engendered an atmosphere in which Lesotho's agriculture hardly excited much intellectual and political interest. A situation was thus created in which Lesotho's agriculture was virtually written off as irrelevant to the country's long-term development

(Mphahama, 2010). Attention then shifted to water resource development culminating in the huge construction programme under the Lesotho Highlands Water Project.

But neglecting agriculture in a country with virtually no non-agricultural sector of note has turned out to be a huge mistake. Current rates of unemployment are put at 45% and they have never seemed better (CIA, 2008). On top of this, the current food/humanitarian crisis facing the country has been so devastating that it has sent the development community into a panic. The Highlands Water Project has also failed to deliver the expected employment and income relief and has rather created enormous social and political tensions, including displacement of populations.

The production problems of Lesotho's agriculture, or the supply-side questions, are complicated by the constraints in both internal and external marketing (Mphahama, 2010). The country's negative terms of trade for most agricultural and industrial products are well known. The internal road network and other infrastructure continue to pose serious constraints to the easy movement of agricultural produce in the country, creating a situation where consumers are motivated to purchase from South Africa. At the same time, internal cost structures escalate domestic production costs with the result that export production of produce with excellent production prospects in the country seems unattractive.

Expectedly, the average growth rate of the economy has slowed in the 1990s, from an average per capita growth rate of 5.1% in the first half of the decade to an average per capita growth rate of 4.7% in the second half, up to 1998. The 2003 Human Development Report of the United Nations Development Programme (UNDP), ranked Lesotho 137th out of 175 countries (UNDP, 2003). Since then, the country has fallen further to rank 156th out of 182 countries (UNDP, 2009). Further, recent developments in the Republic of South Africa, including technical changes in mining and the demands of democratic structures mean that Lesotho must look internally for solutions to its economic problems. This realization has recently created immense interest in the formulation of an internal marketing policy targeting a range of agricultural commodities. Over the years, the government has been promoting efforts to diversify away from foodgrains and invest in sub-sectors in which the country has comparative advantage. Because of its climate, abundant water supply and pest-free status, horticulture has been identified as a sub-sector with enormous comparative advantage (Mphahama, 2010). There is also evidence that demand for horticulture products is growing especially in the light of increasing awareness about the role of horticulture products in boosting resistance against diseases.

On the basis of previous work in Lesotho the following four main factors should be considered in programmes to boost horticulture production.

- the key physical production factors such as the infrastructural setting and land tenure arrangements;

- human capacity situation including the capacity of the extension systems and the availability of relevant skills for policy analysis and leadership;
- the credit system; and
- the marketing system.

Recent studies in a number of countries, including Ethiopia (Feleke *et al.*, 2003; Alemu *et al.*, 2003) suggest that these factors are important and that institutional constraints have an impact on farmers' responsiveness to policy changes. The fact that 15 years after the potential for commercial profitability of asparagus production was established (EFP, 1995), little or nothing is happening in that direction leaves little doubt that institutional constraints could be important in Lesotho. Further, the recent experience with the introduction and popularization of *Paprika* in 1999 is quite instructive and demonstrates that Lesotho farmers are indeed receptive to new ideas. In that particular instance, Lesotho farmers enthusiastically embraced the new crop which was however not suited to the country's agro-ecological conditions and thus not likely to have sustained positive impact on the farming system and farmers' welfare. This experience suggests that a well-structured institutional arrangement, integrating the full spectrum of research, extension and knowledge networks, should be able to recognize these limitations and insulate farmers from exploitation.

It is necessary to examine these factors, among many, and determine the extent to which they engender or inhibit agricultural development and thus form a basis for recommending interventions that strengthen the relevant institutions that cater to them. On the basis of international experience, it is possible to posit that institutional development would contribute to more effective utilization of the physical, human, and financial resources placed at the disposal of Lesotho under on-going or prospective technical cooperation agreements.

The purpose of this study was to identify the key institutional factors constraining the production and marketing of horticultural products in Lesotho. More specifically, the intention was to:

- investigate factors that affect horticulture production and marketing in Lesotho;
- make policy recommendations on the basis of the results.

8.2 Study area

Geographically, Lesotho is an enclave, being completely surrounded by only one country, South Africa, to make it one of only three such entities in the world (the others are the Republic of San Marino, an enclave in Italy, and the Vatican City, an enclave in the city of Rome, also in Italy). The country is located between latitude 28°S and 31°S, with Longitude 27°E and 30°E, and has a temperate climate with cool dry winters and hot wet summers (Baffour, 2003). Its peculiar mountainous terrain has earned it several nicknames, some of which are: 'The mountain kingdom' or 'the roof of Africa' or 'the kingdom in the sky'.

The mountains actually cover approximately 65% of the total land area, justifying the identification of the country with its mountain topography. It is often said that the lowest points in Lesotho are higher than most other places in the world, averaging about 1,500 metres, with the highest point of 3,300 metres, *Thabana Ntlenyana*, being the highest in Southern Africa. The country has a total land area of about 30,340 square kilometers.

The country is divided into 10 districts, namely: Butha-Buthe, Mokhotlong, Leribe, Teyateyaneng (also known as Berea), Thaba-tseka, Maseru, Mokhotlong, Mohaleshoek, Qacha's nek, and Mafeteng. These districts are distributed across the different agro-ecological zones of the country. For purposes of this study, one district, namely Thaba-tseka was chosen from the mountains, Mafeteng and Butha Buthe were selected from the foothills, while one, Teyateyaneng was selected from the lowlands (Lesotho Bureau of Statistics, 2006). The distribution of the selected districts is shown in the Map of Lesotho presented in Figure 8.1.

Figure 8.1. Map of Lesotho. Districts where the study was conducted are marked dark gray (Lesotho Bureau of Statistics, 2006).

8.3 The data

The variables examined in the study are presented in Table 8.1. Research examining institutional constraints to smallholder development as part of other studies have generally included these variables.

The signs of the coefficients show the direction of influence of the variables on the dependent variable. It follows that a positive value indicates an increase in the likelihood that there will be a change to the alternative option from the baseline to the alternative (Gujarati, 1992). Hence, in this study, a positive value implies an increase in the probability of increasing the production and marketing of horticultural products.

a. Gender (GNDR): Gender is clearly an important factor in horticulture production and marketing especially in a country like Lesotho where gender-based stereotyping is the norm. Decision making roles are normally divided between men and women depending on the nature of the economic or social activity involved. In general, the legal system regards women as minors and do not have the power to make important decisions in the household

Table 8.1. Definition and description of variables examined in the study.

Variable	Variable description	Anticipated sign +/-
Dependent		
Production satisfaction	dummy: satisfied = 1, 0 otherwise	
Independent		
Gender	dummy: male = 1, 0 female	+/-
Age	continuous	+
Years of education experience	continuous	+
Market information	dummy: have access = 1, 0 otherwise	+
Access to production skills	dummy: have access = 1, 0 otherwise	+
Visits by extension personnel	categorical: yes = 1, 0 no	+
Extension quality service	dummy: good = 1, 0 otherwise	–
Grading to market standards	dummy: meet standards = 1, 0 otherwise	+
Access to title deeds	dummy: have access =1, 0 otherwise	+/-
Transport	dummy: yes=1, 0 otherwise	+
Membership of farmer association	dummy: yes=1, 0 otherwise	
Storage	dummy: yes=1, 0 otherwise	+
Contractual markets	dummy: yes=1, 0 otherwise	+
Road infrastructure	dummy: yes=1, 0 otherwise	+

in relation to resource allocation. But the situation becomes complicated when a woman is either widowed or has a non-resident spouse who is probably employed in South Africa in a variety of income generating activities not readily available in Lesotho. In such cases, women may make decisions when it comes to growing of crops, but under clear delegation.

b. Age (AGE): This variable is the actual age of the household head/respondent measured in years. According to Bembridge (1984), age determines the behavioural patterns of a household. Younger farmers are expected to be more energetic in doing arduous farm tasks than older farmers who are likely to avoid the more arduous operations and settle for those that are less physically demanding. Age is also associated with experience and the length of time over which an individual has been accumulating capital for investment in farm operations. Younger farmers are expected to be less technically experienced as well as have less capital at their disposal.

c. Years of education (YRSEDU): Bembridge (1984), confirmed the importance of education in decision-making processes with implications for capital accumulation and adoption of innovative practices in production and marketing. In agricultural production, education plays a significant role in the extent to which farmers process information about new inputs and methods and the adoption of improved agricultural techniques. The absence of education is therefore expected to have a negative impact on production and marketing of horticultural produce. It is therefore hypothesized that there is a positive correlation between education and horticulture production and marketing.

d. Production satisfaction (PRDNSATISF): This dependent variable measure whether a farmer is satisfied with his production or not. The variable is for production rates, participation and level but with emphasis to production satisfaction of the farmer, and this variable explains the production information of the farmers which is notoriously unreliable at times as farmers tend to inflate for prestige purposes or deflate to evade taxation. But when they are not required to state how much they have produced in a season, they are more likely to be honest as to whether or not they are satisfied.

e. Contractual market (CONTRCTMKT): This dependent variable measures whether or not the farmer has access to market contract or not. The relationship between market contract in horticulture production and especially marketing is an important one and at the same time hard to attain because, for a farmer to have access to a market contract, there are certain qualifications a farmer has to meet or have. For instance like taking part in the formal market or having access to capital or credit. On the other hand, contracts ensure the availability of a guaranteed market for the farmers, thus promoting market participation in horticulture production because it is through contracts that farmers are assured of readily available inputs, ready and accessible market, support and credit and loans to buy inputs at lower prices.

f. Market information (MKTINFOR): Information in farming business is an important determinant of communication. The variable, access to market information was measured by the farmers' ability to access market information and their ability to interpret it. To capture this variable, farmers were interviewed on communication networks that are accessible to them like radios, TVs, etc. The communication could either be on the availability of markets or inputs being sold at a lower price for the farmers. Access to information has been set as a dummy variable, where a household with access to information takes the value of one and a household that has no access to information takes a value of two. Access to information was expected to positively influence production and market participation; implying that households with access to information would be more likely to participate in both.

g. Extension visits (EXTNVISITS): Contact between the extension officers and the farmer is important, and this variable which is on its own an important source of information for farmers (Enki *et al.*, 2001). Denoted by one if farmers are being visited and two if otherwise; extension visits have a positive effect on farming.

h. Quality service of extension (QUALITYSERV): Access to extension services is an important variable in the farming sector because through this service, farmers gain access to farming advice and farming knowledge. New ways and techniques of farming are also provided by the extension service. Farmers were asked to rank it from excellent, very good, satisfactory, poor and very poor. The better the service provided by the extension, the better the quality of farming business there will be. In this study, it is hypothesized that the quality of extension service provided to the farmers is poor. This variable is analysed as categorical.

i. Grading to market standard (STDGRADING): In this study, there are grading standards which small-farmers have difficulty meeting and are therefore excluded from profitable markets. According to Kherallah and Kirsten (2002), there are regulations imposed by markets to meet consumer demand and create market niches. These regulations are trickling down to the production level thereby affecting the structure and characteristics of the market downstream.

j. Title deeds (TITLEDEEDS): This variable represents serious constraints especially when it comes to land. Farmers without title deeds to land but are in the farming business, are highly constrained as the land could be re-possessed from them at any time. This variable is therefore hypothesized either negative or positive for those farmers who have acquired title deeds.

k. Transport (TRANS): Transport ownership was hypothesized to be a huge constraint because many farmers did not have their own means of transporting the produce to the markets. According to the interview findings, many famers used public transportation when others used hired transport which was confirmed to be costly and unreliable as owners of the cars were sometimes not available. In addition, availability of transportation helps reduce long market distance constraint.

l. Road infrastructure (RDINFR): Road infrastructure is measured by the accessibility of road networks that are adequate to farmers to be able to travel to the nearby or furthest market and their conditions. The poorer the road condition, the harder it is for farmers to travel and transport their products and according to the findings of this study, road infrastructure is one constraint that is hindering the marketing process.

m. Membership of farmer association (ASSMEM): This variable was deemed important because in Lesotho there is a well-developed system of traditional cooperation which small farmers draw upon to address labour bottlenecks and other production constraints. The *letsema* has been in operation for centuries and entails farmers working in groups to address a problem by collective action. It was also observed that some farmers join the Districts Farmer Association (DFA) which serves other objectives including extension. Through this Association small farmers are able to access inputs and credit. The work of Ostrom and others show that customs and social conventions designed to induce cooperative solutions can overcome the collective action difficulties and help achieve efficiency in resource use (Nabli and Nugent, 1989).

8.4 The model and analytical framework

In order to conclude as to the extent to which horticulture production was constrained, a suitable response or dependent variable should be specified and examined for possible variations under alternative scenarios. The most popular and intuitively appealing variable in this regard is farm revenue. The impact of key institutional variables on the farm revenue can then be examined as a basis for judgment one way or another. But farm revenue is an inconclusive guide in an environment where farming is not completely oriented to the market, making it difficult to accurately monetize the benefits from farming. Also, there is the notorious tendency of traditional households to give unreliable information about their production performance depending on their perception of the purpose of the investigation. However, when people are asked to simply state whether or not they are satisfied with a particular situation, experience shows that they tend to provide more reliable information. For this reason, the study decided to model production satisfaction as a binary choice variable such that when a farmer reports satisfaction with the previous year's production it is scored one (1) or zero (0) otherwise. Specifying such a model is no different from the approach taken by D'Haese *et al.* (2003) in analyzing how participation decisions are influenced by a set of institutional factors in the former Transkei region.

Of the large number of variables obtained through the sample survey, the institutional factors can be identified as: standard and grading, land access, transport availability, possession or otherwise of title deeds to land cultivated, extension services, and availability of markets. The hypothesis to be tested is that the probability that farmers will be satisfied with the outcome of their production activities will depend on several elements in the environment of the farmer. For instance, where the farmer has access to land and other productive resources, extension services, title deeds, etc., the chances are that the farmer is likely to perform at levels that he/she finds satisfactory. But this attribute as well as the possible other factors influencing it are unobservable which makes the problem one that is amenable by any of the qualitative choice models such as probit, logit or tobit models (Greene, 2000). The probit model is chosen in this particular study. The probit model was necessary to avoid selection bias in the sample (Yúnez-Naude and Taylor, 2001).

To proceed, the model of production satisfaction can be stated in general terms as follows:

$$\mathrm{Y} = \mathrm{PS} = f(X_1, X_2, \ldots X_n) \tag{1}$$

Where:

Y is the dependent variable that captures what the small producers think about the results they are achieving in their horticultural production, and the X's in the model represent the set of institutional factors already mentioned above.

Such a model can be specified as follows:

$$y^*_i = \beta_1 + \beta_2 x_{2i} + \ldots + \beta_k x_k + \mu_i \tag{2}$$

But the handicap is that y^*_i cannot be observed in reality but can only be inferred. This means also that its exact determinants can only be estimated on the basis of the dummy variables constructed for this purpose which can be defined as:

$$y_i = 0 \text{ if } y^*_i < 0 \text{ and} \tag{3}$$

$$y_i = 1 \text{ if } y^*_i \geq 0 \tag{4}$$

From the foregoing equations, it can be deduced that:

$$\mathrm{Prob}\,(y = 1) = \mathrm{Prob}\,(u_i > -\beta' x_i) = 1 - \mathrm{F}(-\beta' x_i) \tag{5}$$

which assumes that F is the cumulative distribution function for the error term u. Under the assumption that the error term, u, is normally and independently distributed, i.e. ($IN(0, \sigma^2)$, we can define a probit model as:

$$F(-\beta' x_i) = \int^{-\beta' xi/\sigma}_{\infty} 1 / (2\pi)^{1/2} \exp(-t^2/2)dt \quad (6)$$

The econometric software Stata-10 is able to calculate the probit coefficients and estimate maximum likelihood ratios based on which model validity can be ascertained. Marginal effects of the independent variables were also calculated and interpreted.

8.5 Results and discussion

The summary statistics of the variables comprising demographic and some production/ marketing data are presented in Table 8.2. In terms of the demographic characteristics of the sample, the summary statistics suggest that the majority of the farmers were male, aged about 60 years on average (ranging from 38 to 82 years). All surveyed household heads had had some education, with average years of schooling of 5.7 years (the raw data showing that years of schooling ranged from 1-15 years). Household size averaged about 8.02 persons, surprisingly large although the skewness suggests that more households had fewer than the average household size.

It was deemed necessary to summarize data relating to productive asset ownership and perceptions about production and marketing activities of the households. The indication from the results is that farm sizes are generally small, property rights are limited and that few persons that had access to land actually owned them (based on the possession

Table 8.2. Descriptive statistics of sample households in Lesotho.

Variable	Mean	Std. deviation	Skewness
Gender	0.61	0.490	-0.458
Age	60.7	8.794	-0.500
Education	5.7	2.561	1.279
Household size	8.02	2.365	-0.118
Land holding	3.2	1.234	0.525
Property rights	0.35	0.479	0.639
Title deeds to land	0.39	0.490	0.458
Satisfied with production	0.28	0.451	0.995
Non-farm employment	0.42	0.496	0.329
Revenue per ha	4,870	>2 million	1.295
Land access	0.48	0.502	0.081
Market distance	3.9	1.713	1.215
Transport problems	0.49	0.502	0.041

of title deeds to the land). Importantly, only about a quarter of the surveyed households expressed satisfaction with the results they obtain from their farming activities in respect of horticultural production. There was also evidence that the households experienced serious problems with marketing of produce where nearly half the sample had serious transport problems and farmers were about 4 kilometres away from the nearest market.

Due to suspected multi-colinearity, the software dropped all but the 7 variables presented in Table 8.3. The variable for membership of farmer associations was also dropped, making it impossible to assess the influence of this variable which is intuitively considered useful and will be investigated separately. The probit model to determine the impact of the set of institutional factors on the attitude of the farmers to the profitability of the horticulture sector is presented in Table 8.3. The intention was to show the probability that the farmer would be satisfied with the production performance of his/her horticulture farming as a result of a number of institutional influences in the farming environment. The analysis imposed revenue per hectare as an offset which could still be influential despite the difficulty in directly modelling that variable in a society where there are valid concerns over its reliability when based on farmer's recall. The results are presented in Table 8.3.

According to the results, the probit model was highly significant and suggested very good fit, with a Wald X^2 of 514.42 and log likelihood of -1,119.79 (P=0.000). This would mean that the modelled variables, with the exception of standards/grading, had strong enough influence to determine whether or not the farmers would be satisfied with the way the outcomes of their farming enterprises. The marginal effects are displayed in Table 8.4 and suggest that there was probably a 9% higher probability of a farmer being satisfied with horticulture production if s/he had land access than if access was non-existent. In general,

Table 8.3. Log-likelihood estimates and goodness-of-fit measures for the identified market and institutional factors.

Variable	Coefficient	Standard error	z-value	*P*-value
Standard/grading	-0.118	0.315	-0.38	0.707
Land access	0.822	0.295	2.79	0.005
Transport facilities	3.401	0.407	8.36	0.000
Title deeds	4.840	0.304	15.93	0.000
Market distance	0.175	0.087	2.01	0.045
Transport problems	-1.182	0.323	-3.66	0.000
Extension visit	3.102	0.509	6.09	0.000

Number of observations: n = 100; Wald chi^2 (7) = 514.42; Log likelihood = -1,119.7855; Prob > chi^2 = 0.0000.

Table 8.4. Marginal effects of the institutional factors implicated in horticulture production in Lesotho.

Variable	Discrete change (dy/dx)	Standard error	z-value	*P*-value	x
Standard/grading	-0.012	0.032	-0.38	0.704	0.43
Land access	0.091	0.043	2.10	0.036	0.48
Transport facility	0.836	0.086	9.74	0.000	0.22
Title deeds	0.905	0.040	22.40	0.000	0.39
Market distance	0.018	0.012	1.71	0.088	3.935
Transport problems	-0.131	0.050	-2.63	0.009	0.49
Extension visits	0.854	0.097	8.84	0.000	0.09

Marginal effects after probit: y = Pr(production satisfaction) (predict) = 0.05027315.

it would seem that possession of title deeds, access to transport facilities, and extension visits had the highest probability of impacting the attitude of the farmer that when these factors were non-existent.

The results with respect to title deeds (as a proxy for property rights), transportation, and extension visits deserve some elaboration. As is well known, traditional tenure systems are the norm in Lesotho where the land is held by the monarch or traditional power élite in trust for the population. At the same time, land is in short supply and most Basotho lack access to land for farming. Of the 30,355 square kilometres of land area, only 9% is suitable for arable agriculture (Kingdom of Lesotho, 2006). As it is, as many as 55% of the population do not have access to land for productive purposes (Kingdom of Lesotho, 2006). In the face of customs that frown at alienation of land outside the community, the scope for acquiring land to start or expand agricultural production is quite limited ((Qhobela, 2001). For this reason, land holdings are generally low, with the average farm sizes at about 2 hectares. According to Pule and Thabane (2004), while it is difficult to find clear evidence of tenure insecurity, the need for reform in the land ownership arrangements cannot be over-emphasized. The results of this analysis does suggest that this tenure arrangement is probably one of the most serious constraints faced by farmers and about which they expressed the most concern, with the indication that there was as much as 90% chance of dissatisfaction with the production results if the land tenure arrangement does not change for a more liberal system that allows for farmers to increase their landholding and expand production of horticulture.

The situation with respect to transportation is again understandable in view of the poor state of the rural infrastructure in the country. The state of the infrastructure is not unrelated to the nature of the terrain and topography of the country. Lesotho is a mountainous country with its lowest point being on average more than 1000 metres above sea level. While the only modern city, namely Maseru, is situated in what is known as the low-lands, the rural areas and farming areas are in the foothills and mountains where access difficulties are often extreme. Despite a small land area of about 30,000 square kilometers, travel within the country can be quite stressful and dangerous due to the high 'elevations and few safe roads' (McNeil, 1996); one of the most charismatic monarchs of the kingdom died in a ghastly road traffic accident in the country on 15 January 1996 (McNeil, 1996) and several high government officials (including at least one government minister) have also met the same fate within the last decade. The few paved roads are narrow and often poorly constructed and pose serious danger to road users. It is therefore understandable that this study would reveal strong influence of transportation on the way farmers felt about the profitability of their farming activities.

The extension service of Lesotho is typical of the extension services in much of sub-Saharan Africa where they face serious constraints of staffing and facilities as well as philosophical orientation. There are frequent criticisms of the agricultural extension service in Lesotho and for its failure to drive the necessary change in the farming populace, especially the rural and resource poor (Mokone and Steyn, 2005). Expert opinion attributes the problem to lack of the requisite skills for dealing with small farmer problems and the absence of the infrastructure for managing the extension service, especially through training, providing the right type of incentives to motivate staff, among other problems. As a result of these problems, farmers are not receiving the guidance they need to apply new production methods, adopt improved inputs and practices, and identify profitable enterprise opportunities, especially in the horticulture sub-sector.

8.6 Conclusion

The current food shortages and attendant high prices have reminded policy about the unresolved problems in many Southern African countries. The virtual collapse of the agricultural sector in the small mountain kingdom of Lesotho is a cause for serious concern for a country that continues to depend disproportionately on South Africa. As domestic calls for improved service delivery become more and more strident in South Africa, its ability to continue to carry its smaller neighbour is questionable. With the increased devastation caused by the HIV/AIDS pandemic, there is a natural anxiety to gain better understanding about the causes of the problem and identify areas of flexibility on which remedial actions can be anchored.

This study was designed to contribute to building better understanding about the institutional constraints to horticulture production. Data was collected from smallholders and gardeners in 5 of the 10 districts of the country and covered a wide range of demographic, production and marketing variables some of which were subjected to econometric analysis to determine the probability that farmers' perception about the profitability of horticulture production would be influenced by a set of institutional factors. There is evidence that more analysis is required on the existing data as well as new information that need to be obtained to have a more objective basis for making definitive statements about the role of institutions in the current state of the horticulture sub-sector in Lesotho.

However, there is enough information to conclude that property rights, especially in respect to land ownership and distribution, is a crucial factor in the way farmers see the potential of the horticulture production in the country. It is equally clear that farmers consider that the extension service and the condition of their infrastructure in general, and in particular the physical infrastructure, deserve some attention. These findings are consistent with views that are widely held both in the country and among the international development community. Policy to address them should therefore be part of a comprehensive national development effort linked to the on-going national vision process. In the case of the land ownership question, it is necessary to recognize the important customary dimensions and proceed with caution in order to bring about change that is at once sustainable and also popular.

The promotion of homestead gardens has been proposed at various levels. The contention is that this would contribute immensely to combating the widespread poverty, growing unemployment, HIV/AIDS, and weak and declining agricultural performance in the country. Policy support for this will be crucial because homestead food production does offer the possibility for marginal households with limited access to land to grow some food for home consumption and also for sale. It is also possible to undertake year-round production of the basic staples of vegetables on such gardens if support is provided to these households for water supplementation such as through water harvesting.

References

Alemu, Z.G., K. Oosthuizen and H.D. Van Schalkwyk (2003). Is increased instability in cereal production in Ethiopia caused by policy changes? Paper presented at the 25th Triennial Conference of the International Association of Agricultural Economics held in Durban, South Africa, 16-22 August, 2003.

Baffour, A. (2003). Lesotho – Africa's best kept secret: pure majesty. Available at: http://findarticles.com/p/articles/mi_qa5391/is_200305/ai_n21330240. 24/06/2008.

Bembridge, T.J. (1984). A systems approach study of agricultural development problems in Transkei. PhD Thesis. University of Stellenbosch, RSA.

Central Intelligence Agency (2008). Facts about Lesotho. The CIA World Factbook. US Department of State (CIA), Washington, DC, USA.

D'Haese, M., N. Vink, G. Van Huylenbroeck, F. Bostyn and J. Kirsten (2003). Local institutional innovation and pro-poor agricultural growth: the case of small-woolgrowers associations in South Africa. Garant: Antwerp, Belgium.

EFP (1995). Commercial study: 'the production of high value horticultural crops. Consultancy Contract No.:94/109 by Exotic Farm Produce Limited.

EIU (2002). Country profile – Lesotho. The Economist Intelligence Unit: London, UK.

Enki, M., K. Belay and L. Dadi (2001). Determinants of adoption of physical soil conservation measures in central highlands of Ethiopia: the case of three districts of Shewa. Agrekon 40, 293-315.

FAO/WFP (2003). Special report FAO/WFP crop and food supply assessment mission to Lesotho. Rome, Italy: Food and Agriculture Organisation of the United Nations.

FAO/WFP (2004). Special report FAO/WFP crop and food supply assessment mission to Lesotho. Rome, Italy: Food and Agriculture Organisation of the United Nations.

Feleke, S.T. R.L. Kilmer and C.H. Gladwin (2003). Determinants of food security in Southern Ethiopia. Paper presented at the 2003 American Agricultural Economics Association Meetings in Montréal, Canada.

Ferguson, J. (1994) The anti-politics machine – 'development', depoliticization, and bureaucratic power in Lesotho, Minneapolis, University of Minnesota Press, USA.

Gill, S. (1993). A short history of lesotho from the late stone age until the 1993 elections. Morija Museums and Archives, Maseru, Lesotho.

GOL (1996) Agriculture: a strategic programme of action – Lesotho sectoral roundtable consultations. Ministry of Agriculture, Cooperatives and Marketing, Maseru, Lesotho.

GOL (2005). The National Vision 2020. The Government of Lesotho, Maseru, Lesotho.

GOL/LCN/WB/USAID/EU (1996). Pathway out of poverty – an action plan for Lesotho. Government of Lesotho, Maseru, Lesotho.

Greene, W.H. (2000), Econometric analysis. New Jersey, USA: Prentice Hall.

Gujarati, D. (1992). Essentials of econometrics. MacGraw-Hill: New York, NY, USA.

Kherallah, M. and, J. Kirsten (2001). The new institutional economics: applications for agricultural policy research in developing countries. Markets and Structural Studies Division, Discussion paper No. 41, International Food Policy Research Institute, Washington, DC, USA.

Kingdom of Lesotho (2006). Kingdom of Lesotho report. International Conference on Agrarian Reform and Rural Development (ICARRD), Brazil, 7-10 March 2006.

Lesotho Bureau of Statistics (2006). Population and housing census. Ministry of Finance and Development Planning, Maseru, Lesotho.

McNeil, D.G. (1996). King of tiny land circled by South Africa dies in car plunge. New York Times, 16 January 1996.

Mokone, G. and G.J. Steyn (2005). Evaluation of performance of extension workers in Lesotho. South African Journal of Agricultural Extension 34, 275-288.

Moyo, S., P. O'Keefe and M. Sill (1993). The Southern African environment. Earthscan Publications: London, UK.

Mphahama, L. (2010). Institutional constraints to horticultural production in Lesotho. Dissertation. Department of Agricultural Economics and Extension, University of Fort Hare (Mimeo).

Nabli, M. and J.B. Nugent (1989). Collective action, institutions and development. In: M. Nabli and J.B. Nugent (eds.) The New institutional economics and development. Amsterdam, the Netherlands: Elsevier Science Publishers.

Pule, N.W. and M. Thabane (2004). Lesotho's land tenure regimes: experiences of rural communities and the calls for land reform. Journal of Modern African Studies 42, 283-303.

Qhobela, S.C. (2001). Land reform and poverty alleviation: Lesotho's experiences during the last two decades. Human Sciences Research Council. Pretoria, Republic of South Africa.

Sen, A. (1981). Poverty and famine: an essay on entitlement and deprivation. Oxford, UK: Claredon Press.

UNDP (2003). Human development report 2003 – Millennium Development Goals: A compact among nations to end human poverty, United Nations Development Programme, New York, USA.

UNDP (2009). Human Development Report 2009. United Nations Development Programme, New York, USA.

Van Schalkwyk, H.D. and A. Jooste (2002). The external agricultural trade environment of Lesotho with specific reference to trade agreements. Compiled as part of a larger Study investigating reforms to Lesotho's Agricultural Marketing Act., Maseru, Ministry of Trade, Industry and Cooperatives.

Van Schalkwyk, H.D. and M.L. Mhalanga (2002). The enhancement of rural marketing in Lesotho, Agricultural Policy and Capacity-Building Project, Ministry of Agriculture, Cooperatives and Land Reclamation, in collaboration with the Department of Marketing, Ministry of Industry, Trade and Marketing.

World Bank (1986). Poverty and hunger: issues and options for food security in developing countries. World Bank Policy Study. The World Bank, Washington, DC, USA.

Yúnez-Naude, A. and E. Taylor (2001). The determinants of nonfarm activities and incomes of rural households in Mexico, with emphasis on education. World Development 29, 561-572.

Part V
Institutional issues in natural resource management and implications for smallholder development in Southern Africa

9. The land question in smallholder development in South Africa

Ajuruchukwu Obi

Abstract

This chapter provides a review of the political, social and economic issues surrounding the South African land question as it has evolved from the arrival of the European explorers-turned colonialists to the inception of democratic rule and land reform. Land remains a highly emotive issue in South Africa and the bulk of government strategies to restructure the agricultural sector since 1994 have revolved around land. How these strategies are being implemented and how they fit into the broader and more comprehensive land reform programme are interesting discussion points. This chapter reviews the land reform programme and how the various components, namely land redistribution, land restitution and tenure reform, are being implemented and what obstacles are in the way of achieving the target goal of transferring 30% of land previously owned by the white population to the black people as a way of redressing past wrongs. Some proposals for a land tax that have been made in the past are also reviewed in the context of the on-going reform and the difficulties in implementing it on a national scale.

9.1 Introduction

From the beginning of creation, land has constituted an important issue in man's existence, even pre-dating organized agriculture. Modern governance arrangements modified the conception of land as a mere productive resource, but that was already begun by the time of colonial rule. An institutional analysis of the land question is easily an attempt to understand the results of the introduction of politics into the ownership and utilization of land and how that process led to a view of land as distinct from the other productive resources such as labour and capital. This is an important discussion for a country like South Africa, particularly in the light of the history of uneven development that characterized its recent past, but is equally important for other countries in the region, including those like Lesotho and Swaziland where traditional tenure systems still predominate.

The central role of land in the South African context is also understandable. In some way, the liberation struggle was pivoted on the gross inequities in the distribution of land between the black and white populations of South Africa. For a society where alternative opportunities for employment outside the farm were limited, insufficient access to land meant that a large part of the population went without work and thus wallowed in abject poverty. The enthusiasm with which the land reform programme was launched is therefore understandable even if little excuse can be made for the inadequate preparation around the *modus operandi* for implementing the programme. For instance, while the land redistribution to achieve the 30% black ownership of land is now hinged on the 'willing buyer, willing seller' modality, the approach was not discussed in earlier documents articulating black positions for a negotiated settlement of the political crises in South Africa based on which the new constitution was drafted and ratified (Lahiff, 2005 and FW De Klerk Foundation, 2007). As Lahiff (2005) noted in a policy brief of the Programme for Land and Agrarian Studies (PLAAS) of the University of the Western Cape, this modality for land redistribution, which is a variant of the market-assisted approach, was not presented as an option in the ANC documents entitled: *Ready to govern* (produced in 1992 as the Constitutional Talks were picking up), and the ANC Manifesto, *The reconstruction and development programme (RDP)* which was issued in 1994.

In fairness to it, the *RDP* document had made no secret of the ANC's determination to fundamentally alter the pattern of land distribution and ownership in the country as part of a programme of agricultural restructuring and democratization. There was a clear awareness of the fact that, despite the enthronement of democratic rule, the effects of decades of apartheid rule were still present in the form of well-laid infrastructural networks and know-how built over many years with substantial state subsidization (Matabese, 1993; Mbongwa *et al.*, 1996; Kinsey and Binswanger, 1996.). Much of this insight obviously came from the conclusions of the *Land redistribution options conference* organized by the World Bank and the Land and Agriculture Policy Centre of Witts University during 12-15 October 1993 in Johannesburg. But by failing to clearly specify the model for land redistribution and put it out in the public domain and allow discussions and debate on its operability and possibly expose its contradictions, a unique opportunity was lost. If such debate would have been held, it perhaps would have been possible to identify the multiple factors, including the inherent short-comings of the willing-buyer-willing-seller modality. It can also be argued that the conclusions of the *Land redistribution options conference*, which included the introduction of the 'market-assisted land reform programme', came too late in the run up to the transfer of power to the ANC government and therefore did not allow enough time for full examination.

As Lahiff (2005) suggested, a willing-seller-willing-buyer model in the South Africa that was emerging from an apartheid regime was a rude joke since the white owners clearly did not have any 'willingness' to part with their land. At the same time, the 'willingness' of blacks to buy the land was meaningless in the absence of the financial means to do so. In the first place, the white owners were under no obligation to sell their land on concessional terms to specified prospective beneficiaries. On their part, the prospective black buyers depended on the state's approval of their application to secure loans to make the purchase. In the meantime, an army of consultants drawn largely from the white community was ostensibly providing support to prospective black buyers to develop business plans based on which loan applications were evaluated. All the elements that could scuttle and slow down the process were therefore in-built within the policy. In addition, the policy did not make adequate provisions for optimal beneficiary selection. At the same time, the black population that was entering the mainstream economy as emerging farmers lacked the skills and experience for effective and productive participation in a formal economic system, while the crucial post-settlement support necessary to overcome this disadvantage was either completely absent or so badly structured that it was irrelevant.

The recommendation of the 2005 *Land Summit* for establishing 'inclusive partnerships and collaboration across government departments and civil society' is incontestable. According to the 1979 World Conference on Agrarian Reform and Rural Development, equitable distribution of land does not answer all the questions; the redistributed land has to be utilized efficiently for the ultimate objectives of rural development, human resource mobilization, increased production and poverty alleviation to be achieved (FAO, 1979; De Janvry, 1984). This is where inclusive partnerships and inter-departmental collaborations and coordination come into the equation and calls for development of area-based plans that are at once integrated, inclusive, participatory and flexible.

This chapter provides a more formal review of the relevant literature on the evolution of the land question in South Africa. However, the review recognizes that the South African land question cannot be viewed in isolation since it is inconceivable that a phenomenon of such enormity could have failed to draw from broader global experiences and lessons. Even if the South African experience ultimately turns out to be unique, there is intellectual value in having a framework for comparison and for pronouncing one way or the other. Furthermore, there is a need to make some contribution to the on-going debate on various aspects of land in the country which imposes the obligation to present a balanced discussion.

In view of the foregoing, this chapter will examine the economics and politics of land and how it has evolved as a formidable tool not only for economic development planning but also for political management and organization of society. Several instances are known worldwide where land has either created the motivation for political action of one type or the other, or has become an important outcome of such actions. The extent to which such experiences are systematic obviously has relevance for policy. For this reason, the different roles that land can and do play, within a multifunctional conception of agriculture, are reviewed. Attention will then be turned to the way and manner land entered the politics of South Africa and how this has governed and mediated political, social and economic life in the country ever since. Within this framework, the legislative developments that have charted the course of land ownership and distribution in the country are reviewed. Early remedial efforts, even during the heydays of apartheid dominance, will also be highlighted as precursors for the new legislations and policies that are now trying to achieve redress and right the wrongs of the past. In this regard, such issues as the series of debates on a possible land tax in the country and the taxing of capital gains will be touched upon.

9.2 Land, economics and politics

As indicated, this review will comprehensively examine the land question in a broad socio-political and economic context. The global experience will be discussed before attention is directed to the South African experience to examine what role land has played in the economic history of the country as well as how it has featured in the process of agricultural development of the country. This will be useful in order to properly situate the whole issue of land price formation and the interaction between agricultural restructuring the value of agricultural lands. The international literature on land in general, and agricultural land in particular, has devoted ample space to the role of land in the development process and how policies about its distribution and use have influenced some of the most interesting political developments in history.

There is a lot of interest in the role of land as a factor of production and what determines access to land for agricultural and other purposes as well as the process by which prices are formed in the agricultural land market. There is equally a rich literature on the land market in general but for purposes of the present study, the review will be limited to the agricultural land markets and how they function. Although relevant research will consist mainly of studies accomplished within related geographical settings, there is value in establishing theoretical and methodological patterns, irrespective of where such work is done.

9.3 Land in South African economy and politics

In a very fundamental sense, South Africa's history has been more about land than much else. Although much of the recent discussion on the land question trace the discriminatory policies to the Land Act of 1913, the events that built up to it date back several centuries to the arrival on the South African shores of Jan van Riebeeck. This section will briefly lay out the historical details as they relate to land and how South African land policies have evolved since then.

9.3.1 From the conquest to 1913

The arrival on 6 April 1652 of the Dutch East Indian Company led by the legendary Jan van Riebeeck marked the beginning of the land problems in South Africa (DBSA, 2005; Plaatjie, 2005). Purportedly en route to more distant lands in East Asia, the group made what was reported as a temporary stopover but found the conditions conducive to farming activities and in 1654 the first piece of land was obtained for farming purposes. Accounts vary as to the initial reaction to the use of land by the Dutch. By one account, the Dutch settlers had met an indigenous population whose primary economic activity was trading in merchandise brought in by the seafarers (Van Schalkwyk, 1995). There may have been some black indifference to the acquisition of land which they did not need anyway. But according to Plaatjie (2005), the indigenous black population was already actively farming on the land long before 1652 and there were clear links between man and the land as a productive resource.

Whatever the initial reaction, Wilson and Thompson (1969) recorded in their historical account of those early days the development of serious tensions between the settlers and the indigenous population. There was a clear sense of unease at the rate at which land was being grabbed by the settlers. The first recorded violent conflict over land may have occurred in 1659 when the indigenous population made a serious attempt to chase the settlers out of their land. This effort proved unsuccessful (Wilson and Thompson, 1969).

Basking in their successful subjugation of the black uprising, the settlers began to make incursions inland. From the coast, the conquerors radiated northwards and eastwards, capturing land in areas of present day Kimberley and the Orange Free State. In the course of the 19th century, it was no longer in doubt that white presence on the land was a permanent feature of the South African agricultural scene. Gill (1993) described some of the conflicts over land that spread into the heartland of the present Free State Province and commonly known as the *Lifaqane* wars fought during 1818-1824.

Alongside the aggrandizing land conquests came another major development with implications for black agricultural participation. This was the discovery of diamonds and gold in several parts of South Africa where mines were springing up to exploit these resources. With more discovery and exploitation of these resources, the mining interests were consolidating and gaining more economic power which attracted more international investment that facilitated improvements in the conditions at the mines. Expectedly, mine employment became more attractive. It was also at this time that the Land Banks were established, from 1907, to concentrate more land in the hands of the white settlers (Murray, 1997). This latter development gradually stripped the African peasant farmers of their productive resources, turning them into wage earners on white-run large farms or in the mines (Keegan, 1986; Jeeves and Crush, 1997; Murray, 1997).

With more settled and commercial-oriented agriculture and opening of the mines, wage employment in both agriculture and mining became important and the period witnessed a sharp decline in the incidence of raids and inter-group conflicts for which men were expected to play important defensive/combative roles for their communities. With considerable spare time at their disposal, the men were naturally attracted to the emerging opportunities for short-term contractual employment in the construction of rail-roads and in the diamond and gold mines springing up in South Africa. Kimble (1979), Swallow and Borris (1988), and Gill (1993) have advanced similar reasons for the decline in agricultural productivity in neighbouring Lesotho following the boom years as the nineteenth century drew to a close. Similar spillover effects may have been inevitable in the other countries that share borders with the country given the dominant influence South Africa has always had in the region.

There is evidence that the increasing agricultural activities brought with them unexpected affluence among the Dutch settlers and, along with this development, increasing envy of the others both within and without who were beginning to develop some interest in what was going on in the territory. The most outstanding such new entrants were the English who apparently saw themselves being marginalized by the new economic prosperity. According to Plaatjie (2005), the war that is variously described as the South African War and the Anglo-Boer War fought over the period 1899-1902 was about land and brought devastating consequences on the territory's agricultural economy.

The phenomenon of the 'poor white man' is no doubt one of the clearest manifestations of the extreme hardship that was unleashed on the immigrant Dutch farmers, or Afrikaaners, whose agricultural economies were virtually wiped out (Plaatjie, 2005). The English had by this time destroyed every Afrikaner farm they could not take and created so much misery among the Afrikaner population that they became virtually the most destitute people within the territory that later became the Union of South Africa. It was clear that the war had indeed given rise to a fundamental restructuring of the agricultural sector in terms of the ownership and control of land and the racial structure of economic power in ways that would reverberate into the next century and possibly to the present day. Murray (1997) has contributed a very graphic description of the role of Lord Alfred Milner, in attempting to 'break the back of (any) lingering spirit of Boer rebelliousness by sprinkling the countryside with English-speaking yeoman farmers'.

During this period, the practice of allocating city areas on racial lines began to take hold. Kassier and Groenewald (1990), cited by Van Schalkwyk (1995), recall the discriminatory practice of allocating land to the immigrants from Europe in a manner that ignored the existence of the black population. The phenomenon of the 'black reserves' emerged at this time as well, representing geographical areas demarcated for the exclusive residence of the black population. Within these reserves, land tenure arrangements followed a different pattern from what prevailed in the white areas, with more traditional communal systems of ownership being kept in place.

The black population was forbidden to seek or use land outside those reserves. In the Orange Free State, at least, a formal legal system had been established, codified in the Orange Free State. barring the indigenous black population from owning or leasing land outside the black reserves (Davenport and Hunt, 1974). Of course, this was not uniformly applied in the rest of the country, with some provinces allowing limited black access to land as a 1905 Supreme Court interpretation of the Pretoria Convention of 1881 revealed (Davenport and Hunt, 1974). The rest of the territory was designated 'white area' (Meredith, 1988) unless otherwise reserved as 'state land'. The state land could be allocated to either blacks or whites but the blacks were obliged to conduct themselves in a manner approved by the state in order to retain their allocation (Meredith, 1988).

By this time, the conflictual environment had stabilized sufficiently for a system of land trading to evolve. According to Kassier and Groenewald (1990), land in the Natal area was being traded on the London Stock Exchange by 1860 when some 15 speculators had acquired about 275,000 ha of land in the area. With the growth of a land market, the need to farm profitably was becoming quite urgent. Some of the factors that motivated the black population to seek wage employment as peace returned to this volatile area have been mentioned earlier in this chapter. To these, Groosskopf (1932) adds the growing poverty among a black population that had lost its land to the whites and now had to live in overcrowded black areas where even the raising of animals was hampered by overgrazing and thefts.

There are numerous accounts of how the African peasantry held up in spite of the embattlement it faced in the hands of racism and deliberate attempts to stifle it. Evidence has been found to support the claim that a prosperous black farming class was able to exploit the burgeoning consumer markets in Johannesburg and Kimberley where mining incomes was creating wealth and effective demand beyond their wildest imagination. The mine workers and urban entrepreneurs in the newly formed towns and cities needed to be fed, and the story is that the food came, not only from the white but also the black farms (Plaatjie, 2005). This was the situation up to the formation of the Union of South Africa in 1910 and unified system of governance came into existence with administrative headquarters in Pretoria.

With the new administration stationed in Pretoria, a new emphasis was placed on centralization of policies on property rights. The in-coming government under Lord Alfred Milner took the first definitive step in this direction by setting up the South African Native Affairs Commission to spearhead the process of centralization of land policy for the whole country (Jeeves and Crush, 1997; Murray, 1997). This commission was named the Lagden Commission after its first chairman. According to Van Schalkwyk (1995), this commission became the architect of some of the most repressive legislations which enthroned white economic interests to the detriment of those of the black population. As noted by Kassier and Groenewald (1990), this era defined the eventual structure and character of South African agriculture. As the era drew to a close, the profusion of legislations that would result in a record 87 bills over a quarter of a century began with two of the most influential and historically significant ones, namely the 1912 Land and Agricultural Bank Act and the infamous Land Act of 1913.

As was noted earlier, the establishment of Land banks began in 1907 as institutions to assist the white (largely English) farmers who were rapidly acquiring agricultural land in the wake of the subjugation of both the black and Afrikaner populations. Following the formation of the Union of South Africa in 1910, the next move was to bring all these colonial land banks under one administration. The Land and Agricultural Bank Act of 1912 was the instrument for achieving this. Its distinguishing feature was its exclusive white focus, targeting only the white farmers and related bodies that supported white agriculture.

In 1913, the series of discriminatory actions regarding ownership and distribution of land was finally given a legislative stamp. This was done by the enactment of the Natives Land Act in 1913. This is easily one of the most discussed Acts in the literature on the South African land policies. According to Davenport and Hunt (1974), The World Bank (1993 and 2003), the African National Congress (1994), Van Schalkwyk (1995), Lyne and Darroch (2003), and others, it was this single piece of legislation that effectively restricted black ownership of land to the areas designated as the Native 'Reserves'. A black person could not purchase, lease or rent land outside this area unless approval was given by the Governor-General (of State President) of the Union of South Africa.

The Native Reserves were later formally designated the 'homelands' about which some of the most intriguing political and governance arrangements were also made that allowed some measure of self-government. According to the National Department of Agriculture's (NDA) documentation, the land area involved was about 17 million hectares which was approximately 13.9% of the national land area (Fenyes and Meyer, 2003). Without a doubt, this was a completely different setting in terms of the relationships between man and the land and the kinds of arrangements that were necessary to access this resource for use in any way. While the principal mode of tenure remained customary and involved the traditional power élite exercising complete control over the land and individuals having only usufructuary rights (Lyne and Darroch, 2003), there was less land available for the generality of the people, resulting in widespread landlessness within the 'homelands'.

9.3.2 Post 1913 laws and the consolidation of white power on land

The next 35 years, up to 1948, saw the consolidation of white power on land and the intensification of the discriminatory rules that progressively dispossessed the black population and impoverished them beyond anything imaginable. A combination of stiff legislations and other measures were employed to achieve the aim of developing white agriculture into a powerful ally for systematic discrimination. At the same time, this was an era marked by continuing political rivalry within the white race that revealed its heterogenous character. As has been mentioned earlier, the English and the Dutch settlers (or Afrikaners) has been major adversaries over the land for the better part of a century. Defeat for the Afrikaners meant extreme pauperization which deepened their bitterness and provided the strongest motivation for political participation to influence legislation in favour of the Afrikaners.

In 1922, the Union government enacted the Cooperative Societies Act of 1922 to create an elaborate network of agricultural cooperative system to support white farming. This system aimed to capacitate white farmers through ensuring the prompt delivery of information about techniques and markets. Working side by side with the newly reconstituted Land Bank that delivered soft loans, the cooperative system was instrumental to the strengthening of a competitive advantage enjoyed exclusively by white agriculture that as a consequence faced much lower transactions cost than the black farming system.

In the meantime, the profusion of land-related legislations continued. In 1936, the Native Trust and Land Act of 1936 was enacted to endow the Governor-General with more powers over the land tenure arrangements within the black reserves (Van Schalkwyk, 1995). This was clearly overkill in a situation where the black population had been consigned into the reserves and was still not free to manage their use of the land. The next year, the Marketing Act of 1937 was enacted to provide support to the white commercial farmers in the marketing of their produce. The aim was to give more certainty to the white farmers as the farmers had assurance of marketing a good part of what they produced and therefore had strong motivation to expand production as much as possible.

Given the massive help in subsidies that was coming from the reconstituted Agricultural and Land Bank, this was in no way a tall order for the white farmers. As would be expected, the benefits of the Marketing Act of 1937 completely by-passed the black farmers in the reserves. Mention needs also be made of the measures that aimed to improve soil quality and further enhance the production environment for white agriculture. Notable among this is the Soil Conservation Act of 1946 which aimed to conserve the soil quality in the white areas.

9.3.3 The apartheid era and land laws

The National Party controlled by the Afrikaner population came to power in 1948. This review will be limited to laws enacted during this era that had implications for land ownership and control and leave out those that were more of a political nature except where these cannot be separated. The Groups Areas Act of 1956 is perhaps the signature legislation of the new regime which had come to power on the ticket to redress wrongs done to the Afrikaner population by the English. The Act divided the country along race lines. Nobody was allowed to own property outside his/her own colour-based area. If there was any hope of somehow circumventing the Land Act of 1913 by acquiring land elsewhere from your place of residence, the Group Areas Act effectively put an end to that dream.

The literature provides other strong indications of stiff resistance on the part of the new regime to improve conditions for the black population even in the face of evidence that such improvements were called for. For instance, Government documents reviewed by Van Schalkwyk (1995) suggest that despite recommendations by the Tomlinson Commission of 1955 for improvement of economic and agricultural conditions in the black areas, the government was adamant and rejected these outright. Of particular interest was the recommendation to introduce freehold tenure in the homelands, and drop the idea of one-man-one-plot which limited the scope for the black farmer to expand his holding and improve productivity and output.

Ultimately, the legacy of the era of the apartheid government as far as agricultural development was concerned was the widening of the gap between the white and black populations. As more and more favourable support programmes were channelled towards the white farmers, less and less went to the black farmers within the reserves, resulting in a widening of the gulf between them and the deepening of black poverty and backwardness. The support packages channelled to the white farmers included access to research findings and the most elaborate and sophisticated systems of agricultural extension education and training, including through the instrumentality of the cooperative system already set up by the British administration.

9.3.4 The era of deregulation and limited black empowerment

As the apartheid stranglehold tightened around the black farming population there were signs of resilience and determination to make farming pay within the black homelands. Bayley (2000) draws attention to the struggle among the black farmers to produce some food for subsistence purposes as more Acts were enacted that reinforced their marginalization. Especial mention is made of the 1968 Marketing Act which, while coming up with some new measures to improve the marketing system, still provided for treating different geographical areas differently. There was still a high level of state support to the cooperative sector alongside the emphasis on confining African farming activities to specified geographical areas. For all practical purposes, the Marketing Act was established in the context of systematic disempowerment of black farmers. Most agricultural households in the reserves/homeland areas were reliant on off-farm incomes and food purchases to supplement their own production. Market interventions designed to benefit white farmers were sometimes implemented in a way that negatively affected net sellers in homeland areas.

However, in the early 1980s, a stark reality was beginning to dawn on the policymakers in South Africa. This review has highlighted the considerable amount of state subsidy that went into the agricultural sector over the years (Van Schalkwyk, 1995). It would have been too much to expect that these support measures, in the way they were structured and delivered, would have no macro-economic and environmental consequences. For one thing, along with the subsidies, the farmers faced state-established producer prices that far exceeded world prices for the commodities such as maize (Van Schalkwyk, 1995). The farmers were also being provided technical support and information about latest research findings on mechanical and biological technology which meant that they could continue to make profits as they expanded hectarage.

And this they did as a rational economic behaviour. Brand *et al.* (1992) noted that large stretches of land came under maize as farmers strove for more and more profits. The upshot naturally was the South African agricultural sector, while generating some of the highest surpluses on the continent, became increasingly environmentally and economically unsustainable. Van Zyl (1989) concluded that for that period, the substantial surpluses being generated did not benefit the country in any meaningful way but rather resulted in a welfare loss for the country as a whole.

Given this realization, it was inevitable that some rationalization would be put in place. Van Schalkwyk (1995) shows that the deregulation of the agricultural sector was actually part of an economy-wide financial sector liberalization that began in the early 1980s to mitigate the macroeconomic effects of the past agricultural support policies. It must be borne in mind that this era also witnessed the intensification of the global condemnation of the repressive regime in South Africa which was being expressed in stifling economic sanctions and almost complete curtailment of contacts with the regime. Without a doubt, therefore, the forces that led to the South African government to consider agricultural restructuring were multiple, coming from both their own policy actions and the impact of the global fight against apartheid.

Along with the financial and economic reforms, the government obviously saw that the apartheid system was unsustainable and that sooner rather than later, there would be accounting to do. From the mid-1980s, a number of locations within the former homelands were identified for agricultural development. The aim was to establish a core of black farmers who would 'emerge' into commercial agriculture. Qwaqwa was one of the former homeland areas identified for agricultural development by the previous government (Jordaan and Jooste, 2004). The idea was to select a set of former employees of the government agricultural support parastatal known as Agriqwa and settle them on land expropriated from white commercial farmers. According to Jordaan and Jooste (2004), approximately 55,000 ha of land were expropriated between 1979 and 1986, divided up and equipped with the necessary infrastructure for modern and sustainable farming. Farm sizes vary from 250 ha to 1000 ha depending on the potential and type of farm.

Another systematic study of the changing policy environment at this time was conducted by Claasen (2000) to analyze the situation of the group of 114 black farmers who participated in the land settlement scheme in the former black 'homeland' of Qwaqwa in 1989. These

114 farmers were mostly selected from Agriqwa, a non-profit government corporation, which was founded for the sole aim of establishing these 'emerging' or 'beginning' farmers. The agency provided financial and technical assistance to the newly settled farmers each of which had been allocated between 350 and 450 ha of farmland. Infrastructure for improved modern farming was also installed and farmers received frequent advisory visits from officials who provided them with every assistance needed to turn farming into a profitable endeavour. Value adding activities seemed to be popular and profitable among emerging farmers. This experiment seemed to debunk the notion of the black farmer as not being responsive to economic incentives and was actually so steeped in tradition that he would be unable to be weaned from a subsistence mode of production. It was clear that these emerging farmers possessed the will and ingenuity to make a success of agriculture in the area (Claasen, 2000).

Similar schemes were initiated in the other former black areas of the country. A review of the post-1994 agricultural development revealed the phenomenal and innovative actions taken in the North West Province to irrigate communal arable lands that were managed under a mixture of tribal tenure systems and freehold with a highly sophisticated irrigation infrastructure. Golder Associates (2004) has completed an assessment of the scheme undertaken during that era and produced a comprehensive document that provides insights into the scheme in terms of its objectives, scope, operational details and problems especially with the inception of pluralistic democratic dispensation and the integration of the former 'homelands' into the rest of South Africa.

According to Golder Associates (2004), the irrigation scheme was established on land that existed under communal tenure arrangements where user rights were allocated by the tribal leadership on a hereditary basis. No payments were made for the acquisition of user rights to land. The development within the Taung area of the province possesses characteristics that are considered typical of what prevailed in the erstwhile Bophutswana Republic, as this former 'homeland' was known. The particular scheme analyzed comprises 4,000-5,000 ha of communal arable land which has been partitioned among 411 farmers who, as explained earlier, do not have to pay anything for the user rights in keeping with the precepts of the communal tenure systems prevailing in the traditional black areas of the country.

When the scheme took off in the period 1980-1989, candidate farmers were selected from the rural communities by tribal/communal leaders in cooperation with the public agencies created for agricultural development within this self-governing homeland. When the scheme began, participating farmers were each allocated 1.7 ha of land within a circular piece of land described as a 'circle' which was served by a rotary irrigation infrastructure or 'pivot' that consisted of pipe-fed irrigating 'machinery' on wheels that delivers water on participating fields according to a predetermined format. Today this scheme has been expanded and each participating farmer is allocated 10 ha of land within the circle on which three other farmers (making 4 farmers in all) own same-sized parcels and sharing one pivot for irrigating their fields. An additional area comprising 1000-1,500 ha arable land is now being cultivated with sprinkler irrigation (Golder Associates, 2004).

Contemporary discourse on the land question within South Africa's troubled past has largely ignored this limited, even if half-hearted attempt to empower the black population and integrate them into the nation's agricultural economy. Not many commentaries have seen this effort in charitable terms. Some commentators have actually gone as far as suggesting that the apartheid regime, realizing that the day of reckoning was in the horizon, needed to have in place arrangements that would not be lost on the historians of the apartheid era.

Whatever the intention of the apartheid government was in establishing the agricultural development programmes in the former homelands at this time, the fact remains that it happened. And its significance and practical implications are even further heightened by what has happened to the schemes with the advent of democratic rule. In his study, Claasen (2000) tracked the progress of the Qwaqwa scheme up to the post-apartheid era and has made a number of interesting revelations. For instance, it was found that after 1994, official policies regarding the previous homelands changed dramatically. The support agency Agriqwa was dissolved and replaced by another one named Agri-Eco.

It was not long before the restructuring and rationalization of Agri-Eco implied the end of financial and agricultural assistance to the emerging farmers in Qwaqwa (Claasen, 2000). As a result of this process of restructuring, the emerging farmers now have to compete independently in a free market environment. These emerging farmers have suddenly been exposed to a competitive market environment without having much experience or preparation for it. The result has been very serious managerial problems for these farmers. The study also found that a number of farmers are experiencing serious financial problems. The refusal of commercial banks and agricultural cooperatives to grant production loans for planting purposes, underlines the extent of the financial difficulties experienced by emerging farmers in the area. A major conclusion of the study is that the success of similar projects is largely dependent on basic support structures in training and agricultural extension work from government and development agencies.

Without question, Government needed to embark on location- and commodity-specific programmes (infrastructure and training) designed to increase the capacity of black farmers to compete meaningfully in a deregulated market. More importantly, government should devote efforts towards facilitating:

- competition along the marketing chains;
- the provision of agricultural information;
- a legally secure framework within which agricultural trade can take place;
- the maintenance of the physical marketing infrastructure.

9.3.5 The era of land reform and agricultural restructuring

Without an iota of doubt, realism had begun to dawn on the apartheid regime in South Africa regarding the role the black population could play in the country's agriculture. Even before democratic pluralism was ushered in, discussion has started on what the nature of the rural restructuring would be (Van Zyl and Van Schalkwyk, 1993; Williams, 1993; World Bank, 1993). According to Van Zyl and Binswanger (1996), the system of racial discrimination was clearly unsustainable and was hurting the economy in very fundamental respects. For one thing, the exclusion of a large segment of the population from meaningful economic participation was preventing the emergence of entrepreneurship in the small scale sector and within the rural economy. The development of such entrepreneurship would go a long way in addressing employment creation and stimulation of the rural economy. In the views of Van Zyl and Binswanger (1996), the political consequences would be quite catastrophic if not addressed decisively and with the minimum delay. An urgent need to restructure the agricultural sector and embark on a redistribution of land was identified and seen as the way to avoid an imminent and 'debilitating pattern of civil disorder and violence...' (Binswanger and Deininger, 1993).

The literature has extensively reported on the various twists and turns in policies and strategies affecting the agricultural sector during the apartheid era (Van Schalkwyk *et al.*, 2003) and changes that have taken place since the new regime in South Africa assumed political power in 1994. Nothing better describes these changes more than Summers and Vinod's (1993) assertion that 'communism is the longest route from capitalism to capitalism'. As has been highlighted earlier, the move towards freeing up the sector pre-dated democratic elections given the realization that the excessive regulation and paternalism was costing the economy dearly. According to the World Bank (1994) and IMF (2000), economic growth grew generally at around 1% per annum towards the end of apartheid rule. Between 1986 and 1992, GDP growth rate averaged 1.03% while employment grew at -0.23% (The World Bank, 1994). Since then, economic growth has averaged above 3% per annum. While questions remain about what the ultimate lessons of the land reform programme would be (Hart, 2003), there are indications from a number of macroeconomic indices that conditions have been changing for the better over the years. For instance, in the second quarter of 2005, the annualized rate of real growth stood at 5% (South African Reserve Bank, 2005).

Bayley (2000) catalogued a number of the most obvious reasons for the deregulation of the agricultural market of South Africa, including broader macro-economic reforms, especially the financial sector reforms that took place in the 1970s, featuring the real depreciation of the Rand in the 1980s and interest rates adjustments. In addition to these factors, Bayley (2000) also highlighted the following developments:

- Realization by the practitioners themselves, including farming and industry leaders, that the control system was not sustainable.
- Dissatisfaction with the operations of the marketing boards which were expressed in a number of legal challenges.
- Despite food self-sufficiency being achieved, the agricultural sector was performing very poorly in terms of its contribution to national wealth (GDP) and employment growth.
- Rising domestic food prices in the 1960s which pointed to the need for market liberalization.

The marketing control structure was also seen as mechanisms for the reinforcement of the marginalization of the black farming population, providing high level of state support to the cooperative sector, promoting a monopolistic paradigm, and using statutory support to encourage the inflation of land prices (Bayley, 2000). There was also a feeling that the government used marketing controls as a vehicle for rent seeking (Bayley, 2000).

The first hint towards liberalizing the agricultural market was the enactment of the 1968 Marketing Act (Bayley, 2000). But critics claim that the associated reforms had very little teeth and could be easily reversed. Critics also had various problems with the management of these early reform measures which were tagged *ad hoc* (Bayley, 2000). In the opinion of several commentators (Van Schalkwyk *et al.*, 2003), much of the amendments made to the earlier control Acts were merely intended to close existing loopholes in the control apparatus and tighten government's grip even more. Producer prices of farm products began increasing quite uncontrollably largely because input costs were also going up quite fast. According to FAO data, rate of price increases in South Africa far exceeded what was observed in many other developing countries of comparable levels of development (Van Schalkwyk *et al.*, 2003). In the face of the deteriorating price situation, researchers (Van Schalkwyk *et al.* (2003), warned of an impending financial disaster in the agricultural sector. By the mid-1980s, the doom-day was already upon agriculture, as several farmers began to experience far-reaching financial setbacks and insolvencies (Van Schalkwyk *et al.*, 2003).

The deregulation of the agricultural market came only with the establishment of pluralistic democracy in the country. In the 1996, The Marketing of Agricultural Products Act of 1996 was passed. The purpose of the new Marketing Act relates to its four central objectives:

1. increasing market access for all South African producers who use the marketing system;
2. enhancing marketing efficiency for agricultural produce;
3. ensuring that export earnings arising from the agricultural sector are optimized; and
4. providing a basis for making agriculture more viable.

There was a clear departure from the excessive controls of the past and the a clear commitment to use agriculture as a vehicle for broad-based development of the country and alleviating poverty for the generality of the population, especially the black South Africans who had been excluded from the mainstream of the country's agricultural economy over a long time. The literature makes clear distinctions between the old regime of controls and the new marketing policies (Table 9.1).

From the standpoint of modelling an asset or commodity markets, the significance of deregulation of the agricultural sector rests largely in the new institutions that arise from such a policy development and influence the length of the marketing channel one way or the other, and invariably affect marketing as well as other transactions costs. Ruijis (2002) has examined these outcomes in respect of cereals marketing in West Africa. But a common theoretical thread can be found within the broad field of institutional economics (North, 1989 and 1991; North and Wallis, 1994; and Eicher, 1999, among others).

The fact that this element is often taken as given provides strong justification to examine it explicitly at this point. The academic and development literature is replete with the

Table 9.1. Distinctions between the Marketing Act of 1968 and the Marketing of Agricultural Products Act of 1996 (Van Schalkwyk et al., *2003).*

1968	1996
Interventionist system	Market deregulation
Increased productivity	Increased marketing efficiency
Reduction of marketing margins	Optimum export earnings
Increased consumption and food self-sufficiency	Food security at household level
Maximum commercial producers on land	More emphasis on small-scale farmers
Economic farming units; minimum farm size	Increased sustainability of agriculture
Non-participative and bureaucratic introduction of intervention	Participative, transparent and all-inclusive
Stabilizing product prices	Producers must themselves stabilize income
Intervention inclusive of single channel; pools, surplus removal, fixed prices, quotas; price support; promotion; general and special levies, registration, records and returns.	Limited to levies; export control; pools; registration; records and returns
Requested by producers or introduced by Minister	Requested by any directly affected group of persons or firms
Consultation not always necessary although certain quantified producer support required	Consultation process prescribed by Act inclusive of all directly affected groups
No political process to approve levies apart from Minister	Levies need to be approved by both parliamentary portfolio committees and the Minister
No maximum period and no interim testing of intervention	All statutory measures to be introduced for fixed period and tested at least every two years

substantial contribution that asset ownership and its equitable distribution make for economic growth. Such thinking is not new any longer with respect to asset ownership but did have a long-running battle with the strong and contrary views of earlier growth models such as those of Kaldor and Kuznets. It has now been established on the basis of international experience that permanently higher levels of growth can be achieved if the productive assets of the society, principally land for an agrarian society, are redistributed in a way that is just and equitable. Aghion *et al.* (1999), Bardhan *et al.* (1999), and Piketty (1999), are some of the development literature that have clearly demonstrated this in a range of developing countries as part of efforts to develop new theories of growth. This thinking is also consistent with the standard position that secure property rights contribute to productivity growth through inducing those who enjoy such rights to exert more labour and management effort and deploy higher levels of investment to enhance the fertility of their land (Feder and Feeny, 1991). The role of land reform in economic development has generally been examined against this background (De Janvry, 1984; Cusworth, 1992; Delgado, 1995; Christiansen, 1996; Greenberg, 2002; and others).

There are more empirical studies, including those undertaken by the World Bank, which provide cross-country evidence on this matter, the most notable being Deininger and Squire (1998), Deininger and Olinto (1999) and World Bank (2003). Using cross-country regressions, these studies have demonstrated that when the distribution of land ownership is unequal, economic growth proceeds at a much slower pace, if at all. A comparison of agricultural growth rates in China and South Africa over the period 1980-1993 illustrates this point more sharply. While Chinese agriculture based on a smallholder model operated on small farms of less than 2 ha on average generated a rate of growth of 5.3% p.a. during that period, South Africa's agriculture based mainly on 60,000 commercial farmers who controlled 102 million ha of land, grew at 1.8% over the same period (Eicher and Rukuni, 1996). What this poor growth performance meant for rural living condition in South Africa is an all too familiar picture. This can be contrasted with the very significant contributions that a more equal distribution of land seemed to have made on human development indicators in China, consistent with positive economic performance described earlier (Burgess, 1998).

Household level effects have also been shown to be quite important. A large amount of studies have also established the important positive impact of a more equal distribution of land on household level welfare and food security, among other favourable outcomes (Blanchflower and Oswald, 1998;). In India, it was shown that land reform positively impacted on the poverty situation in the country and that, at least in one State of the country, namely the West Bengal, productivity increases resulted from reforms in the tenancy arrangements (Besley and Burgess, 1998; Banerjee *et al.*, 1998). Despite very strong criticisms by several organized groups, it has been shown that the reforms in the Philippines have produced some significant changes in the investment patterns of beneficiaries of land reform who were shown to devote more resources to the education of their children than households not affected by land reforms (Deininger *et al.*, 1999). Similar results have also been obtained from Brazil where as much as a 5-fold increase in the income of beneficiaries of land reforms was observed (Buinainain *et al.* 1999). How the beneficiaries of land reform are selected remains an important aspects of a land reform programme (Van Rooyen and Njobe-Mbuli, 1996).

As has been noted earlier, intensive discussions among several stakeholders preceded the launch of the Land Reform Programme in South Africa. In its own contribution, the World Bank undertook a modelling exercise that took account of various factors, including land use patterns, net farm incomes, household size, and agricultural income shares. The purpose was to assess the extent of income and employment generation that the land re-distribution process would deliver (World Bank, 1993). The findings of the modelling exercise were that both rural employment and household incomes would increase as a result of the re-distribution process. The logic of this finding is obvious. With more rural people expected to have access to this vital productive resource as a result of the various components of the land reform process, it seemed obvious that farming would become more widespread and lead to a stimulation of the rural economy through creating higher purchasing power among producers. The increased supply of food in the rural areas would also have a positive impact on food prices and hence real incomes of consumers. The World Bank assessment also anticipated an increased incidence of part-time farming which will generate a multiplier effect by increasing the size of part-time jobs (World Bank, 1993).

The market-assisted land reform has gained in popularity in recent times. This is because it has been found to be a more cost-effective option for promoting equity in the redistribution of assets in developing countries (The World Bank, 2003). Many land reform programmes are taking place today within an environment in which markets have been liberalized and subsidies to the farm sector have been discontinued as part of an economy-wide structural adjustment programme. In such a situation, it is preferable to adopt land re-distribution measures which avoid elements of coercion or the favouring of one group over another. The argument of the World Bank is that when markets are assisted to work in distributing land, more land is transferred to poor people who otherwise would be unable to compete in a completely free and unfettered market (The World Bank, 2003). Such an arrangement is also reputed to avoid the emergence of bloated bureaucracies and non-performing farming units.

With the foregoing in mind, Van Zyl and Binswanger (1996) enumerated a set of design criteria for an effective market-assisted land reform programme, including:

- Adopting self-targeting strategies to ensure that only willing buyers and willing sellers are matched.

- Employing financial grants in such a way as to motivate buyers to seek profitable opportunities and provide the necessary incentives for sellers to relinquish their land in exchange for commensurate cash reward.
- Gives flexibility to the beneficiaries in the choice of enterprise as well as level of investment, etc.
- A decentralized institutional structure that incorporates mentorship after start-up.

Implicit in the theory behind the market-assisted land reform as seen by the World Bank, and the design criteria for the South African programme (see Aliber, 1996; Van Zyl and Binswanger, 1996; Lund, 1996 and others) is a recognition of the centrality of efficient utilization of the re-distributed land. The willing-buyer-willing-seller framework foresees a situation where only people genuinely interested in retaining the land in its current use are attracted to buy the land. The RDP was very emphatic on the need to use land productively for agricultural and other productive pursuits. According to the RDP, the land reform programme was to be a mechanism for building 'the economy by generating large-scale employment, increasing rural incomes and eliminating overcrowding' (African National Congress, 1994). Again, this question was so important that the RDP endorsed the instrument of a land tax which would be used to 'free up underutilized land...and promote the productive use of land' (African National Congress, 1994). But, interestingly, the RDP also asserts that 'land is the most basic need for rural dwellers' (African National Congress, 1994).

In terms of actual programme content, the recent document released by the Department of Land Affairs in preparation for the 2005 Land Summit provides a comprehensive account of what is being done on the land reform programme in the country (DLA/DoA, 2005). This account is all the more authentic for the fact that it is being presented more than a decade after the scheme has been operational and therefore includes also the lessons learnt from the implementation and some of the remedial actions taken to correct mistakes. A number of official documents and academic publications have also been produced in recent times to provide insights into the background and implementation of the land reform programme in South Africa. As has been indicated elsewhere, the land reform programme is being implemented through three main elements, namely the land redistribution programme, restitution, and tenure reform. These three components of the land reform programme are briefly reviewed in the next few sections.

Land redistribution

This component of the land reform programme is considered the 'flagship' of the programme (DoA/DLA, 2005). At first, the scheme defined the rather ambitious goal of transferring 30% of the nation's land (about 24 million hectares of agricultural land) to black ownership by 1999. The expectation was that about 3 million people would benefit from this programme. The bulk of this land is held under commercial large-scale agriculture dominated by the white population while the rest is state land taken over during the period of apartheid rule for various state programme including military and conservation programmes. The new constitution that provided a democratic alternative for South Africa recognized existing property rights (the so-called 'property clause'). This meant that current owners of the candidate land for redistribution would have to be compensated for the land that would be taken from them. The World Bank estimated that the programme would cost between R22-26 billion to implement because market-related prices would have to be paid for the land, on a willing-buyer-willing-seller basis (DoA/DLA, 2005). Annually, this was expected to average between R1.5-1.7 billion.

But, with just about 1% of the land transferred in about 5 years into the programme (DoA/DLA, 2005), the lack of realism in the goal was revealed and in 1999, the programme was reviewed. With the target re-distributable land unchanged at 30%, the duration was extended to 15 years. In addition, steps were taken to facilitate the process which was designed to be almost fully market-driven. One of such steps was the increase in the level of the cash grants provided to prospective land buyers to enable them acquire land and put them into productive use. Initially, an amount of R15,000 was approved per beneficiary household under the Settlement and Land Acquisition Grant (SLAG) established at inception in 1994. This amount was adjusted to R16,000 per beneficiary household in 1998. The grant amount was set at the same level as the housing grant provided by the government.

The idea was that South Africans who were historically disadvantaged by the apartheid laws and were as a consequence landless and poor would receive a cash grant in that amount and use that to purchase agricultural land, develop and operate same. However, the reality was different. Farmland prices were, however, above levels that would allow an individual to economically go it alone. According to Lyne and Darroch (2003), an arrangement was worked out whereby individual households pooled their small cash grants to raise enough money to purchase a reasonable sized farm which they would now operate communally.

In time, the unsustainability of SLAG programme was demonstrated quite starkly by the frequent conflicts within the farming groups which were loose cooperatives, close corporations, and similar bodies. One of the problems revolved around the difficulty in forging a common purpose among individuals who had never engaged in farming and lacked the experience for any kind of profitable enterprise. Their sheer numbers also often proved too unwieldy to be managed effectively (DoA/DLA, 2005). According to Lyne and Darroch (2003). and others, by 2000, only a small part of the transferable land had been transferred and the outlook for the future was quite bleak. As part of an overhaul of the entire land reform programme, the new Minister that assumed office following the 1999 elections suspended SLAG and launched a review of the programme.

The result of the review is the on-going Land Redistribution for Agricultural Development (LRAD) which was established in 2000 and began activities in August 2001. According to a review by Swanepoel *et al.* (2004), the main element of LRAD is that it enlarged the size of the grant per recipient. Any applicant, not just the very poor as defined under SLAG, qualified for a minimum amount of R20,000 upon meeting the requirement to make an own contribution of R5,000 (Swanepoel *et al.,* 2004). A maximum grant of R100,000 can be obtained but requires an own contribution of R400,000 (Swanepoel *et al.,* 2004).

A special feature of the LRAD programme is that the own contribution of R5,000 to qualify for the minimum grant amount of R20,000 can be in the form of applicant's labour valued at going rates. The idea of the own contribution is to secure the commitment of the applicant. It continues the legacy of its predecessor of being limited to previously disadvantaged individuals, invariably from the black population. A minimum age-limit of 18 years was set in line with national norms. Applicants do not have to have agricultural experience although it is considered essential for the success of the farm business. It is required that the applicant submits a clear business plan which applicants can receive help from the Department of Agriculture to prepare.

These modifications have, however, done little to significantly improve the pace of the process. For instance, by February, 2005, the programme had managed to re-distribute no more than 3.5 million hectares to some 168,000 households. In terms of total land re-distributed, this means that only about 14.6% of the target has been attained in 11 years (6 years since the programme was revised). If current rates do not change, the programme may be able to achieve no more than 30% of its target overall by 2014. The achievement in terms of human numbers is even more worrying, at a mere 4.8% of target, fuelling growing concerns about the rural pauperization as the only visible outcome of the decade-long agricultural restructuring programme. Unemployment rates are still high in the country; on the basis of the broad definition of unemployment, jobless rates in the country stood at about 31% in 1993 (on the eve of the inception of majority rule in 1994), and have been deteriorating ever since, to about 38% by 1997, rising to about 39% in 2005, and not much different today (UNDP, 2007).

Land restitution

The other component of the land reform programme is accessLand Restitution by which presently dispossessed persons or groups (including communities) that could establish pre-1913 ownership of land can have such ownership restored. According to the White Paper on the Land Reform Programme (DoA/DLA, 2005), Land Restitution involves 'returning land, or compensating victims for land rights lost because of racially discriminatory laws passed since 19 June 1913'. Official estimates put the number of affected persons at more than 3.5 million who were forced out of their original land and compelled to settle in 'scheduled' areas which where later designated 'homelands'. Three sets of legislations provided the legal backing for the actions, namely:

- Native Land Act No. 27 of 1913;
- the Development Trust and Land Act No. 18 of 1936; and
- the Prevention of Illegal Squatting Act No. 52 of 1951.

These actions went under different names as follows: the 'black spot removal', 'removal of labour tenants', 'removals from mission stations', 'removals for the sake of forestry requirements', and 'internal removals'. The new Constitution of South Africa that came into force in 1996 provides in section 25, sub-section 7, that:

> *'a person or community dispossessed of property after 19 June 1913 as a result of past racially discriminatory laws or practices is entitled, to the extent provided by an Act of Parliament, either to restitution of that property or to equitable redress.'*

Already foreseen during the process of negotiation for the establishment of a multi-racial democratic system in the country, this provision arose from the Restitution of Land Rights Act No. 22 of 1994. To drive the process, a body known as the Commission on the Restitution of Land Rights was established and vested with the following tasks:

- To promote equity for persons or groups dispossessed by the policies of the past, especially those who are as a result landless and can be classified as poor, including the rural poor.
- To facilitate a developmental orientation to the solution of the problem by encouraging the relevant stakeholders to organize in the framework of viable development initiatives.
- To implement the restitution process in a manner that would promote reconciliation within the country.
- To contribute towards an equitable redistribution of land rights in the country.

In addition, a Land Claims Court was established to deal with ratifications or adjudication in respect of claims.

As would be expected, this scheme made a slow initial start. Having gone into operation in 1995 as in the case of the other components, only 41 claims had been decisively settled by 1998, three years after. It was obvious that things were not going well. The next general elections were coming in 1999 and President Nelson Mandela was soon leaving office as he had pledged. At the same time, the disgruntlement of the NGO sector, including the 'landless' people's groups represented by the Landless People's Movement (LPM), provided grounds for worry against the backdrop of the developing situation in Zimbabwe where the land reform programme had clearly gone out of hand.

Five main concerns were highlighted by the secretariat of the Commission on the Restitution of Land Rights as follows:

- duplication of functions between the Department of Land Affairs and the Commission;
- improperly defined lines of accountability;
- a dualistic approach to policy and procedures;
- the slow pace of the mandatory judicial process for settling claims;
- lack of integration of the restitution process into the broader land reform programme.

A review was subsequently conducted in the course of 1999 and led to the enactment of an amendment to the Restitution Act. The powers of the Minister of Land Affairs received a boost in the amendment as claims could now be finalized at the ministerial level rather than in the Land Claims Court. Within a year of the review, the pace of land claims settlement had picked and the process was moving quite smoothly. An assessment conducted in 2001 showed that the review process had been effective in unshackling the process from its bureaucratic straightjackets. Table 9.2 presents results of the assessment.

Table 9.2. Land restitution outcomes as at March, 2001 (Department of Land Affairs and Department of Agriculture, 2005).

Land restoration	
Households awarded land	16,764
Land cost	R191,270,645.00
Hectares of land restored (Court process)	173,805
Hectares of land restored (Ministerial approval)	109,421
Total beneficiaries receiving land	100,584
Financial compensation	
Households awarded compensation	10,921
Financial compensation awarded (Land Claims Court)	R21,860,330.00
Financial compensation awarded (Ministerial Approval)	R287,043,658.10
Restitution total	
Claims settled as at 31 March 2001	12,094
Total households	27,685
Total restitution beneficiaries	164,661
Total restitution cost	R500,174,633.10

Land tenure reform

The third key component of the land reform programme is the reform of land tenure systems in the communal areas to enhance the accessibility of land within the former independent homelands where traditional systems of land tenure were in force previously and largely prevail today. It is necessary once again to delineate the specific areas of the country where the 'independent homeland administrations' were established during the apartheid era. Parts of four of the present nine provinces were affected. These include the Eastern Cape, Limpopo, Kwazulu-Natal, and North-West provinces. According to the old administrative structure, these areas were known as the Ciskei, Transkei, and Kwazulu. Today, agricultural production still occurs in areas designated 'communal areas'.

The past practices featured control by the traditional power élite which administered the allocation of land to the members of the community (Jeppe, 1980). As has been noted earlier, the communal tenure system in the South African communal areas mirrored the practices in other parts of Africa where the traditional authorities remain the trustees in land matters and wield considerable powers. What has changed in South Africa, though, is that the powers of the traditional authorities have been scaled down or officially eliminated without a corresponding and clear modern governance arrangement to replace them. This has meant a considerable degree of confusion, while opportunities for corrupt manipulation of the process still exist for the traditional leaders. According to Lyne *et al.* (1997) and Lyne and Darroch (2003), the main problem has been the increasing unpredictability of the whole system. It is understandable that where clear rules and guidelines are absent, the scope for arbitrariness will be high. In a system characterized by land scarcity due to inequitable sharing of a limited supply of land within the former homeland area, economic opportunities are severely limited for those whose livelihoods depend almost exclusively on the agricultural sector. The highly constrained land rental market due to uncertainties as to whether those who rent will eventually return land to original owners is another consequence of the imprecise system of rules on land ownership and distribution within the former independent homelands of the country.

In order to protect the residents of these former homelands against the abusive and corrupt practices of the traditional leaders, an interim law was passed in 1996. This law is referred to as the Interim Protection of Informal Land Rights Act 31 of 1996. But ultimately, the need is for the people to have ownership of the land by legal changes that introduce more secure land rights in these communal areas. The progress in this regard seems to have now come in the form of the recently enacted Communal Land Rights Act (CLARA) of 2004. The main emphasis of the new Act is to facilitate the transfer of communal land to tribes as well as to individuals and communities. While this is no doubt a welcome improvement on the current situation of landlessness among many residents of the former homelands, it is not surprising that a number of groups will find that this threatens their existing privileges and powers (Lyne and Darroch, 2003). For instance, while the CLARA was still in draft form being circulated for discussion and debate, the Inkatha Freedom Party which draws strong support from rural Zulu tribes people, protested that the Act would disempower traditional leaders (Lyne and Darroch, 2003).

In the area of land tenure, the government has two other related concerns, namely the issue of labour tenancy, and the frequent problems of eviction of farm workers from commercial farms. In the first case, namely the labour tenancy issue, this is an arrangement whereby black farm workers use their labour to pay for usufructuary rights to a portion of a white farmer's land. This was an arrangement that was worked out to circumvent the apartheid legislation incorporated in the Natives Land Act of 1913 that prevented blacks from owning land in areas designated white areas. Even with apartheid laws annulled and a land reform programme running into its second decade of implementation, some of these practices still remain and continue to be a source of considerable worry for the government which has enacted a specific legislation, the Land Reform (Labour Tenants) Act 3 of 1996. The aim of this Act is to make the land rights of commercial farm workers and labour tenants more secure on the commercial farms.

The issue of eviction of black farm workers from white farms is clearly a labour relations question for which a comprehensive set of labour legislations exist outside the agricultural sector. Since the end of apartheid rule, workers' rights have been strengthened nationally. But it would seem that the agricultural sector fits into a special category probably due to the relatively more powers a farm owner would generally be expected to have arising from the nature of farming, land, etc. The government is of the view that, as part of the overall programme to redress past wrongs, such employment insecurities should be discontinued. In pursuance of this, the government has put in process a legal framework in the form of the Extension of Security of Tenure Act 62 of 1997 (ESTA). The idea of ESTA is to protect farm workers from unlawful evictions from the farm which have implications for both their employment and accommodation as many of them live and work on the farms. Many farms have also built schools for the children of the farm workers. Eviction in this case would no doubt have far-reaching consequences for the entire family of the farm worker. Hence the need for protection of land rights to ensure that loss of job and accommodation can only happen on legal grounds.

A review carried out for the National Department of Agriculture (DoA) by Swanepoel *et al.* (2004), suggests that there are serious production problems associated with communal land use. According to the review, while the provincial departments of agriculture in the affected four provinces deploy a large part of their budgets and human resources towards agricultural development in the communal areas, production falls far short of potential. Findings reported by the Eastern Cape Department of Agriculture show that out of a potential maize yield of over 4 tonnes per ha, farmers realize only about 200 kg per ha. In

Kwazulu-Natal, the actual production is about one-sixth the proven potential. While as much as 38% of the national beef herd is kept in the communal areas where crop production remains an adjunct to livestock production, extremely low animal off-take denies both farmers and consumers the benefits of this important sub-sector.

Some of the factors contributing to the poor production performance in communal areas have been identified and summarized as:

1. Continuous shallow ploughing of the same land creates an impermeable layer of soil which hampers the flow of moisture to crops grown under rain-fed conditions.
2. Forced fallowing of large tracts of land as a result of inadequate supply of labour, inputs and equipment.
3. Sub-optimal resource use as a result of over- or under-investment of resources which leads to low productivity.
4. The use of community-based models of empowerment which lead to serious internal conflicts that consume an inordinate amount of resources and time and contribute to low production and productivity.

For any tenure reform programmes to lead to agricultural development, including the integration of the black population to the country's agricultural economy, it is imperative to take a holistic view of the problem. Some of the issues highlighted above will certainly be crucial, but further detailed studies to ascertain the problems and what strategies are needed to contain them, will be necessary.

9.4 Agricultural taxation and the rural land tax debate

Some of the longest running debates in the land market of South Africa concern the issues of taxation of agricultural land. In the era of land reform in the post-apartheid South Africa, the main concern has been on how the mechanism of taxation can be used for increasing the availability of land in the market to increase the volume of land transactions by encouraging land owners not to hold on to unproductive land. The next few subsections examine some of the salient issues related to the debate.

9.4.1 General considerations on taxes

It is necessary for a more informed view, to present the wider evidence on the importance of taxes in general as reflected in the economics literature. A number of researchers have worked on the broad subject, notable among them being Stiglitz (1986), Stallman and Jones (1997), Mankiw (1998), etc. It generally considered that taxation is good from the point of view of generating resources to finance national development and redistribute the national wealth in a more equitable manner. A tax system is therefore evaluated on the basis of its capacity to carry out these functions while not imposing corresponding hardships on the society. On the basis of the review of the various works on the subject, the following attributes of taxes can be summarized:

- *Economic efficiency*: it has been said that taxation is a re-distributive mechanism. This means that it must achieve this without interfering the efficient allocation of other resources to make the economy productive. If this is not the case, then the purpose of taxation would be defeated. At the same time, the tax system must leave the consumer choices undisturbed, which is to say that consumer should be free to consume after the tax regime is imposed what they consumed before the imposition of the taxation. In the opinion of Stallman and Jones (1997), the condition of economic efficiency requires that the tax system be as broad-based as possible, as opposed to having a narrow base, so that to the extend practicable, inefficiencies are kept to a minimum.
- *Competitiveness*: it is required that a tax system does not put a country at a disadvantage vis-à-vis other countries in attracting investors or buyers of its goods and services. This is the whole point of making the tax system competitive so that it is not so high that it discourages foreigners or too low that it generates very little to finance needed development. A tax system should be able to attract new business and retain existing ones and should provide the necessary incentives for tourists and others whose spending behaviour influence economic activities in the country, one way or another.
- *Administrative simplicity*: the implementation of a tax scheme is very important. Often, a tax law may be in place but implemented in a way and manner that hurts rather than helps the economy. Some of the difficulties may be that the system is so administratively unwieldy that it leaves considerable scope for officials to mismanage the system and obstruct the smooth flow of the rewards to the national treasury. This is the case when a tax system allows a lot of loopholes that creates opportunities for corrupt gain for those who are entrusted with its management and also allows room for the potential tax payers to avoid paying tax. When a tax system is simple, it is easy for the taxpayer to understand it. Such a tax system is also easier and less expensive to administer.
- *Adequacy*: this is easily the most important attribute because it is at the heart of the very reason a tax is imposed, namely to generate sufficient funds to finance the various obligations of the government. An adequate tax system is therefore one that is capable of generating enough revenue to meet the developmental needs. This means that the tax system must be responsive to a number of demographic factors such as population growth rate, labour force growth rate, etc.
- *Fairness and equity*: these are important attributes in terms of the distributional aspects of a tax system. The tax system should not have a discriminatory element that is it must bear equally on groups or individuals in the same circumstances. This is horizontal equity which requires that all entities in the same relative socio-economic circumstance must be taxed to the same degree. The other category is vertical equity which differentiates between groups and individuals in dissimilar circumstances. This raises an important question in respect of the appropriate methodology for evaluating the equity of a tax system – based on the payments received by the tax payers, or the ability to pay.

As Stallman and Jones (1997) observed, there are no ideal tax systems with all the above attributes. The design and implementation of a tax system, regardless of the level of national administration for which it is intended, is a matter of trade-offs and compromises. In discussing taxes in general, it is essential to bear in mind that taxes general specific effects and the field of welfare economics has studied these in some detail. As Mankiw (1998) has observed, taxes generally make both buyers and sellers worse off than previously. The direct or indirect effect of the tax on buyers and sellers is known as the tax burden or the tax incidence. Mankiw (1998) notes that the tax burden is distributed between buyers and sellers and the associated supply and demand elasticities determine the severity of the tax burden. The more inelastic the supply, the greater the tax burden on the seller relative to the buyer. Conversely, the more inelastic the demand, the greater the burden on the buyer relative to the seller. Figure 9.1 illustrates the effects of taxes on buyers and sellers when tax is imposed at a given level within a specific market.

Supply and demand analysis represent a useful and powerful tool for analyzing the various effects that different types of taxes can have on individuals and groups in the society (Frank, 1994). If we assume the imposition of a constant tax per unit of an output, we should be interested to know how this impacts on the equilibrium price of the commodity and the quantity of the output that will be supplied at the ruling price.

Suppose the tax is imposed on the seller, with the original supply schedule represented by S_1S_1 and price at P_{eq} which represents the equilibrium price associated with quantity supplied at Q_{eq}. The imposition of the tax on the seller will lead to an increase in the

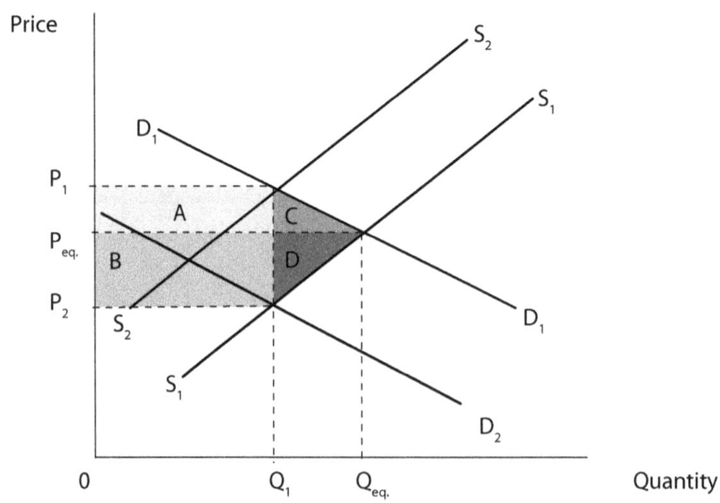

Figure 9.1. Illustration of the effects of taxes on buyers and sellers.

price from the original equilibrium point at $P_{eq.}$ to a new level at P_1. The seller's wish will definitely be that he is able to completely cover the extra costs brought about by the tax and still sell the same quantity. But the market situation suggested by the slope of the supply curve is a competitive one where it is not always possible for a seller to 'eat his cake and have it'. What would normally happen is that a buyer, on whom no tax has been specifically targeted, but who all the same is required to indirectly assume responsibility for the tax burden of the seller, will become less motivated to buy the commodity to the same degree at the higher price.

Thus, the first thing that happens comes from the reaction of the buyer in reducing the demand for the commodity which results in the demand curve D_1D_1 shifting to the left to D_2D_2 where a preference is declared for a lower price P_2. Since that price means that the seller will sell at a loss if the same quantity is supplied, the logical thing will be to reduce the quantity supplied to Q_1 which corresponds to a new intersection point of S_1S_1 and D_2D_2. But this is an unsustainable position because at that price, the seller's costs are not even being covered. This will force the buyer to accept a higher price at P_1 while buying a reduced quantity Q_1. So, a tax imposed on a seller forces the seller and buyer into a compromise in order to accommodate the reduced enthusiasm of the buyer for the higher-priced commodity and the frustration of the seller with higher costs. Thus, there is a sharing of the tax burden here between the buyer and seller – the seller receives a reduced price while the buyer pays a higher price. If we assume the tax burdens for the seller and buyer are t_s and t_b, respectively, their shares of the tax burden can be denoted as follows:

$$t_s = P_{eq} - (P_1 - T) / T \qquad (1)$$

Where:
t_s = the tax burden on the seller;
P_{eq} = the original equilibrium price;
P_1 = the final price received by the seller;
T = the amount of tax imposed per unit of commodity.

In the same way, we can denote the buyer's share of the tax as follows:

$$t_b = (P_1 - P_{eq}) / \mathrm{T} \qquad (2)$$

Where:
t_b = the tax share of the buyer;
P_1 = the final price paid by the buyer;
P_{eq} = the original equilibrium price;
T = the tax imposed on the seller.

It should be noted that both t_s and t_b are proportions or shares of the absolute size of the tax imposed. In that case, both should sum to unity according to the following relationship:

$$t_s + t_b = 1 \quad (3)$$

Developing further the idea of the tax burden, it is important to note that although the buyer eventually pays P_1, the difference between what the buyer was willing to pay (P_2) and what she eventually pays (P_1) does not all accrue to the seller because of the tax ($P_1 - P_2$) which accrues to the government. In Figure 9.1, this corresponds to areas A and B. While the buyer bears the portion B, the seller bears the portion A. As a result of paying these taxes, both the seller and the buyer incur costs equivalent to losses in the producer and consumer surpluses respectively. The loss in consumer surplus is equivalent to the area D, while the loss in producer surplus is equivalent to the area C.

It is usually of both empirical and policy interest which of the parties incurs the most cost from the imposition of the tax. That is, who bears the bulk of the burden of tax given the relation above? In other words, which is greater, t_s or t_b? There is no fixed rule on who bears the most burden of the imposed tax. As indicated earlier, this would depend on the slope of the demand and supply curves (Frank, 1994; Mankiw, 1998). When elasticity is explained in its commonplace, lay meaning as the responsiveness of the quantity bought to changes in price from the differential viewpoints of the buyer and seller, the tax impacts can be a little clearer. According to Frank (1994), if the supply is highly unresponsive to changes in price (vertical to the horizontal or almost so), then t_b (the tax burden on the buyer) is close to zero while t_s (the tax burden on the seller) is close to unity. This means that the seller bears the bulk of the tax. On the other hand, if demand is highly unresponsive to price, that is almost horizontal, then t_s will approach zero while t_b will be almost unity.

In the situations above, a relatively inelastic supply schedule implies fixed supply and absence of close substitutes to which the buyer can turn. On the other hand, a relatively elastic supply curve implies that supply is highly variable and that the buyer can choose among substitutes. In the first case, the buyers bear most of the burden of the tax. In the other case, the suppliers do not have any alternative than to go on supplying the product. As a result, the bulk of the tax burden falls on them. This provides some endorsement to the notion that 'a tax tends to fall most heavily on the side of the market that can least escape it' (Frank, 1994). Similar points are also made with respect to capital gains tax, suggesting that the poorer segment of the economy absorbs the most tax while the wealthier segment has the ability to avoid or delay capital transactions so that capital gains accrue without being realized.

The foregoing statement about who feels the pinch most compels policymakers to be extremely cautious in deciding on any tax to raise revenue for the government. The common approach of politicians is to impose taxes on those who 'can best afford them' (Frank, 1994). The interest on who feels the tax burden also leads to the consideration of some important tax concepts. Distinction is commonly made between legal incidence of the tax and the economic incidence of the tax (Frank, 1994). In the case of legal incidence, the question is who is being targeted by the tax. The tax either targets buyers or sellers. The economic incidence of the tax refers to the shares of the tax burden that is borne by different parties. Regardless of the legal incidence of the tax, that is, where the tax is placed, the burden of tax will be shared in the same way. It is the slopes of the supply and demand curves that determine how the tax burden is distributed across the population. When the slopes of the demand and supply curves are the same, the tax burden is more less shared equally between the parties involved in the transaction.

The other costs associated with taxation are referred as deadweight loss or social costs of the tax. When some external economic instrument leads to an increase in the price of the commodity above its equilibrium level, it is said that a deadweight loss has occurred. This is usually associated with the introduction of the value- added tax (VAT). There is always an intrinsic value attached to a commodity by the consumer which they try to compare with the price on offer. According to Mankiw (1998) when the value attached to the commodity by the consumer exceeds the price on offer, they buy, otherwise they don't. When the consumer buys because of a positive assessment of value relative to product price, there is a 'gain from trade' (Mankiw, 1998). When the consumer does not buy because of a negative assessment of the price relative to the intrinsic value, there is a loss from trade (Mankiw, 1998). Figure 9.1 can be used to analyze this situation in the same way as the previous analysis related to the imposition of a tax on the seller of a commodity.

9.4.2 Agricultural land tax in the global/historical context

The debate on the pros and cons of a land tax in South Africa has generally drawn from economic theory, international experience and the unique demographics of the country itself. Appeal was usually made to international and historical experience to show that governments the world over have always relied upon land to generate the necessary tax revenue to finance the affairs of state (Franzsen, 1992). These motives were shown to be invariable, whether one was looking at ancient China during the Huang Ti and Hsai Dynasties (2700-2200 BC), Egypt of the Pharaohs, Mesopotamia, or the Roman Empire (Franzsen, 1992). On the basis of a historical assessment by Woolery (1989) as cited by Franzsen (1992), perhaps the most famous land tax advocacy of the modern era is credited to Henry George whose idea was that governments could lead themselves out of poverty by heavy taxation of land over and above what was then deemed tolerable. Bird and Stock (2002) show that in the 1990s, both developing and industrialized countries raised the proportions of their local taxes attributed to property and land taxes. For instance, about 40% of the locally derived taxes in the developing countries came from property and land taxes while the figure for industrialized countries was about 35%. The transition economies performed poorly in this regard, at a mere 12% (Bird and Stock, 2002).

In general, it has been established that land taxes make it possible for local governments to exercise fiscal responsibility without correspondingly distorting the economy (The World Bank, 2003). The firm belief is that a tax on land compels landowners to make effective use of their land, especially where the tax is based on the ownership of the land rather than the use to which it is put or the output from the land. International experience (Brueckner, 2000; The World Bank, 2003) equally suggest that land tax is an ideal revenue source for local government financing since it is directly tied to a source of local production rather than distortionary transfers from the central government treasury. It has also been shown that land taxes can induce local development as local authorities realize an obligation to deliver much-needed services to those who pay for them.

In an age of decentralization of governmental structures to the local level, many observers (among them Skinner, 1991; Strasma, 1993; Boadway, 2001 and Eaton, 2001), think that the ability to raise their own finances will make the process more effective. Enquiries into the role of land taxation in economic development has been going on for a long time (Van Sickle, 1925). The fact that local beneficiaries of the local taxes directly bear the tax burden is also considered an important advantage of an effective land tax especially if the local authorities have the full and complete discretion to set the tax rates, identify the taxable land/taxpayers, assess them and carry out the necessary collection of the due taxes (The World Bank, 2003).

The arguments (Wald, 1970; Holland, 1970; Harris, 1976; Woolery, 1989; and Franzsen, 1992) in favour of a land tax put a lot of emphasis on the attributes of land which predisposed it to much easier tax assessment and administration, including the fact that:

- land is not concealable, which must be frustrating to those who have a strong propensity to avoid other forms of taxes based on less tangible indicators of wealth;
- land is locationally stable and therefore convenient to access and assess;
- the identity of the taxpayer for a land tax is easily established without complication;
- land is an asset with perpetual positive value upon which a tax can be assessed.

9.4.3 The South African experience with land and agricultural taxation

Very few issues have dominated the agricultural policy environment of South Africa for as long as the questions of agricultural taxation and rural land tax. But for all practical purposes, they have been more important by the heated debate they have generated than by their actual impact since systematic taxation specifically targeting the agricultural sector has yet to be implemented in the country. The questions of agricultural taxation and the land tax have also been the most contentious, being known to generate a great deal of emotional reactions from the farming community and others (Franzsen and Heyns, 1992; Van Schalkwyk, 1995.).

Figure 9.2 provides a simple framework for analyzing the incidence of a land tax and its social costs. As is well known from economic theory, the supply of land is fixed in the pure case which gives us a supply curve that is perfectly perpendicular to the horizontal axis. One of the most famous cases for a land tax was made by Henry George on the grounds that it would help to re-distribute national wealth and alter the obscene picture of 'monstrous wealth and debasing want' (Mankiw, 1998). Whether or not such a redistribution of national was feasible given the characteristics of land and the economic realities can be examined by recourse to economic analysis.

Figure 9.2 presents a negatively sloping demand curve for land in relation to a vertical supply curve depicting an inelastic supply of land. If we assume an equilibrium price of land, P_{eq}, which prevails at the point where the negatively sloping demand curve intersects with a vertical supply curve, economic theory predicts that the introduction of a land tax would lower the price of land to a level P_s which is below the equilibrium level. The difference between the equilibrium price and the market price of the land accrues to the government as tax revenue from land. According to Van Schalkwyk (1995), the decline in land price as a result of the imposition of a land tax would happen because it would have meant an additional cost to the farming community. The options of discounting the extra tax-induced production cost by, for instance, raising product prices or land rents do not

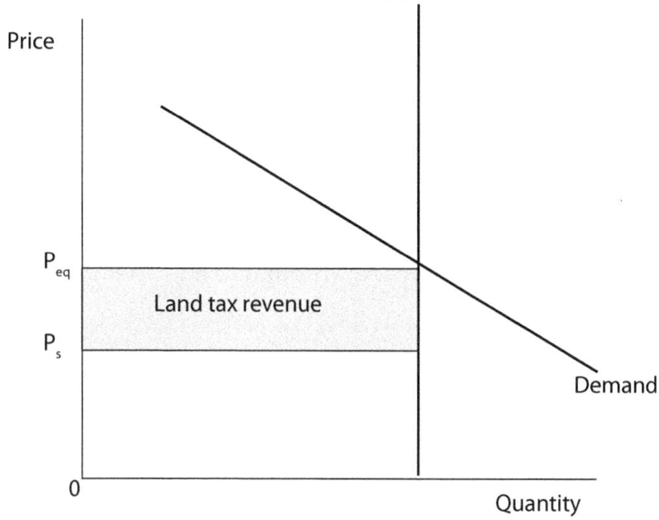

Figure 9.2. Illustration of the special case of land tax incidence.

seem quite feasible. But the foregoing scenario is true only in the short run. In the long-run, a lot of changes are expected to take place within the physical environment of the land market, including with respect to the infrastructures associated with land use. This will have implications for land productivity (that is, agricultural output from a unit of agricultural land), and therefore translate into a change in the supply of land from one period to the other. In such circumstances, a given unit of farmland can be priced higher than previously to take account of the productivity-enhancing improvements. This consideration is also discussed generally in relation to land quality issues and is easily handled where land quality indices are available.

According to Nieuwoudt (1995), such physical infrastructures that can influence the actual supply of land include irrigation equipment, development of pastures for livestock feeding, fencing, orchard development for horticultural production, etc. Since land tax also falls on improvements such as these, the prospects of higher taxes my discourage such investments and so lead to situations where agricultural development stagnates or farmland prices do not rise over time. To that extent, a rural land tax may turn out to be inimical to agricultural progress.

The fact that the country's commercial agricultural sector evolved as a highly sheltered sector that enjoyed a high degree of protection from successive governments is probably an important factor in the difficulty successive government's have had in successfully selling the idea of taxes. As has been very well-discussed in the literature, for most of the apartheid era in South Africa, and especially in the 1980s, agricultural policy defined a single aim of attaining 'self-sufficiency in respect of food, fibre and beverages and the supply of raw materials to local industries at reasonable prices' (Van Schalkwyk, 1995).

In order to achieve this aim, the government extended a wide variety of subsidies and state support schemes to motivate increased farm investment and production. Van Schalkwyk (1995) summarized this farmer support environment to embrace such incentives as setting of domestic prices above world market prices, provision of sophisticated extension services that facilitated access to improved technologies and other forms of preferential treatment. This was also a period during which the international community had imposed a set of economic sanctions that called for strong internal response towards self-preservation. Thus, it would obviously have seemed anathema to turn around and start taxing the same farmers.

But agricultural taxation in one form or the other is not a completely unfamiliar variable. According to Franzsen (1992), prior to the establishment of the Union of South Africa in 1910, a wide range of taxes had been levied in different parts of the vast area that now constitute the Republic of South Africa. As early as 1677, some form of agricultural taxation had existed in the Cape of Good Hope (area now embracing the Western Cape Province) in the form of what was known as 'agricultural income tax payable in kind' and administered by the Dutch East India Company better known by its Afrikaans acronym,

VOC (*Vereenigde Oost-Indische Compagnie*). This tax was also described as a 'tithe' to the extent that it represented a proportion of the income earned in the same way as the one paid in Christian Churches, or 'zakaath' as in Islam. To this tax was added two more forms known as 'transfer duty' in 1686 and 'recognition fees' that took effect in 1714. Citing his earlier research (Franzsen, 1990), Franzsen (1992) revealed that another form of tax known as the 'stamp duty on transfer deeds' also came into effect in 1714. The first violent resistance of the indigenous Khoisan and amaXhosa against European settler occupation and expansion took place between 1799 and 1803 and one is left to imagine whether there was a link between the introduction of these taxes and local reactions (Morris, 2004).

Outside the Cape Province, taxes that affected the farming communities were also in effect. For instance, in three of the political entities that later came under the umbrella of the Union of South Africa, namely the Natal Republic, the Transvaal (then known as the South African Republic) and the Republic of the Orange Free State, an elaborate system of taxation was introduced to bring a broad spectrum of the citizenry within the tax net (Franzsen, 1992). In these three Republics, taxes as diverse as transfer duty, stamp duty on transfer deeds, hut tax, land tax, and property tax were imposed on the inhabitants (Franzsen, 1992). An indication of the comprehensiveness of the tax system was the existence of a tax on absentee landlords whose uninhabited dwellings were taxed, and the hut tax which embraced subsistence farmers and the peasantry (Franzsen, 1992). In a review of the literature on the land tax in South Africa in the period 1652-1994, Van Schalkwyk (1995) adds that another purpose of the hut tax was to create a strong income need among peasants that would motivate them to seek work in gold mines and sugar cane fields as general labour.

As it became clear in the early 1990s that land reform was inevitable in South Africa and was only a matter of time, conversation and debate ensued on the implications of introducing a land tax in the country. One of the major concerns around the success of the land reform programme was the potential cost to the government of implementing what would obviously be a monumental undertaking in view of the large number of potential beneficiaries and the amount of the land to be re-distributed. Experts reasoned that a tax on land would provide government with enough resources to finance the land reform programme (Franzsen, 1992; Van Schalkwyk, 1995; Katz Commission, 1998; DLA/DoA, 2005).

Evidently, Van Zyl and Vink (1992) did not share the enthusiasm of the those who think land taxation will work and actually saw ample scope for immense administrative overhang in a land tax in South Africa. Their conclusion that the introduction of a land tax in South Africa would contradict international experience may have drawn largely from the uniqueness of the structure of the agricultural sector with its sharp dichotomies and the background of far-reaching state support applied disproportionately on the commercial, large-scale agriculture sector that was largely white-owned.

In any case, the debate raged on into the era of land reform which came into effect in 1994. In May 1995, a Land Tax Sub-Committee of the Katz Commission, sat for the first time to investigate the possible implications of a land tax in South Africa (Katz Commission, 1998). The White Paper on Land Policy has identified a number of clear advantages of the rural land tax for South Africa, among which the following have been emphasized:

- Intensification of land use which would lead to landowners releasing surplus land into market to avoid paying tax on idle land.
- Elimination of land speculation by adjusting the tax to align with the eventual use of the land and therefore discourage the tendency to acquire land cheaply purportedly for agriculture and then turn into non-agricultural use which might have higher returns. It is expected that such a graduated tax in this case will reduce the farmland prices and will therefore benefit small-scale farmers for whom the high price of farmland is a strong disincentive.
- The problem of absentee landlords will be eliminated by a rural land tax, so that only genuine farmers will acquire land. Again, this will bring land prices down.
- When the rural land tax is combined with the removal of existing rules on sub-division of farmland, it will become easier to re-distribute land from large holdings to medium and small holdings and this will be beneficial to small-scale farmers who can then acquire only the size of farmland they can operate.
- It will be an additional source of budget revenue for the government without corresponding administrative costs since it is an easy tax to administer.

The recent Land Summit held in Johannesburg in July 2005 came up with the recommendation that a rural land tax should be introduced in the country and the Minister of Agriculture and Land Affairs, by her pronouncements, appears favourably disposed to such a policy regime. It would seem that whether or not a land tax is introduced in South Africa is only a matter of time.

Despite its relative popularity, a number of issues make the introduction of land taxes a carefully considered option. These issues relate to the fairness and equity as well as effectiveness. In this regard, policymakers and researchers outline a number of requirements for administering a land tax, including having a comprehensive official record (that is, a cadastral record) of available land portions/plots, their sizes, values, ownership statuses, production parameters and marketing data, specifically with respect to input and output costs and prices (The World Bank, 2003). There are also frequent concerns with the basis for the land tax, i.e. whether it will be based on the land area, its value, or some other criteria and how to deal with the technical tax administration questions of identification, assessment and collection. According to Bird (1974), all these require a clear legislative position on property rights and an administrative structure that routinely collects and updates information on land and its ownership and sets and implements guidelines for assessment, collection and enforcement of land tax.

Questions are also often raised about what the optimum tax structure should be to ensure that collection rates are good and discontent is kept at a minimum. Hamid (1983) and Hoff (1991a and 1991b) have shown that the introduction of a land tax in poor societies with inadequate infrastructures for insurance in the presence of excessive farm risks can exacerbate inequalities and lead to land being concentrated in a few hands. It is therefore generally considered that a 'simple, possibly flat, tax that may be waived for very small landowners' may be more manageable and lead to better collection rates, as examples from Kenya and Indonesia show (Bird and Stock, 2002). How much discretion the local authority should have in the administering of the land tax system is also a major issue. If the implementation of a land tax will have the effect of strengthening fiscal responsibility of the local authority, then it should have the opportunity to build up the necessary in-house capacity in that area.

But pressures to deliver services within a tight budget constraint may lead to a revenue maximization behaviour on the part of the local authority that may hurt rather than help the investment climate. For instance, too high a tax rate may lead to investors leaving the particular local area in favour of areas with more favourable tax regimes. Conversely, too low a rate may put the local area at a serious disadvantage in terms of revenue maximization. For these reasons, it has been suggested some central government intervention in setting the broad bands of tax rates within which the local authorities can then exercise discretion. According to Boadway (2001), both minimum and maximum rates help in avoiding two of the most serious pitfalls of tax administration, namely tax competition and tax exporting. Tax competition can arise when better-off local governments use lower tax rates to lure businesses away from worse-off local governments unless there is minimum rate below which they cannot go. On the other hand, if a maximum tax rate is not established and enforced, there is a tendency to impose high tax rates on businesses with the intention of passing on the tax burden to non-residents. According to Boadway (2001), this latter situation delinks those who pay the taxes from those who benefit from them.

References

African National Congress (1994). The reconstruction and development programme. The African National Congress (ANC). Pretoria, RSA.

Aghion, P., E. Caroli and C. Garcia-Penalosa (1999). Inequality and economic growth: the perspective of the new growth theories. Journal of Economic Literature 37, 1615-1660.

Aliber, M. (1996). Benefit-cost analysis of land redistribution: conceptual issues. In: J. van Zyl, J. Kirsten and H.P. Binswanger (eds.) Agricultural land reform in South Africa. Cape Town, RSA: Oxford University Press, pp. 563-588.

Banerjee, A., J. Dolado, and R. Mestre (1998). Error-correction mechanism tests for cointegration in a single-equation framework. Journal of Time Series Analysis 19, 267-283.

Bardhan, P., S. Bowles, and H. Gintis (1999). Wealth inequality, wealth constraints and economic performance. In: A. Atkinson and F. Bourguignon (eds.), Handbook of Income Distribution Volume 1. North Holland: Amsterdam, the Netherlands, pp. 541-604.

Bayley, B. (2000). A revolution in the market. The deregulation of South African agriculture. Oxford Policy Management, UK. 142 p.

Besley, T. and R. Burgess (1998). Land reform, poverty reduction and growth: evidence from India. London School of Economics, London, UK.

Bird, R.M. (1974). Taxing agricultural land in developing countries. Cambridge, MA, USA: Harvard University Press.

Bird, R.M. and E. Stock (2002). Land and property taxation around the World: a review. Paper presented at the World Bank Land Workshop, April 3-6, 2002, Budapest, Hungary.

Binswanger, H.P. and K. Deininger (1993). South African land policy: the legacy of history and current options. World Development 21, 1451-1475.

Blanchflower, D.G. and A.J. Oswald (1998). What makes an entrepreneur? Journal of Labour Economics 16, 26-60.

Boadway, R. (2001). Intergovernmental fiscal relations: the facilitator of fiscal decentralization. Constitutional Political Economy 12, 93-121.

Brand, S.S., N.T. Christodoulo, C.J. Van Rooyen, and N. Vink, (1992) Agriculture and redistribution: A growth with equity approach. Development Bank of Southern Africa paper, Midrand, RSA.

Brueckner, J.K. (2000). Fiscal decentralization in developing countries: the effects of local corruption and tax evasion. Annals of Economics and Finance 1, 1-18.

Buinainain, A.M., J.M. Da Silveira, H.M. Souza and M.M. Magalhaes (1999). Community-based land reform implementation in Brazil: A new way of reaching out to the marginalized? Paper presented at the GDN Launch workshop, Bonn, December 5-8, 1999. Available at: www.gdnet.org.

Burgess, R. (1998). Land, welfare and efficiency in rural China. Unpublished paper. London School of Economics, London, UK.

Christiansen, R. (1996). Overview of land reform issues. In: J. van Zyl, J. Kirsten and H.P. Binswanger (eds), Agricultural land reform in South Africa. Cape Town, RSA: Oxford University Press, pp. 367-389.

Claassen, J.H.D. (2000). Management Profile of Emerging Farmers in the Qwaqwa Area – A Geographical Survey', University of the North, Qwaqwa Campus (mimeo).

Cusworth, J. (1992). Zimbabwe: issues arising from the land resettlement programme. In: Dudley, N., J. Madeley and S. Stolton (eds.), Land is life – land reform and sustainable agriculture. London, UK: Intermediate Technology Publications, pp. 89-101.

Davenport, T.R.H. and K.S. Hunt (1974). The right to land. David Philip Publishers: Cape Town, RSA.

De Janvry, A. (1984). The role of land reform in economic development: policies and politics. In: C.K. Eicher and J.M. Staatz (eds.) Agricultural development in the Third World. Baltimore, USA: John Hopkins University Press, pp. 263-274.

Deininger, K and L. Squire (1998). New ways of looking at old issues: inequality and growth. Journal of Development Economics 57, 257-285.

Deininger, K and P. Olinto (1999). Asset distribution and growth: new panel estimates. The World Bank, Washington, DC, USA (mimeo).

Deininger, K., F. Lara, M. Maertens, and P. Olinto (1999). redistribution, investment, and human capital accumulation: the case of agrarian reform in the Philippines. Paper presented at the World Bank's Annual Conference on Development Economics, April 2000, Washington, DC, USA.

Delgado, C.L. (1995). Africa's changing agricultural development strategies: past and present paradigms as a guide to the future. Food, Agriculture and the Environment Discussion Paper #3, International Food Policy Research Institute, Washington, DC, USA.

Department of Land Affairs and Department of Agriculture (2005). Land and agrarian reform in South Africa: an overview in preparation for the land summit, 27-31 July, 2005', Ministry of Agriculture and Land Affairs, Pretoria, South Africa.

Development Bank of Southern Africa (2005). Development report 2005 overcoming underdevelopment in South Africa's second economy. Midrand (Johannesburg), Development Bank of South Africa (DBSA), Human Sciences Research Council (HSRC), and United Nations Development Programme (UNDP).

Eaton, K. (2001). Political obstacles to decentralization: evidence from Argentina and the Philippines. Development and Change 32, 101-127.

Eicher, C.K. (1999). Institutions and the African farmer. Issues in Agriculture No. 14, Consultative Group on International Agricultural Research (CGIAR), Washington, DC, USA.

Eicher, C.K. and M. Rukuni (1996). Reflections on agrarian reform and capacity building in South Africa. Staff Paper No. 96-3, Department of Agricultural Economics, Michigan State University, USA.

Feder, G. and D. Feeny (1991). Land tenure and property rights: theory and implications for development policy. The World Bank Economic Review 5, 135-153.

Fenyes, T. and N. Meyer (2003). Structure and production in South African agriculture. In: L. Nieuwoudt and J. Groenewald (eds.) the challenges of change: agriculture, land and the South African economy. University of Natal Press: Natal, RSA.

Food and Agriculture Organisation of the United Nations (1979), Report of the World Conference on Agrarian Reform and Rural Development, FAO, Rome, Italy.

Frank, R.H. (1994). Microeconomics and behaviour. New York, USA: McGraw-Hill Inc, pp. 51-55.

Franzsen, R.C.D. (1990). Regskritiese ondersoek na hereregte in Suid-Afrika. University of Stellenbosch, RSA.

Franzsen, R.C.D. (1992). Introductory comments on land taxation: an historical and legal survey. In: Franzsen, R.C.D. and C.H. Heyns (eds.), A land tax for the new South Africa? The Proceedings of a Conference on 20 March 1992, The Centre for Human Rights, University of Pretoria, pp. 3-17.

Franszen, R.C.D. and C.H. Heyns (eds.) (1992). A land tax for the new South Africa? University of Pretoria, Pretoria, RSA.

F.W. De Klerk Foundation (2007). Land reform: a contextual analysis. The FW De Klerk Foundation, RSA.

Gill, S.J. (1993). A short history of Lesotho from the late stone age until the 1993 elections. Morija Museum and Archives, Lesotho.

Golder Associates (2004). Resuscitating plan for Taung irrigation scheme. A consultancy report submitted to the North West Provincial Department of Agriculture, Conservation and Environment.

Greenberg, S. (2002). Land reform and transition in South Africa. Paper Presented at the Development Studies Seminar 2002/14 (RAU Sociology), University of Witwatersrand, Johannesburg, RSA.

Groosskopf, J.F.W. (1932). The poor white problem in South Africa: rural impoverishment and rural exodus. Stellenbosch, South Africa: University of Stellenbosch.

Hamid, N. (1983). Growth of small-scale industry in Pakistan. Pakistan Economic and Social Review 21, 37-76.

Harris, C.L. (1976). Local responsibility and land taxation: lessons from United States experience. In: Wong, J. (ed.), The cities of Asia: a study of urban solutions and urban finance. Singapore 411, University Press.

Hart, T.G.B. (2003). Land reform: what have we learned. South African Journal of Agricultural Extension 32, 15-27.

Hoff, K. (1991a). Introduction: agricultural taxation and land rights systems. The World Bank Economic Review 5, 85-91.

Hoff, K. (1991b). Land taxes, output taxes, and sharecropping: was Henry George right? The World Bank Economic Review 5, 93-111.

Holland, D.M. (1970). The assessment of land value. Madison, USA: The University of Wisconsin Press.

IMF (2000). South Africa: taxing capital gains. International Monetary Fund (Fiscal Affairs Department), Washington D.C., USA, pp. 2-14.

Jeeves, A.H. and J. Crush (1997). Introduction. In: A.H. Jeeves and J. Crush (eds.) White farms, black labor. Pietermaritzburg, RSA: University of Natal.

Jeppe, W.J.O. (1980). Bophuthatswana – Land tenure and development. Department of Development Administration, University of Stellenbosch, RSA.

Jordaan, A.J. and A. Jooste (2004). Strategies for the support of successful land reform: a case study of Qwa Qwa emerging commercial farmers. South African Journal of Agricultural Extension 32, 1-14.

Kassier, W.E. and J.A. Groenewald (1990). The agricultural economy of South Africa. In: C. Csaki, T. Dams, D. Metzger, and J. van Zyl (eds.), Agricultural restructuring in Southern Africa. Windhoek, Namibia: International Association of Agricultural Economists.

Katz Commission (1998). Report of the land tax sub-committee of the commission of inquiry into certain aspects of the tax structure of South Africa. Eighth Interim Report of the Commission of Inquiry into Certain Aspects of the Tax Structure of South Africa, Pretoria, RSA.

Keegan, T.J. (1986). Rural transformations in industrializing South Africa – The Southern Highveld to 1914. Johannesburg, RSA: Ravan Press.

Kimble, J. (1986). Towards an understanding of the political economy of Lesotho: the origins of commodity production and migrant labour, 1830 – 1885. Ma Dissertation, National University Of Lesotho.

Kinsey, B. and H. Binswanger (1996). Characteristics and performance of settlement programmes: a review. In: J. van Zyl, J. Kirsten and H.P. Binswanger (eds.), Agricultural land reform in South Africa. Cape Town, RSA: Oxford University Press.

Lahiff, E. (2005). Debating land reform, natural resources and poverty – from a willing-seller-willing-buyer to a people-driven land reform. PLAAS Policy Brief No 17.

Lund, S. (1996). An overview of the land reform pilot programme. In: J. van Zyl, J. Kirsten and H.P. Binswanger (eds.), Agricultural land reform in South Africa. Cape Town, RSA: Oxford University Press.

Lyne, M. and M. Darroch (2003). Land redistribution in South Africa: past performance and future policy. In: L. Nieuwoudt and J. Groenewald (eds.) The challenges of change: agriculture, land and the South African economy. Natal, RSA: University of Natal Press.

Lyne, M., M. Roth and B. Troutt (1997). Land rental markets in Sub-Saharan Africa: institutional change in customary tenure. In: Rose, R., C. Tanner and M.A. Bellamy (eds.), Issues in agricultural competitiveness – markets and policies. Aldershot and Brookfield, International Association of Agricultural Economists, pp. 58-67.

Mankiw, N.G. (1998). Principles of economics. New York, USA. The Dryden Press/Harcourt Brace College Publishers.

Matabese, N. (1993). Current state of land distribution. Proceedings of the Land Redistribution Options Conference, 12-15 October, 1993, Land and Agriculture Policy Centre, Johannesburg, RSA.

Mbongwa, M., R. Van Den Brink and J. Van Zyl (1996). Evolution of the agrarian structure in South Africa. In: J. van Zyl, J. Kirsten and H.P. Binswanger (eds.), Agricultural land reform in South Africa. Cape Town, RSA: Oxford University Press.

Meredith, M. (1988). In the name of apartheid: South Africa in the post-war period. New York, USA: Harper and Row.

Morris, M. (2004). Every step of the way: the journey to freedom in South Africa. Pretoria, RSA: Department of Education.

Murray, M.J. (1997). Factories in the fields: capitalist farming in the Bethal district, C. 1919-1950. In: A.H. Jeeves and J. Crush (eds.), White farms, black labor. Pietermaritzburg, RSA: University of Natal.

Nieuwoudt, W.L. (1995). The impact of a land tax on future investments. South African Journal of Economics 63, 85-90.

North, D.C. (1989). Institutions and economic growth: An historical introduction. World Development 17, 1319-1322.

North, D.C. (1991). Institutions. Journal of Economic Perspectives 5, 97-112.

North, D.C. and J.J. Wallis (1994). Integrating institutional change and technical change in economic history: A transaction cost approach. Journal of Institutional and Theoretical Economics 150, 609 624.

Piketty, T. (1999). Theories of persistent inequality and intergenerational mobility. In: A. Atkinson and F. Bourguignon (eds.) Handbook of income distribution. Amsterdam, the Netherlands: North-Holland, pp. 429-476.

Plaatjie, T.K. (2005). AgriBEE a panacea or anathema: a historical approach. Paper presented at the conference on Introducing Black Economic Empowerment in the Agricultural Sector: Discussing the Framework, organized by Trade Conferences International, 13-14 April, 2005, Indaba Hotel, Johannesburg, South Africa.

Ruijis, A. (2002). Cereal trade in developing countries: a stochastic equilibrium analysis of market liberalization and institutional changes in Burkina Faso. PhD Thesis, Groningen University, Groningen, the Netherlands.

Skinner, J. (1991). If agricultural land taxation is so efficient, why is it so rarely used? The World Bank Economic Review 5, 113-133.

South African Reserve Bank (2005). Quarterly bulletin September 2005, Research Department, South African Reserve Bank, Pretoria, RSA.

Stallman, J.I. and L.L. Jones (1997). Property tax and tax reform in Texas: an equity concern. Department of Agricultural Economics Faculty Paper Series, Texas A&M, College Station, Texas, USA.

Stiglitz, J.E. (1986). Economics of the public sector. New York, USA: W.W. Norton & Company.

Strasma, J. (1993). International experience with market-based measures, including land taxes, for land redistribution. Proceedings of the Land Redistribution Options Conference, 12-15 October, 1993, Land and Agriculture Policy Centre, Johannesburg, RSA.

Summers, L.H. and T. Vinod (1993). Recent lessons of development. The World Bank Research Observer 8, 241-254.

Swallow, B.M. and B. Borris (1988). Cooperative agricultural development and food production in Lesotho. In: Prah, Kwesi, K. (ed.) Food security issues in Southern Africa. Roma, Institute of Southern African Studies, National University of Lesotho.

Swanepoel, F., A. Strobel, A. De Lange, E. Nesamvuni, and A. Nyamande-Pitso (2004). The evaluation of empowerment policies, strategies and performance within the agricultural sector. Integrated Report, Pretoria, RSA, National Department of Agriculture.

UNDP (2007). Country programme outline for South Africa 2007-2010. Executive Board of the United Nations Development Programme and the United Nations Population Fund (UNDP/UNFPA), New York, USA.

Van Rooyen, J. and B. Njobe-Mbuli (1996). Access to land: selecting the beneficiaries. In: J. van Zyl, J. Kirsten and H.P. Binswanger (eds.), Agricultural land reform in South Africa. Cape Town, RSA: Oxford University Press.

Van Schalkwyk, H.D. (1995). Modeling South African agricultural land prices. PhD Thesis, University of Pretoria, RSA.

Van Schalkwyk, H.D., J. Groenewald and A. Jooste (2003). Agricultural marketing in South Africa. In: Nieuwoudt, L. and J. Groenewald (eds.), The challenge of change – agriculture, land and the South African economy. Pietermaritzburg, RSA: University of Natal Press, pp. 119-135.

Van Sickle, J.V. (1925). Taxation of land in Austria. The Journal of Land and Public Utility Economics 1, 215-225.

Van Zyl, J. (1989). Interrelationships in maize markets in Southern Africa II: welfare aspects of the farmer support programme. Development Southern Africa 3, 360-384.

Van Zyl, J. and H.D. Van Schalkwyk (1993). The working of the South african land market and its role in a rural restructuring programme. Proceedings of the Land Redistribution Options Conference, 12-15 October, 1993, Land and Agriculture Policy Centre, Johannesburg, RSA.

Van Zyl, J. and H.P. Binswanger (1996). Market-assisted rural land reform: how will it work? In: J. van Zyl, J. Kirsten and H.P. Binswanger (eds.), Agricultural land reform in South Africa. Cape Town, RSA: Oxford University Press.

Van Zyl, J. and N. Vink (1992). An agricultural economic view on land taxation in South Africa. In: Franzsen, R.C.D. and C.H. Heyns (eds.), A land tax for the new South Africa? The Proceedings of a Conference on 20 March 1992, The Centre for Human Rights, University of Pretoria, RSA, pp. 32-59.

Wald, H.P. (1970). Basic design for more effective land taxation. In: Taylor, M.C. (ed.), Taxation for African economic development. London, UK: Hutchinson Educational.

Williams, G. (1993). Setting the agenda: a critique of the world bank's rural restructuring programme for South Africa: a summary. Proceedings of the Land Redistribution Options Conference, 12-15 October, 1993, Land and Agriculture Policy Centre, Johannesburg, RSA.

Wilson, M.H. and L.M. Thompson (1969). A history of south Africa to 1870. Cape Town, RSA: David Philip.

Woolery, A. (1989). Property tax principles and practice. Taoyuan, Taiwan: Land Reform Training Institute.

World Bank (1993). Options for land reform and rural restructuring: a summary. Proceedings of the Land Redistribution Options Conference, 12-15 October, 1993, Land and Agriculture Policy Centre, Johannesburg, RSA.

World Bank (1994). South Africa. The World Bank/International Bank for Reconstruction and Development, Washington, DC, USA.

World Bank (2003), Land policies for growth and poverty reduction. International Bank for Reconstruction and Development/The World Bank, Washington, DC, USA.

10. Institutional factors in natural resource management in the Eastern Cape Province of South Africa[1]

Mafabia Mokhahlane and Ajuruchukwu Obi

Abstract

This chapter examines the institutional factors governing the existing patterns of communal rangelands utilization in the Eastern Cape Province of South Africa. The discussion covered various institutional factors particularly those relevant to natural resource management at the communal level. Informal institutions such as customary laws and traditional institutions seem to have gone out of fashion in many places while in a number of communities indigenous institutions still operate but competitions between them and State structures has resulted in conflicts and confusion over definition, responsibilities, management procedures and enforcement mechanisms. A key finding is that lack of government support to communities has resulted in the poor management of resources. The consequence of this has been deterioration of the rangelands and impoverishment of the communities. The indication is that transaction costs and good governance structures are important in the common property institution and high levels of trust and cooperation between members are necessary to reduce these transaction costs. Furthermore, collective action on communal lands enables communities to share ideas on how best to manage their resources to ensure sustainability.

10.1 Background

In most of the South African rural systems, livestock farming is a tradition and it serves as an enterprise playing a vital role of enhancing food security thereby alleviating poverty (Coetzee *et al.*, 2005). Poverty, according to World Bank (2004), makes people less powerful because they have access to fewer resources for sustaining themselves. Most farmers, however, remain poor because of poor livestock management resulting from lack of knowledge on how best they can manage their livestock and because of the high population rate. On the other hand, despite various policies related to natural resource management and tenure reforms, degradation of rangelands remains a big challenge.

Besides enhancing food security, livestock farming for small-scale farmers is also an important means of generating income for the improvement of the livelihoods of rural

[1] The Masters dissertation on which this chapter is based was supervised in part by Professor Gavin Fraser at the Department of Agricultural Economics, University of Fort Hare and his contributions are gratefully acknowledged. Dr Sikhalazo Dube led the project under which this research was funded and his generous support is equally gratefully acknowledged.

people (Masiteng *et al.*, 2003). Livestock, in most of the rural communities, graze on communal rangelands, which, according to Duvel and Afful (1997), are areas used by animals and people for different purposes such as grazing, collecting fuel wood, water, medicinal plants and wild fruits. The form of tenure in most of the communal rangelands is a common property under which non-members are excluded from utilizing the resources within the rangelands (Shackleton and Cousins, 2000).

Despite the fact that communal rangelands continue to support rural farmers and their livestock with natural resources, most of them are perceived to be degraded and the main argument behind this degraded land is overstocking and excessive harvesting of wood. Excessive harvesting changes vegetative structure, leading to loss of certain plant species thereby disturbing the soil surface (Shackleton, 1993). In addition, degradation occurs because of the free rider problem which is common on communal rangelands. This indicates that there is no motivation for managing the grazing lands and as a result, the rangelands become free for all those who want to use them (Ainslie, 2002).

10.2 Problem statement

The chiefs and their headmen have commonly controlled communal rangelands, were responsible for allocating land to people, and imposed rules as to how to utilize the resources. However, during the apartheid era, the responsibility of controlling rangelands went to the government and various laws and permits regarding natural resource management were established wherein offenders would pay for the misuse of resources (Twine, 2005). For example, Van Averbeke (2000) indicated that in the Ciskei and Transkei former independent homeland areas, the control over the allocation and enforcement of land rights on traditional communal tenure was transferred from the chief and his community to the government. This indicates that local institutions became ineffective in regulating the system of communal tenure, because the chief no longer had a say regarding the management of their resources.

Several programmes and schemes were initiated during the apartheid era. The aim of these programmes was to improve land usage, for example, the betterment planning commenced in 1955 with the intention of conserving the environment in order to improve agricultural production. In most of the African communities, the land was divided into residential and agricultural land. People who previously occupied agricultural land were moved to residential areas (De Wet, 1995). Instead of improving land management and ensuring sustainable livelihoods, the betterment planning programme, however, worsened conditions because more poverty, deterioration of land and fewer agricultural activities were experienced (Letsoala, 1987). Agricultural activities decreased because natural resources were isolated from the homesteads and the joint operation by the communities regarding farming was broken. Furthermore, most of the communities failed to understand the scheme because of the approach used.

The Eastern Cape Province was not an exception from these various schemes and programmes by the government. For example, the culling programme in the Eastern Cape was one of the common programmes for livestock improvement. Undesirable breeds were culled in order to breed animals of higher quality (Ainslie, 2002). Most of these programmes however, had little impact on the production system and on the conservation of resources (Cousins, 1999). The main reason for this limited impact was the lack of active participation of the farmers and the inadequate attention paid to the prevailing cultural and socio-economic circumstances.

Besides the betterment planning and other schemes, new initiatives such as Broadening Access to Agriculture Thrust (BATAT) implemented between 1995 and 1998, were also put in place but this system was not able to put ideas into practice. The idea was that extension services should help farmers with communal rangeland management practices. However, these extension services were not able to do so because farmers were only trained in commercial cattle farming and, as a result, they lacked an understanding of communal rangelands and stockowners' objectives and practices. BATAT system focused more on technical aspects than social aspects, promoted commercial grazing and improved breeding stock. This resulted in community conflict because the focus was only on livestock owners without taking into account other rangeland users (Peden, 2005).This then called for the institutional capacity building in communal areas to regulate natural resource use. Issues such as destocking and control of animal movement were then proposed. The goal of these new sets of rules and regulations for livestock farming was to influence communal rangeland usage to avoid overexploitation of resources. These institutional arrangements on communal rangelands represent a set of collective actions and are the result of co-operation to the extent that the community is able to share information on how best they can utilize resources within their rangelands (Omamo, 2005).

Several studies have been conducted to assess the impact of these rules and regulations on the management of natural resources. For example, Ainslie (1999) carried out a study on managing common property natural resources in Tyefu location at Peddie in the Eastern Cape Province and discovered that there was a conflict between Tyefu and the neighbouring areas regarding utilization of natural resources on communal rangelands. The neighbouring communities utilised the rangelands even though they were not open to them. The rationale behind this was that without any institutional arrangements put in place for the management of natural resources, the land became misused and the outcome was degradation. In addition, Ashley and Ntshona (2003) discovered that in those rural communities depending on the State forestry reserves for various natural products like medicinal plants and building materials in the Eastern Cape Province, government's privatisation programmes caused a major threat.

Institutions, according to Vink (1989), have developed in various ways. For instance, there are those that have developed because of explicit human design and this process is called 'institution building' while other institutions emerged when community representatives met and negotiated types of institutions they would favour. The supply of these institutional changes include among others, the cost of collective action or cost of reaching an agreement that is suitable for members of the community. This, in turn, is determined by factors such as cultural and traditional knowledge, and equal control among the communities in terms of decision-making.

To date, no systematic assessment of how these institutions influence the use of rangelands in the Eastern Cape Province has been conducted especially in the light of the new arrangements that emerged in the wake of democratic rule in South Africa and the opening up of the economy to the outside world to which it was closed for many years under apartheid rule. The conflicts and confusions in land use in this province between communal and freehold tenure system and the new arrangements introduced to enhance tenure security need also to be understood. For example, rangelands in the north-eastern parts and central parts generally are communally owned, while those in the western part of the province, are largely commercially owned (Hoare, 2002). The present study is an attempt to put together information about how these different systems interact and what their impacts have been on the management of natural resources and what the implications have been for livelihoods in the smallholder agricultural systems of the province.

10.3 Land degradation in the Eastern Cape

According to the Eastern Cape Department of Agriculture (CSIR, 2004), the Eastern Cape Province is one of the three most degraded provinces in South Africa. During 1988-98, the area of land used for crops and grazing decreased. Land degradation was among other contributing factors. Other factors included the droughts of the mid-1980s and early 1990s, stock theft in the communal areas, increased production costs, lack of support for communal farmers and the collapse of the agricultural infrastructure (Eastern Cape Provincial Government, 2006).

Some of the areas with the highest soil degradation include Herschel, Mount Fletcher and Middledrift. The major contributors in the communal rangelands are deforestation and loss of vegetative cover from overgrazing and overstocking with livestock. This makes communal rangelands more degraded than commercial rangelands (Eastern Cape Provincial Government, 2006).

Land degradation can be classified into agricultural practices that include overgrazing, deforestation and overexploitation of vegetation. These factors are categorized under the direct causes of land degradation and are all common in the Eastern Cape Province (Meadows and Hoffman, 2002). However, the severity of land degradation is greatest in the communal areas under which grazing is the dominant form of land use. The first observable form of degradation is loss of vegetative cover and change in species composition. Bush encroachment is also severe in the rangelands of the Eastern Cape and communally held lands are likely to be more adversely affected than commercially farmed areas (Meadows and Hoffman, 2002).

In the Eastern Cape, about 51% of the land consists of open areas of unmanaged natural vegetation and this includes the forest, woodland, grassland, shrub land and low fynbos. About 1.61% of the province is classified as degraded and 0.12% eroded. Roughly 8.2% of land is being cultivated which is a higher percentage of land while 1.12% of the land is used for forest plantations (CSIR, 2004).

Figure 10.1 illustrates land degradation in the Eastern Cape Province. In general, the extent of degradation in the Eastern Cape Province is categorised into insignificant, light, moderate and severe. According to the Eastern Cape State of Environment Report (CSIR, 2004) citing Hoffman *et al.* (1999), about 47% of the land in this province is moderately degraded followed by 36.5% which is light and 11.1% severe, while 5.5% is insignificant. It is evident from these percentages that almost half of the land in this province is moderately degraded.

10.4 Existing institutional factors for natural resource management

There are different institutional factors governing the use of communal rangelands and these rules differ with communities but they have the common feature of managing mobility within the rangelands. These rules can be either formal or informal and in some areas, they are used interchangeably while in some, people are still adhering to their informal rules when utilizing rangelands (Ainslie, 1999). However, when effective, these institutions can bring fruitful results on the management of natural resources in communal rangelands. This section discusses both formal and informal institutional factors for natural resource management. Different studies done by other scholars will be reviewed wherever appropriate.

10.4.1 Informal institutional factors

Informal institutional factors are commonly from the communities themselves and they include among others, religious beliefs, traditional laws and social norms. The chief and his committee, which in most cases consists of elderly members of the community, normally set the rules.

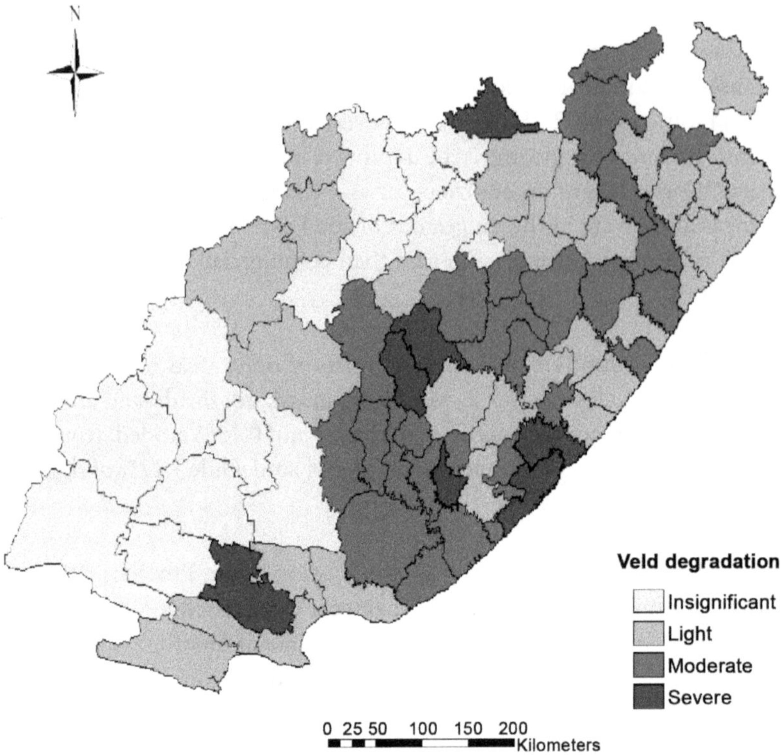

Figure 10.1. Land degradation in the Eastern Cape Province Eastern Cape State of the environment report (CSIR, 2004).

Traditional and customary institutions

The assumption of these institutions is that existing local rules and regulations could be important instruments for the management and development of rangelands once appropriate legal and institutional frameworks are provided to communities (Nothard *et al.*, 2005). The common feature of these institutions is that rural communities have full control over their resources and continue to use traditional mechanisms and rules to define access and resource use by all community members.

The role of traditional institutions is to provide rules and regulations on how the resources should be utilized (Tshabalala, 1995). Those in charge, such as the chief and the headmen, are responsible for issues like land allocation and grazing control. However, traditional institutions differ with communities. For example, Timothy (1998) indicated that rangelands in Ethiopia traditionally belong to a clan. This system of ownership allows the community to work together in conserving resources within their rangelands thereby reducing the risk associated with overstocking. Each clan or sub-clan has its traditional boundary and all clan or sub-clan members jointly can use the existing resources in their area and protect the natural resources. Under ordinary circumstances, members of a clan are permitted to enter with their livestock into the territory of the other clan for grazing.

Although in some other areas traditional institutions still play an important role in natural resource management, Twine (2005) argued that in other areas, especially in the rural communities of southern Africa, laws imposed by the chiefs are no longer effective. Traditional institutions for natural resource management have been weakened and communities are no longer able to impose some rules regarding the use of rangelands. For example, Uma *et al.* (2003) in their studies discovered that traditional institutions in African countries have deteriorated and this has resulted in more degradation of land which caused poverty and inappropriate settlement. Based on the above arguments, it can be concluded that traditional institutions are indeed crucial to the management of resources because the community knows their land better and can develop regulations, which can accommodate all the community.

A traditional community institution is based on mostly face-to-face relationships among social units and is faced with the challenge of practicing rangeland improvement activities, which require the mobilization of financial and labour contributions from tribal members (Messer, 2001). African rangeland regimes, on the other hand, consist of various institutional and water use rights, land rights, rules enforced by customary authorities, self-enforcing social institutions or conventions, and externally or internally enforced contracts (Coetzee *et al.,* 2005).

Besides institutional diversity, farmers in the rural areas of eastern and southern Africa have centralized governments with political structures in which chiefs perform administrative, governmental and legal functions. In contrast, most pastoral groups in eastern and western Africa have customary governments with relatively democratic elders' councils holding legislative and legal authority, or neither minimal customary government. The chiefs or elders' councils have authority to enforce rules (Chapeyama, 2004). This indicates that traditional authorities continue to play a role in Natural Resource Management (NRM) with varying degrees of authority and power. In some other countries, chiefs are asserted powers as chairpersons of sub-district NRM structures thereby diverting some of the benefits they obtain to build their own power base (Kowero *et al.*, 2003). This shows that the specific activities of traditional authorities are restricted, as some chiefs have engaged with the new governance structures thereby enjoying the benefits while others are feeling their power eroding.

Although many farmers are changing their ways, some continue to manage their livestock in their traditional way. However, their traditional system of management is no longer able to cope with the shortage of pasture and instead is adding to the problem of land degradation.

Traditional management knowledge is gradually being lost as more of the younger generation of pastoralists is attracted to urban areas (Twine, 2005). Under community ownership, local institutions may keep their traditional roles of managing the resources, deciding how to allocate resources between pastures and deciding on the nature of the rights to be allocated to members and non-members. These opportunities may empower local institutions and provide them with the ability to organize collective action and sustain the livelihoods of their communities (Ngaido and McCarthy, 2004).

Customary or indigenous law, on the other hand, is a source of law that provides a set of legal rules, particularly for the allocation and use of communal land. It can be distinguished from western or general law in that it is generally unwritten but meaningful to rural communities. Customary law is more closely attached to people's culture, hence territorial in nature (Cousins, 1995). Fisher (1993), cited by Payne (1997), argued that under customary law, land is regarded as belonging to the community at large not an individual. This system has an authority where people agreed to its existence and its enforcement and effectiveness is largely dependent upon the respect and power by the traditional authority structures charged with its implementation (Corbett and Daniels, 1996).

Customary law is seen as a body of rules established over the years with the intension of resolving disputes between individuals (Ankomah and Fox, 2000). Under customary law, the power of rights is regularly partial and some of these rights may not be easy to implement (Ortmann, 2000). Most of the rules from customary laws are immemorial rules coming from the general background of what was the issue of general information, which was then retained in the memories of the chief and his people until they became part of the immemorial rules (Roy, 2005). In addition, indigenous peoples' institutions administer customary law, and the validity of such laws and their contents, including the related procedures, is generally known about, at least by the older members of the community.

Bennett and Barrett (2003), in their study on the management of communal rangelands in the Eastern Cape, discovered that at Koloni, communal rangelands are grazed during summer while the arable land is used during winter as a communal forage reserve. However, the case is different with Guquka, in that the arable land is used in all seasons. The conclusion that can be drawn from the customary system is that it forms a pillar that supports peoples' decisions regarding the use of their rangelands.

Cultural norms

Culture reflects peoples view on what is happening in the real world and where groups of people join to adapt to their natural and social environment. Mazrui (1980), cited by Zenani and Mistri (2003), also define culture as a community and the way they are living based on their values and beliefs. Therefore, culture plays various roles depending on the community. That is, culture functions as the people's vision towards reality and includes in it customs, tradition and religion.

Culture plays a very important role in the management of resources because it conveys knowledge and information in a society, which then applies when utilizing their resources (Stevens and Treurnicht, 2001). Culture also acts as an obstacle to the use of natural resources. For example, Nkosi (2005) indicated that in some of African rural communities, when there is death in a family, no one is allowed to till the land or perform any activity on rangelands. People mourn until that person is buried and after a funeral, in some cultures, people are not allowed to go to rangelands or till the land until a ceremony of cleansing of the family is done.

Religion and livestock farming

Different people and different religions can interpret the perception of religion differently. Gardner (2002), cited by Zenani and Mistri (2003), defines religion as that which provides a way of coming across a sustaining innovative power, whether it comes from God, a remarkable existence in life, or simply the source of all life. Religion can also be associated with the way in which communities live and in most of African traditional religions, most communities are integrated with their natural resources. Furthermore, African communities still believe that land is a gift from God, particularly on communally owned lands. Religion on African land is also connected to the fact that ancestors are buried in it. That is, without land, there would be no home for the dead. The land is valued as a resource for livelihood for it produces food and water for people and their animals (McCall, 1997). For example, in KwaZulu-Natal, when people go hunting or are looking for medical plants on their rangelands, they burn their traditional stick so that they are able to communicate with their ancestors. By doing this, they believe that their ancestors will bless them with whatever they want on rangelands (Nkosi, 2005).

This shows that, in some African communities traditional religion still plays a significant part in people's life. Besides respecting their land, many people still slaughter goats, cattle and sheep as offerings to God and ancestors. These offerings assume three principal forms, namely, offerings of thanksgiving for national, tribal and family successes, offerings in cases of deaths, chronic sickness and epidemic diseases and finally offerings of prayer in terms of hardship, poverty and drought (Quail Report, 1980).

Apart from various offerings, there are certain days under which, because of religious beliefs, people do not do anything on communal rangelands, like collecting fuel wood, water for domestic purposes or thatch grass. These days are Sundays and Saturdays and, as a rule, people have to respect these days. Attitudes toward religion and religious beliefs vary widely. For example, according to Islamic religion, the land belongs to those who care for it and this land can be divided into various categories, e.g. land reserved for God and owned by religious organization or land owned collectively by the community (Payne, 1997). Furthermore, in other cultures like Xhosas, cattle play a crucial role of connecting the living and the dead that is they are used for ancestral spirits. This shows that cattle are not just slaughtered for joyful times but also in relation to religious beliefs (Soga, 1937). Religion as an institution, therefore, can also affect the way rangelands are used because there are people who adhere strongly to their religious beliefs.

10.4.2 Formal institutional factors

Besides informal institutions for the management of communal rangelands, there are formal institutions. These institutions include, *inter alia*, government policies, farmers' organisations and markets. These types of institutional arrangements are mostly used in the rural communities of South Africa for the management of rangelands

Policies for natural resource management

In the Eastern Cape Province, the management of rangelands for livestock farming is under the responsibility of the Department of Agriculture. The provincial department consists of the following: Applied Research and Training, Extension, Land Administration, Land-use Planning, Finance, Engineering, Youth and Gender, Veterinary Services and Administration. Bisho is the head office, but there are regional offices in Umtata, East London and Port Elizabeth, which consist of sub-regions, wherein there are district offices from which extension services work. There are also non-governmental organisations, such as environmental agencies, development forums, community organisations, common property associations, farmers' unions and co-operatives (Scogings *et al.*, 2000). Some of the legislative and other mandates that are applied for the improvement of livestock farming are (Department of Agriculture, 2003/2006):

- Animal Identification Act, Act 8 of 2001;
- Livestock Improvement Act Meat Safety Act, Act 7 of 2001;
- Animal Health Act, Act 3 of 2002;
- Livestock Improvement Act, 1997, Act no 25 of 1997;
- Limitation and Animal Protection Act (Ciskei), 1986, Act 20 of 1986; and
- Livestock Brands Act, 1962.

Environmental Conservation Act of 1982, Act 100 of 1982

Environmental Conservation Act, Act 100 of 1982, was the first attempt in South Africa to address environmental protection in a joint way. Its aim was to provide co-ordination in all activities that might have an impact on the environment. However, this Act did not provide for a complete environmental control measure and this resulted in it being reviewed in 1989. The Act became the Environmental Conservation Act of 1989. There were improvements observed in this new Act, which included, *inter alia*, circulation of regulations and policies for environmental conservation by the Minister of Environmental Affairs (Brauteseth, 2000).

The aim of the Environment Conservation Act of 1989 of South Africa was to establish, maintain and improve the environment, which would in turn contribute to the welfare of people, promotion of natural resources as well as improving literacy in the communities through environmental education. All of these were to be achieved through the implementation and co-ordination of integrated environmental monitoring programmes (Swatuk, 1996). During this time, the Department of Environmental Affairs was introduced for co-ordination of environmental policies and decisions among ministries. In this policy, owners of land were compensated for restrictions forced on them and this was done to monitor progress in the management of the environment (De Bruyn *et al.*, 2000).

In addition, the aim of this Act was to prohibit littering and identify any action, which might have an impact on the environment. This Act was responsible for regulating all activities on the environment to protect it from any disturbances, deteriorations or destruction from activities by people (Breedlove, 2002). Only policies concerning the waste management and environmental impact assessment were circulated and this led to yet another criticism of the policy. This resulted in a new policy called the National Environmental Management Act of 1998, the aim of which was to promote the sustainable use of resources and co-operative governance in the management of environment and equitable access to natural resources. This was carried out through the creation of ethics for decision making on issues related to the environment. In this Act, cultural norm as an institution was taken into account for sustainable development (ELAW, 1999).

The Common Property Association Act of 1996

The Communal Property Association Act (CPA) was introduced in 1996 to ensure security of tenure for all, even within common property resources (CPRs) (Makhanya, 1999). It was found that, although many of the CPRs had potential to improve the livelihoods of the beneficiaries, there was a general lack of equity in the rights of ownership, access and usage among the beneficiaries, which threatened the sustainability of many CPRs. The Act guarantees certain rights and protections to individual members and gives associations the right to enforce rules and regulations and to impose and implement them. This Act also enables communities to join and be known as communal property associations in order to acquire, hold and manage property on a basis agreed to by members of a community in terms of a written structure.

National Heritage Resource Act of 1999

The National Heritage Resources Act of 1999 gives communities an opportunity to take part in the management of their cultural and religious resources. According to the Act, traditional resources involve all belief systems such as, graves and burial grounds, cultural groups for collective, cultural and spiritual grounds as well as landscapes and national aspects of cultural meaning (Zenani and Mistri, 2003). The main aim of the National Heritage Resource Act, according to Breedlove (2002), was to introduce a mixed and cooperative system for the management of the national heritage resources and to encourage communities to conserve their resource for future generations. This indicates that traditional resources in South Africa are of cultural importance and as a result, should be taken care of so they remain available for the next generation. Some of these traditional resources are found on rangelands.

Communal Land Right Act of 2004

The Communal Land Rights Act (ClaRA) was enacted in 2004 with the major aim of supporting communities living in communal areas by promoting security of tenure for South Africans living on rural lands, which are commonly known as traditional communities. In these communities, interests in communal land are held by means of formal and informal land rights generally known as 'permission to occupy', which are managed and administered by traditional authorities. This indicates that ClaRA's interest was on how rural poor in South Africa hold land right and how those rights are administered (Kariuki, 2004). The intention of government was to secure property rights to facilitate development and to broaden equality into rural areas under traditional administration. Under this Act, traditional communities were to select their traditional councils as land organization committees. However, the role, which was to be played by these land administration committees regarding individual or household rights to own individual portions, remained unclear with this Act.

10.4.3 Farmers unions and cooperatives in South Africa

Farmers' unions as well as the agricultural cooperatives in South Africa are organised and controlled by farmers to promote their interests. The objectives of the farmers' unions and cooperatives are to support organised agriculture in marketing and processing of agricultural products by the rural communities, which will in turn benefit the entire farming community (Philip, 2003). Generally, farmers join a local farmers' organization, which is regarded as the starting point for organised agriculture and is linked to a provincial agricultural union, which in turn is affiliated to Agri South Africa. Agri South Africa acts as a representative for all farmers at national level in order to support farmers both financially and socially. Farmers have various associations, to mention but few, grain producers, woolgrowers, poultry producers, timber growers, beef producers and vegetable and fruit producers (Agricultural Digest, 2005).

Besides Agri South Africa, there is National Farmer's Union (NAFU). The aim of this union is to promote the interest of farmers from previously disadvantaged farming communities. To achieve this aim, NAFU as a national organization, encourages participation of farmers in decision-making in the improvement of agricultural farming. NAFU in partnership with other broad-based organisations has formed a union aimed at empowering the small-scale farmers to ensure that they play a meaningful role in promoting agriculture.

In the Eastern Cape Province, the National Department of Agriculture (NDA), Agri South Africa and NAFU have developed a strategic plan, which concentrates on the objectives, challenges and opportunities for agricultural development. The main aim is to implement the community group development strategies, which will focus on the establishment of livestock improvement schemes and food production strategies.

Table 10.1 demonstrates some of the agricultural schemes in The Eastern Cape Province. Various institutions are involved for the effectiveness of these schemes and this include among others, Agri-East Cape, University of Fort Hare, National African Farmers Union (NAFU), Farm Africa, Industrial Development Corporation (IDC), Amatole District Council and Agricultural Research Council (ARC). These institutions have joined in order to share some skills on how best they can improve the standard of living of rural communities (Department of Agriculture, 2003).

Table 10.1. Improvement schemes in the Eastern Cape Province (Department of Agriculture, 2003).

Scheme	Objective
The Eastern Cape communal soil conservation scheme, Provincial Notice No. 57 of 2001	To promote the creation of certain soil conservation mechanism in order to maintain and improve the production potential of communal land and prevent excessive soil loss through erosion
Livestock improvement scheme	Promote proper forage and pastoral risk management and promote efficient and effective flock / herd management practices
Farmer to farmer support leadership scheme	To improve the production efficiency and management capacity of emerging farmers
Resource planning scheme	To provide funding for proper identification, planning, utilisation and conservation of natural resources in order to maintain and improve the agricultural production potential of communal or farm land to improve integrated rural planning

9.4.4 Markets for livestock farmers

Markets play an important role in providing incentives for people, including farmers, to increase their income. Fenyes and Groenewald (1985), cited by Coetzee *et al.* (2005), pointed out that there are four primary functions of livestock markets, namely, acting as centres for collection and local produce exchange, providing services, acting as distribution and bulking points for goods imported from other areas. However, most small-scale livestock farmers have market problems. For example, in the Eastern Cape Province, less than 10% of the total livestock from small-scale farmers is marketed (Ainslie, 2002). The argument behind this is poor infrastructure and an information gap that exists in rural communities.

The Commark Trust (2003) indicates that these farmers sell their cattle through local marketing channels, either directly to other rural households for ritual slaughter or to local entrepreneurs. Because of this, farmers tend to sell their animals at a low price. These informal markets are differentiated by a high degree of seasonality, poor market information regarding prices and quality needed. Besides informal markets, farmers also rely on speculators to sell their animals during critical periods of the year. This results in farmers selling their animals below market value because of weak negotiation power and bad timing.

10.4.5 Extension services

The role of extension service is to improve the standard of living in the rural communities. Thus, extension agents act as facilitators, planners, organizers, motivators and educators (Terblanche, 2005). Therefore, there is a need for effective extension services, which helps in training famers on how best they can adopt various management practices because training contributes towards improvement of rural farmers. In addition, Coetzee *et al.* (2005) stressed that most of the rural farmers are illiterate; as a result, training programs should make use of visual aid materials and adequate illustrations and should work towards developing farmer's negotiation skills. Generally, extension services are seen as a means of transforming suitable agricultural practices to farmers to help them change their attitude towards adoption of these innovations.

10.5 Effects of institutional factors on natural resource management

Every development strategy or programme has its own impact for economic development. This implies that institutional factors, both formal and informal also have an impact on the management of natural resources and their impact varies with the community. Their impact depends on how effective the communities perceive them to be. In general, these institutional factors on rangelands act as a constraint for human activities in that they limit some of the activities on rangelands (Twine, 2005). This section discusses some of the effects of institutional factors on the use of communal rangelands. Issues such as conflict and confusion resulting from institutional factors for the management of resources will be dealt with.

10.5.1 Conflicts on the management of natural resources

There is usually a close linkage between tenure and conflict over land. Within a society, competing claims for control and use of land may cause conflicts. Population growth and changing economic factors can increase competition for access to land (Buckles and Rusnak, 2006). Competition is usually regulated by a society's tenure rules, which are developed in response to dynamic social, economic and political relationships. When these tenure rules are unable to adjust sufficiently rapidly to changing circumstances, the chance of conflict arising is increased (Matiru *et al.*, 2000). For example, customary tenure systems usually originated in areas where resources were extensive compared with the population size and, importantly, where there was a shared social consensus between the various holders of rights. When this social consensus breaks down, conflicts arise.

Contrary to the provisions of various acts by the government, the traditional leaders in some African rural communities believe that the land belongs to the people and that it will be re-allocated to members of the community in the traditional manner (Ghee and Valencia, 1990). This is because traditional leaders fear losing authority over the allocation of land to the State and this causes conflict between the traditional leadership and the State. Besides the fear of losing the land, conflict also results from lack of agreement about how to utilize resources on rangelands. This is because individual community members have different views regarding the use of their rangelands and when they fail to collaborate or respect communal governance structures, conflicts results (Warner, 2000). This indicates that lack of understanding and respect causes conflicts regarding land use and management.

Buckles and Rusnak (2006) argued that conflicts could hinder sustainable management efforts particularly if the interests of influential members within the community are such that collective management is not supported. For example, in the Eastern Cape, there is a conflict caused by influential members regarding access to water dams. Those without other sources of water except for the water dams want to fence off the water dams to prevent contamination by animals while those with other alternatives regarding water sources are not interested in conserving communal water because they want their animals to have access to the water dams. On the other hand, newly settled households are in conflict with existing households over grazing rights. The new settlers do not have any arable land but have livestock and the conflict arises when they wish to utilise grazing lands that are already in use (Manona, 1980, cited by Deshingkar and Cinderby, 1999).

McCarthy *et al.* (2000) pointed out that causes of tenure conflicts in African countries are similar and can arise from various situations. Conflicts may arise because of unclear boundaries between farmlands and grazing lands. Other conflicts between individuals, according to Engel and Korf (2005), occur when they fight over boundaries that were once allocated by customary institutions because in this type of institution, the rules and regulations are not written and there are no records kept. These can lead to heavy conflicts because everybody can claim that the land, both grazing and cultivated fields belong to him/her.

Conflicts also arise between the local population and new settlers. The government allocates new settlers the land that has been used in the past for grazing while, on the other hand, those people who are now staying in the urban areas employ paid herdsmen to look after their animals (Tonah, 2002). This conflict results in a situation where the community no longer follows any rules regarding the use of land, thereby, leading to an open access situation. Conflicts also emerge within local communities, if parts of the group want to make use of new statutory laws to gain individual rights to pieces of communal grazing land, whereas the majority still regards the total land as common property (Warner, 2000).

10.5.2 Confusions on the management of natural resources

Besides conflicts, unclear boundaries between indigenous and statutory laws also add to confusion in consideration to natural resource management (Ramirez, 2002). Many traditional rules governing rangeland use are no longer effective as a result this creates confusion because community members do not know exactly which law should they follow when utilizing rangelands. In some instances, people lay claim to land falling under two nearby villages while maintaining commitment to only one headman or chief with regard to resource management. This is a difficult situation and community members have been known to take advantage of this situation by extending their rights of access to natural resources thereby confusing other members (Kirkland *et al.*, 2005).

Cocks *et al.* (2006) indicated that in the Eastern Cape Province, lack of governmental support from the local government for natural resource management creates confusion especially after the former homeland administration was changed into the new Eastern Cape Province. This shows that in this process of changing to the new Eastern Cape, the administration has been taking its time in attending to matters associated with natural resource management.

10.6 The geographical context, experiment and findings

The next several sections present the geographical context of this study and provide a description of the experimental study sites in which the study was conducted. Along with these are detailed descriptions of the experiments carried out to assess the status of natural resource management in the province and the role of institutions.

10.6.1 The geographical setting

The Eastern Cape is located on the south-eastern seaboard of South Africa and is the second largest province with an area of 170,600 km^2 which represents 14% of South Africa's land mass. The majority of the people speak isiXhosa, followed by Afrikaans and English. In terms of climate, the province varies from mild temperate conditions i.e. 14-23 °C along the coastal areas to slightly more extreme conditions of 5-35 °C among the inland areas, with the inland mountain areas experiencing winter snows and summer rainfalls. Mid-2006 population is 6.9 million people.

The major towns and cities in the Eastern Cape (Figure 10.2) are Aliwal North, Bisho, Butterworth, East London, Grahamstown, Humansdorp, King Williams Town, Port Elizabeth, Queenstown and Umtata, among others. Port Elizabeth and East London are the centres of the two largest industrial regions of the province and are both served by well-equipped ports. The major industries in the province include agriculture, textiles and clothing, tourism, wool, timber and transport, with tourism being the major growth industry (CSIR, 2004).

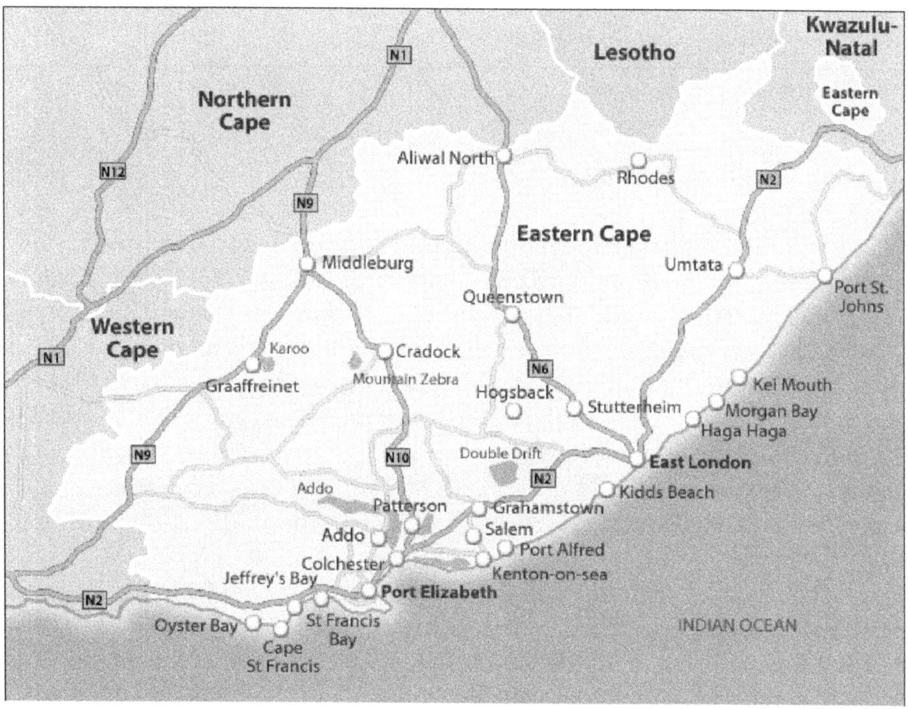

Figure 10.2. Eastern Cape Province. Source: Rainbownation (http://www.rainbownation.com/travel/maps/index.asp?loc=10).

In the Eastern Cape, 64% of land is used for stock farming and this consists of beef cattle, sheep, goats and game. Twenty percent of the land is set aside for crops such as maize, vegetables, pineapples and citrus. Commercial forestry occupies 5% of land use while 1% of land is reserved for conservation.

10.6.2 Experimental study sites

Table 10.2 illustrates the three communities under which the research was conducted. These areas have been selected because of their difference in vegetation types, which made it possible to compare and contrast between different institutional factors and land use patterns.

Table 10.2. Experimental study sites (Eastern Cape Province).

Vegetation zone	Locations	Site
Sweet grassveld	Magwiji	Sterkspruit
Sour grassveld	Tsaba	East London
Valley bush	Lashington	Seymour

Study site 1: Magwiji community

Geographic location. Magwiji is located at the north-west of the Eastern Cape Province. It falls under the Senqu municipality. This community is located on 30°37'60" South and 27°22'0" East with the altitude of 1,507 m above sea level. This community is about 20 km away from Sterkspruit, which is centrally located and serves as the administrative and service centre. Sterkspruit is situated near Herschel in the north-west of the Eastern Cape, which was formerly part of the Transkei. This district is bordered by the Orange River, Free State, Lesotho and the towns of Lady Grey and Barkley East. Sterkspruit has a population of about 165,000 people from 84 villages with Magwiji included. There is little economic variation beyond a few commercial services located in the town (MPCC, 2004).

Climate. The rainfall is relatively high from November to April (401-500 mm) and low from May to October (151-200 mm). Average temperature varies; the highest is in January (20-22 °C) and the lowest in July (9-10 °C). The area is dry with scarce rain during winter and frosty winters with hot summer months.

Rangelands. The natural vegetative cover in the Magwiji community is mostly sweet grass. The sweet grass is palatable for the animals and it is plentiful in summer. In winter, farmers opt for supplementary feeding because the grass is inadequate for the animals. The area is dry and degraded with most of the land being used for communal grazing. Besides grazing, natural resources within the grazing areas are used and these include fuel wood and wild fruits. The rangelands are not fenced and this enables the neighbouring villagers to utilize them.

Livestock. Sheep are the common livestock kept by the community followed by cattle. Livestock graze on communal rangelands where non-members are excluded. Cattle are kept for drought power, for paying *lobola* and for slaughter during funerals. Sheep are kept for wool production and for slaughter during ceremonies. Goats are few in number and are kept for religious purposes.

Study site 2: Tsaba community

Geographic location. Tsaba is located at the eastern part of the Eastern Cape Province. It falls under the Buffalo City municipality. The community is located 33°08'18.6' South and 27°29'38.4' East with the altitude of 176 m above sea level. The main town is East London, which is on the Indian Ocean coast between Buffalo and Nahoon Rivers and is the second largest industrial centre in the Eastern Cape Province. Tsaba falls among the disadvantaged communities surrounding East London.

Climate. The climate varies from mild to warm and humid, tending towards sub-tropical. Most of the rainfall is experienced during the spring and summer months. The average rainfall is 850 mm. Winter temperatures ranges from 7 °C to between 19 and 25 °C. Summer day temperatures range between 28-31 °C. Snowfalls are experienced in winter on the mountain ranges. The windiest months are during spring while between January and March is the calmest months (Eastern Cape Parks, 2007).

Rangelands. The natural resources obtained from the rangelands include wild animals, fuel wood, trees and shrubs, grass and traditional medicines. Vegetative cover is mainly sour field grass. Large parts of the rangelands are covered with invader plants such as blue bush, which is dangerous to the animals. The rangelands are fenced and in good conditions

Livestock. Cattle and goats are the common livestock in this area with cattle dominating. Cattle are kept for draught power and for slaughter during burials. Goats are kept for home consumption and religious purposes. Both cattle and goats are also kept for cash from sales, prestige and customary needs. Farmers use traditional medicines and dosage from the livestock health practitioners to treat animal diseases.

Study site 3: Lashington community

Geographical location. Lashington is situated in the central part of the Eastern Cape. The community is in the Seymour district, which is about 25 kilometres north-west of Fort Beaufort and 20 kilometres west of Balfour. The community falls under the Nkonkobe municipality. Lashington is located 32°38′ South and 27°28′ East with the altitude of 828 m above sea level. The main town is Fort Beaufort, which lies between the Keiskamma and Great Fish Rivers (Eastern Cape Parks, 2007).

Climate. The climate of the Nkonkobe Municipality is semi-arid with an annual rainfall not exceeding 600 mm. Rains are higher during summer and winters are dry with frequent frosts.

Rangelands. The vegetative cover is mainly valley bush, which is suitable for both cattle and goat production. The rangelands are not only used for grazing animals but also for collecting fuel wood and building poles and dung. Invader species on the rangelands, such as black wattle, tend to reduce livestock production.

Livestock. Cattle and goats are dominant in this community. They are kept for different purposes. Cattle are kept for draught power while goats are kept for slaughter for home consumption. Summer months are not favourable for livestock as they are susceptible to diseases because of hot weather conditions, which make it possible for external parasites such as ticks to reproduce.

10.6.3 Methodological questions

The variables used in this study are based on the institutional factors both formal and informal that affect the management of natural resources within communal rangelands. Most of the variables are developed from the Institutional Analysis and Development (IAD) framework.

The variables that directly or indirectly can affect the use of communal rangelands include, *inter alia*, beliefs and norms, collective action choice, conflicts, sanction rights and land security. These variables, according to the FAO (1991), can help a researcher in understanding how a village is organised and how it operates. Favourable social and cultural practices promote cooperation among people while unfavourable practices create barriers for sustainable management among community members. These, in turn, result in conflicts and frictions regarding the way in which resources should be utilised and managed. This indicates that these rules have a significant impact on the way in which the community interacts with its environment.

Besides unfavourable practices, problems and disputes over the use of natural resources can also result in conflicts. For example, Sanginga *et al.* (2007), in their research on the dynamics of social capital and conflict management in multiple resource regimes, discovered that almost all farmers have been involved in conflict over the use and management of natural resources. There was, however, no significant difference in the number of conflicts reported by men and women. These conflicts, which involved a variety of stakeholders and resource users, were grouped into three broad categories: community-level conflicts opposing farmers within same communities; intercommunity conflicts opposing different communities or farmers from neighbouring villages, and supra-community conflicts opposing farmers and communities with higher-level formal institutions or individuals.

Regarding collective action choice, Dasgupta (1995), cited by Adhikari (2001), hypothesized that higher growth rate negatively affects the likelihood of collective action.

The extension service is one of the variables having an impact on the management of rangelands. For example, in their study on the impact of social, institutional and ecological factors on land management practices in mountain watersheds of Nepal, Paudel and Thapa (2003) revealed extension services as the strongest factor influencing the adoption of land management practices. This indicates a positive relationship between extension service and resource management. Table 10.3 illustrates variables used for analysing institutional factors on communal rangelands. These variables were developed from Ostrom's (1990) design principles, cited by Nilsson (2001).

Nilsson (2001) in his research on the management of communal rangelands in Endabeg used these principles as variables to describe institutional factors governing the use of communal grazing land in Endabeg. Some of the results showed that the penalty commonly used to pay damages for minor violations include, *inter alia*, defined amount of local brew, which is perceived as favourable compared to costly legal action. In addition, monetary fines are also used to punish rule breakers. Conflict resolution mechanisms are not directly applicable to grazing on communal grazing of Endabeg because everyone seemed to accept the condition of open access. The normal procedure when a conflict occurs is that the sub-village chairman sits down together with the accused person and a group of respected elders to solve the conflict in a composed manner.

Besides the above institutional arrangements, information on the attributes of the resources and attributes of the community is also of importance. As a result these attributes are also included in an IAD analysis framework The assumption is that these attributes also have an impact on the effectiveness of the institutional arrangements.

Table 10.3. Definition and description of variables for institutional analysis.

Variable	Definition	Indicator	Measure
Use of rangeland	How are the rangelands used?	Utilisation of resources	1 = grazing animals 2 = other uses
Land security rights	Ownership rights	Who owns the land	1 = community 2 = state
Clearly defined boundaries	The boundaries of the resource and specification of individuals, i.e. defined boundaries and rules about who has access to a common resource pool	Existence of clear defined boundaries, e.g. excludability	1 = yes 2 = no
Collective choice arrangements	Collective decision rules. That is, group of people involved in setting rules, e.g. operational rules	Who set the rules? Existence of rules and norms	1 = state 2 = community 3 = both
Monitoring/enforce-ments	Monitoring and enforcing the rules	How are the rules enforced?	1 = state 2 = community 3 = both
Graduated sanctions	Sanctions or penalties given to violators of rules	Fines and charges	1 = yes 2 = no
Conflict-resolution mechanisms	Arrangement for disagreement between people	Conflict resolving mechanism	1 = conflicts 2 = no conflicts
Minimal recognition of rights to organize (indigenous institutions)	The rights of resource users to devise their own institutions	Decision making body	1 = yes 2 = no
Local organisations	A well organized structure with group of people having the same set of goals to achieve	Existence of local organisations	1 = yes 2 = no
Extension services	Services provided by the government to help farmers	Training service and advices	1 = effective 2 = ineffective

10.6.4 About the Institutional Analysis and Development (IAD) framework

According to Wendel (2004), IAD framework developed from various disciplines. This includes, *inter alia*, anthropology, economics, political science and sociology. The methodological framework originated from Ostrom (1990), who emphasised that the communities using a common property are capable of solving their local problems for as long as they are given rights to do so and they can use rules such as graduated sanctions

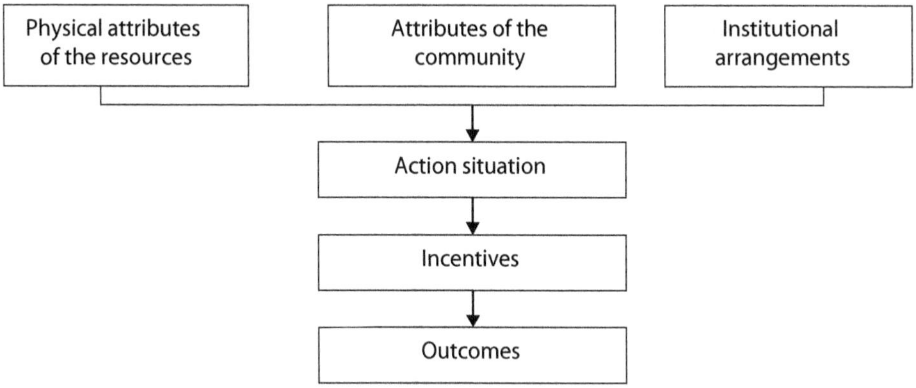

Figure 10.3. Framework for institutional analysis (Adapted from Smajgl et al.*, 2003 and Ostrom, 2003: 46).*

and monitoring to influence the contingent behaviour of individuals (Nilsson, 2001). The IAD framework is used to measure institutions and their effects on individual's incentives and their resultant behaviour. The IAD framework is also used to analyse various situations individuals face in decision-making (Smajgl *et al.,* 2003). The IAD framework is divided into the following:

Physical attributes of the resources

The physical attributes of the resource refers to the structure of land itself. That is, its vegetative cover, size, carrying capacity and quality (ICLARM, 1996). Thus, physical attributes of the resource includes the area's geographical and hydraulic conditions wherein the size of the resource system and number of resource users are crucial (Wendel, 2004).

Attributes of the community

It involves key attributes of the individuals. That is, attributes of resource users and differences between farmers in terms of how they cooperate and whether they have similar or heterogeneous interest. It also involves people's cultural, social and political aspects that influence their behaviour, and how these affect the ability to solve common problems (Wendel, 2004). Attributes of the community also involve issues such as how many people are involved in the utilization of resources and whether they work together in terms of how to use the resource (Nilsson, 2001).

Dependency, trust and differences between resource users in terms of racial and cultural backgrounds are important aspects of the collective action. This is because it is believed that if resource users depend on the resources for generating income, they are likely to contribute to collective activities. However, Tang (1992), cited by Wendel (2004), argued that if resource users do not have any other sources of income, they will not be able to invest financially in the resource and this can affect collective actions. On the other hand, Gopalakrishnan (2005) stressed that trust is determined by people's belief about others due to their various experiences.

Institutional arrangements

Institutional arrangements in the IAD framework have been categorized into eight categories. However, for this research, only those institutional arrangements, which were relevant for the study, were used. These institutional arrangements represented existing informal institutional factors in the study areas. Table 10.4 illustrates institutional arrangements from IAD framework. These institutional arrangements are developed from Ostrom's (1990) design principles.

10.7 Justification of Institutional Analysis and Development (IAD) framework

The main objective of this research is to describe institutional factors and their impact on the use of communal rangelands. The IAD analysis framework facilitates in describing institutional factors and their performance. The framework also helps the researcher to understand the socio-political structures and processes that govern the decision-making environment. Thus, IAD analysis consists of contextual factors, which consist of the user

Table 10.4. The institutional arrangements for the IAD framework (Ostrom, 1990, cited by Nilsson, 2001).

Institutional arrangements	Definition
Clearly defined boundaries	Individuals or households that have rights to withdraw resource units from the CRP must be clearly defined, as must the boundaries of the CRP itself
Collective-choice arrangements	Most individuals affected by operational rules can participate in modifying the operational rules
Monitoring	Monitors who actively audit CRP conditions and appropriator behaviour are accountable to appropriators or are appropriators
Graduated sanctions	Appropriators who violate operational rules are likely to assessed graduated sanctions (depending on the seriousness and context of the offense) by other appropriators, by officials accountable to these appropriators or both
Conflict resolution mechanisms	Appropriators and their officials have rapid access to low cost local arenas to resolve conflicts among appropriators, between appropriators and officials
Minimal recognition of rights to organize	The rights of appropriators to device their own institutions are not challenged by external governmental authorities

group's social, cultural and political aspects, which can have an impact on the way people behave in terms of the decision to comply with institutional arrangements and contribute in collective action (Smajgl *et al.*, 2003).

Wendel (2004) performed a study on institutions and common pool resources in Botswana and used IAD analysis to examine how institutional arrangements affected the probability of cattle owners to govern successfully the grazing land of Matsheng village in Kgalgadi district of Botswana. The results showed that rangeland management in this village is not sustainable. This is because of ongoing degradation of water and land, which hinders cattle owners from investing in resource management. In terms of defined boundaries in Matsheng, about 10% of the communal grazing land has been lost and this worsened grazing pressures on communal land because of dual grazing rights, which allowed everyone to use the rangeland. Monitoring and enforcements in this village does not involve the resource users. Resource users are also not involved in the decision-making process and are not consulted about decision planning.

10.8 Institutional arrangements and institutional performance

In order to find the correlations between the institutional arrangements and institutional performance, Ostrom's (1990) correlation pattern was used. Thus, a qualitative assessment was carried out to determine the strength of institutional arrangements on the management of communal rangelands. Institutional performance helped in determining the effectiveness of institutional arrangements. Each institutional arrangement was classified as strong, weak or absent and the management of communal rangelands was considered strong, weak or absent based on how many design principles were identified and how well they appeared to operate, hence based on institutional performance. The management was considered strong where all the institutional arrangements were strong, weak if between one and three institutional arrangements were weak or absent. With more than three institutional arrangements being weak or absent then management was considered absent (Quinn *et al.*, 2007)

Physical attributes of the resources as well as the perception of the community towards a particular institutional arrangement were used to measure how strong or weak the institutional arrangements were. That is, the strength of institutional arrangements was evaluated based on how the respondents spoke about them. For example, for graduated sanctions, the researcher asked if there were any fines or punishment given to violators of rules. If there was evidence that this rule was present but not fixed, or there was a conflict over it, then the sanctions as a rule was ranked weak. On the other hand, if there was no evidence of conflict and these sanctions were clearly present, then the principle was considered strong. In many cases therefore, conflict was regarded as the determining factor over a principle, or evidence that the rule was regularly broken was used as a sign to determine weakness of that rule.

Nilsson (2001) used Ostrom's correlation pattern to find correlation between the design principles (institutional arrangements) and institutional performance. These design principles are a feature that contributes to maintaining the institutions and their resource base, and gaining user loyalty for the rules (Tucker, 2000). Institutional performance was categorized into robust, fragile and failure. The results showed that institutional performance in Endabeg was a failure because of no clear defined boundaries, weak collective choice arrangements and there was no monitoring and enforcements.

A further analysis was carried out using chi-squared analysis to test whether particular design principles were more likely to be having an effect on communal rangelands. The reason for using this analysis was to examine whether the impact of various institutional arrangements on communal rangelands differ from one another and because the variables were categorical in nature not numerical. That is, to determine whether there was a statistically significant difference between institutional arrangements in terms of their impact on rangelands.

10.9 Expected results

There are various variables that are assumed to have an impact on the use of communal rangelands. The IAD framework was used to describe these variables and their impact on the use of communal rangelands. The following are some of the expected results:

a. *Collective choice arrangements*: The expectation is that the communities are collectively managing their rangelands and that they adhered to the set rules. It is also expected that communities are involved in the decision-making process
b. *Sanctions:* The expectation is that violators of rules are given fines. It is also expected that there are other ways of punishing violators apart from fines.
c. *Conflict resolution management:* It is expected that there are conflicts regarding the use of communal rangelands and some of these conflicts can result in violence. Conflicts are expected to come from those who do not want to abide by the rules. It is also expected that communities are responsible for solving their disputes.
d. *Grazing camps and herding:* It is expected that the availability of grazing camps, as one of the rangeland management practices in the communities, plays a major role in improving the rangeland conditions. Managing livestock movement during grazing is also expected to reduce the incidences of animals grazing at the wrong places especially when the camps are closed.
e. *Extension services:* Extension services play a major role of helping resource users to adopt various rangeland management practices. The expectation is that where there are effective extension services and communities are well informed about the existing institutional factors for managing their rangelands. An effective extension service is also expected to encourage farmers to form more local farmer's organisations and to adapt to more rangeland management practices.

f. *Local farmers organisations:* Communities with a greater existence of local organisations are expected to make higher contributions for management practices and are more likely to establish penalty systems.

10.9.1 Access to land

Figure 10.4 illustrates access to land by new settlers. According to this figure, new settlers are allocated land by the government. That is, 98%, 96% and 90% of the respondents from Lashington, Tsaba and Magwiji respectively have been allocated land by the government.

10.9.2 Grazing lands

In the three communities, all the respondents have access to grazing lands. These grazing lands are used for grazing animals and for other purposes such as collecting fuel wood, medical plants, building poles and thrash grass. Comparing grazing lands in these three communities, Tsaba has good grazing lands followed by Lashington. On the other hand, Magwiji's grazing lands are not in good condition. Rotational grazing is practiced in all the three communities.

In addition, Figure 10.5 shows that about 94% of the respondents from Lashington have access to these grazing lands by virtue of being resident in these community followed by 80% and 85% from Magwiji and Tsaba respectively. Other members have applied from the village committees. The grazing lands are the property of the government that sets the rules on how the communities should utilize the land. The responsibility of the communities therefore is to collectively enforce and ensure that the rules are followed. In

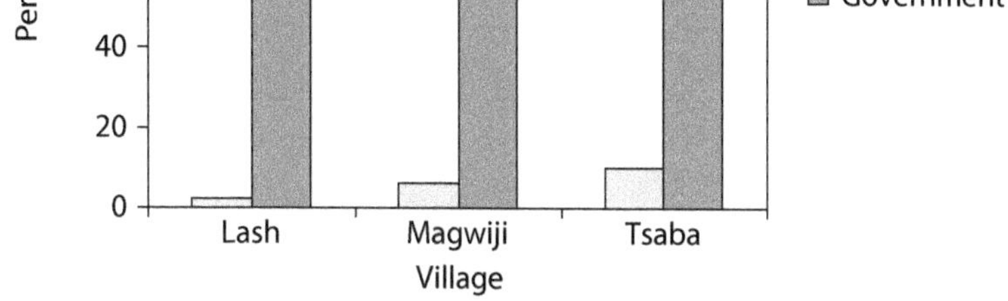

Figure 10.4. Distribution of land to new settlers in Magwii, Tsaba and Lashington.

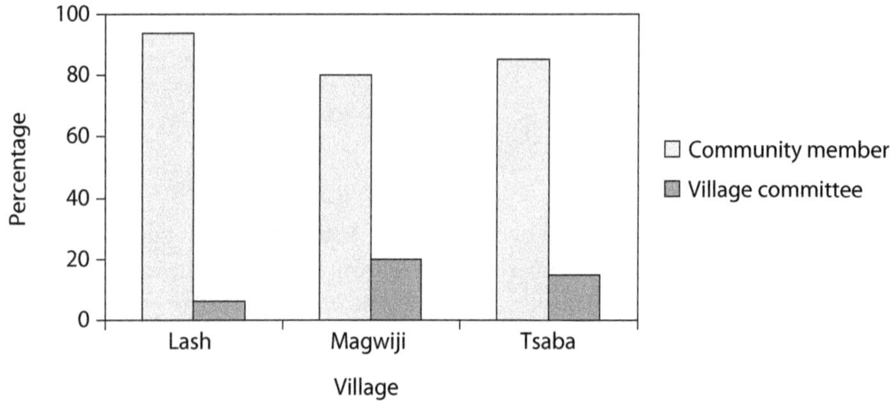

Figure 10.5. Distribution of respondents having access to grazing lands in Magwii, Tsaba and Lashington.

each community, a chairperson works with the chief and his community members to ensure that the rangelands are used effectively and efficiently.

10.9.3 Livestock

Most of the respondents from the three communities keep livestock. These livestock graze on communal rangelands. Table 10.5 illustrates the number of livestock for 2007 from each community. According to the table, Lashington has the most cattle followed by Magwiji and Tsaba. Conversely, Magwiji has the highest number of goats followed by Lashington and Tsaba. Lashington also has the highest number of sheep followed by Magwiji with 940 sheep and Tsaba with no sheep.

Table 10.5. Livestock numbers in Magwiji, Tsaba and Lashington.

Community	No. cattle	No. sheep	No. goats
Magwiji	838	940	1,456
Tsaba	266	0	271
Lashington	2,841	1,017	890

10.9.4 Rules governing the use of communal rangelands

The majority of respondents in the study areas are aware of the rules governing their communal rangelands and the community enforces these rules collectively. However, only 6% of the respondents are not aware of the existence of these rules as shown in Figure 10.6.

In all three communities, some of the common rules on the use of communal rangelands include among others, sanction rights and monitoring, conflict resolution management and clearly defined boundaries. For example, community members are not supposed to graze their animals when a certain piece of rangeland is rested. When asked about the effects of these rules on their day-to-day activities, respondents from the three communities indicated that the rules restrict their activities since they have to abide by the rules.

Rotational grazing used to be a common management practice in all the three communities before fences were stolen. Thus, the communities understood that there were certain times whereby a certain camp would be closed in order to improve the vegetation in that area and when this camp is closed, they were not to graze their animals. On Saturdays and Sundays, most of the community members do not normally perform any activities on rangelands since these are their rest days and to some, it is because of religion.

Management of livestock movement during grazing

The management of livestock movement during grazing is very important because it restricts animals from grazing where they are not supposed to graze. The owner of the animal caught grazing at the wrong place pays a fine depending on the type of animal. In the three communities, lawbreakers pay a fine of R2.50 for sheep and goats and R5.00 for cattle. Sometimes this creates conflict because there are those who do not want to pay. In Magwiji, most of the respondents did not want to respondent to the matter. Those who responded indicated that they do not usually manage livestock movement during grazing.

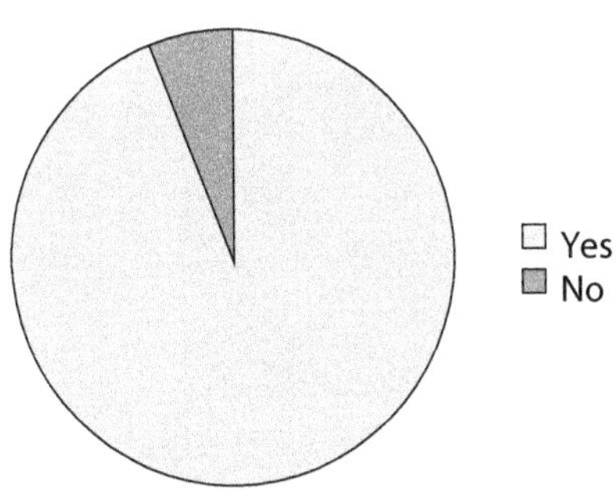

Figure 10.6. Distribution of responses by rules governing the use of rangelands.

However, the case is different with Lashington because 49% of its respondents indicated that they do manage their livestock mobility and only 22% said they do not. The rest did not want to respond. In addition, 54% of people from Tsaba reported that they do not herd their animals.

Fencing of rangelands

One of the major problems on the grazing lands is lack of fencing. Unfenced grazing lands enable the non-members to utilize the resources within the rangelands without permission. In the three communities, lack of fencing resulted in communal rangelands being misused especially by the neighbouring villages. For example, about 86% of the respondents from Magwiji reported that their grazing lands are not fenced and 79% of people from Lashington reported the same case. Rangelands in Tsaba are fenced as a result; there are no threats from the neighbouring villages. Even though rangelands are not fenced in Lashington, threats from the neighbouring villages is not as common as in the Magwiji community with its people complaining about this threat and indicating that this is one of the major factors that has led to the bad condition of their rangelands. Natural landmarks are used as boundaries in the grazing camps for rotational grazing in Lashington and Magwiji since there are no fences.

Local farmer's organisations

Farmer's organisations are important to farmers in that they act as a starting point for joint agriculture. This is where farmers join and form associations under which they are collectively able to market their agricultural products. In all the three study areas, there are local farmers' organisations as shown in Figure 10.7. For example, in Magwiji, about 90% of the respondents belong to the wool association and the benefit they get from this organisation is that it is easier and cheaper to sell their wool.

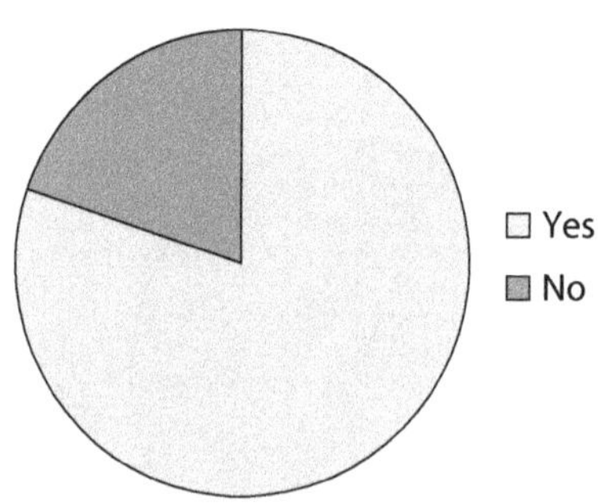

Figure 10.7. Distribution of farmers who belong to farmers associations.

Payment of animal fees by community members

Beside different farmers' associations, farmers have a common fund under which they pay animals fees. This animal fee depends significantly on the community (Chi^2 23.09, $P<0.0001$). For example, Figure 10.8 illustrates that payment is less likely to occur in Lashington with 87% of the respondents saying they do not pay such fees. In Magwiji and Tsaba, 40% and 58% of the respondents indicated that they do not pay any fees while the rest said they pay animal fees. The main purpose of these fees in the three communities is to enable farmers to belong to farmers' associations. These fees are then taken to a common fund to be used for the good of the animals.

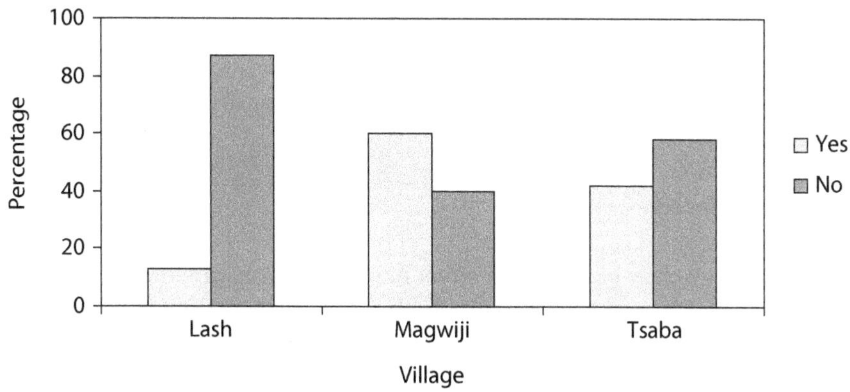

Figure 10.8. Distribution of farmers who pay animal fees in Magwii, Tsaba and Lashington.

10.9.5 Conflicts and the use of communal rangelands

Conflicts exist in the three communities. There are more conflicts in Magwiji as shown in Figure 10.9, where 80% of the respondents reported the existence of conflicts in this community. However, the case is different with the other two villages because only 28% of the respondents in Lashington and 35% in Tsaba reported the prevalence of conflicts. These conflicts mainly arise from law breakers who graze their animals where they are not allowed and violent conflicts over the use of natural resources with neighbouring villages.

10.9.6 Sources of water for livestock

In all three communities, animals use dams and rivers for drinking. The distance between villages and water points is less than 1 kilometre in Lashington and Tsaba while in Magwiji is between 1 to 5 kilometres. Communities face a shortage of water for their animals during low rainfall seasons, especially in Magwiji, and this affects rangelands conditions. Villagers

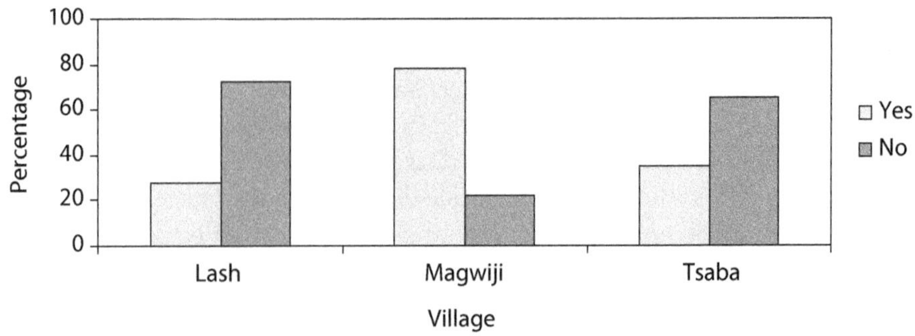

Figure 10.9. Distribution of conflicts in Magwii, Tsaba and Lashington.

from Magwiji complained about distant water points for their livestock. The communities also complained about insufficient water for the animals during dry seasons since they only rely on dams and rivers.

10.9.7 Sources of income

Communities obtain their income from either farm or off-farm activities. Farm income is generated from activities such as selling farm produce, livestock and livestock products while off-farm income is obtained from social welfare grants such as disability grants, pensions and child support grants. Other off-farm income can be obtained from remittances. In the three communities there are no contract markets as a result farmers sell their produce in the nearby towns or to neighbours.

Figure 10.10 illustrates off-farm incomes in the three study areas. According to the figure, about 90% of the community members obtain most of their off-farm income from social grants, mainly pension grants and about 10% of their income comes from remittances. For example, in all the three communities, the income obtained from pension grants per year per person is R9,840.

Apart from off-farm income, average income generated from agricultural activities such as selling of crops and animals per year differs in the three communities (Figure 10.11). For example, Lashington obtain most of its farm income from selling of crops (R469.47) while Magwiji get about R744.90 from selling livestock and Tsaba get R473.91 from selling animals mainly goats. This shows that most of the income generated in these three communities comes from off-farm sources mainly pensions.

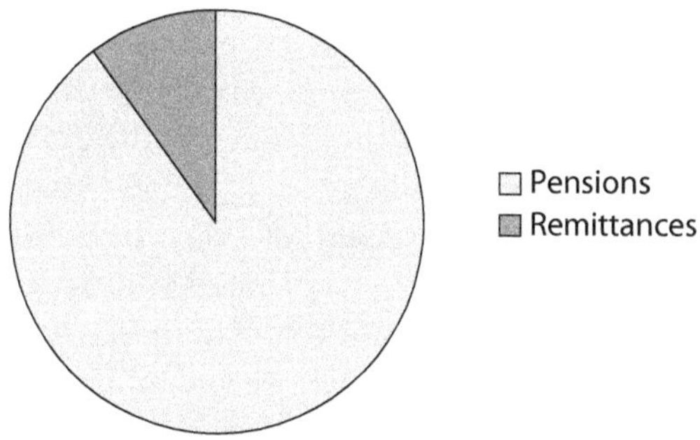

Figure 10.10. Distribution of off-farm income for Magwii, Tsaba and Lashington.

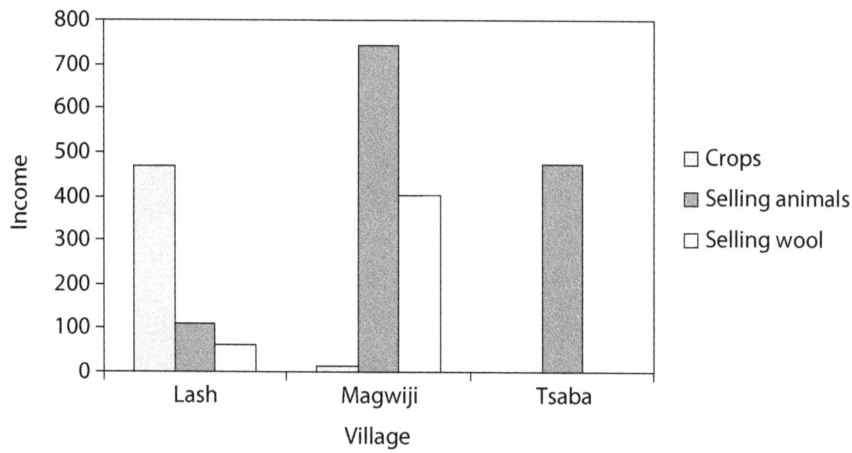

Figure 10.11. Distribution of farm income in Magwii, Tsaba and Lashington.

10.9.8 Household expenditure

Household expenditure is the amount of money spent on clothes, furniture, food and education. It was found that in all the three communities, more money is spent on food than on fuel, transport or education. That is, the average amount of money spent on groceries in Lashington is R4,562.67, while in Magwiji is R4,836.73 and Tsaba is R4,330.87. These results show that farm income cannot cover household expenditure as a result people use their off-farm incomes to cover their expenditures.

10.10 Conclusion and policy implications

It is clear from the foregoing that wide-ranging institutional arrangements have always existed in the region to regulate the use and management of rangelands and these have functioned relatively efficiently over the years. However, new institutions have arisen in recent years. One noticeable effect of this development is the emergence of considerable conflicts and confusion between communities and within groups in the use of natural resources and the extent to which they contribute to improved livelihoods and sometimes pose serious threats to environmental sustainability. Case studies carried out in three communities were reported and confirm the important differences in access to agricultural land and grazing land and asset ownership, including livestock such as sheep and goats. There is evidence that membership of groups has implications for welfare for the smallholder farmers; those farmers who belong to farmer associations seemed to be better off than those who did not. Working with local people to promote group or collective action, perhaps through the introduction of collective innovation principles, may be a useful strategy to pursue as government grapples with the task of saving the environment while addressing the serious poverty in the rural areas of the country.

References

Adhikari, B. (2001). Property rights and natural resources: impact of common property institutions on community-based resource management. Available at: http://www.gdnet.org/pdf/adhikari.pdf.

Ainsle, A. (1999). When community is not enough: managing common property natural resources in rural South Africa. Development Southern Africa 16, 375-395.

Ainsle, A. (2002). Cattle ownership and production in the communal areas of Eastern Cape, South Africa. Research report no.10. Pretoria, RSA.

Agricultural Digest (2005). Agricultural unions. Available at: http://www.nda.agric.za/docs/Digest2000/Digest10.htm.

Ankomah, B. and R. Fox (2000). Property rights and land use regulation: a comparative evaluation. Agrekon 39, 248-253.

Ashley, C. and Z. Ntshona (2003). Transforming roles but not reality? Private and forestry development on the wild coast, South Africa. Available at: http://www.propoortourism.org.uk/Publications/SLSA_ASHLEY_Ntshona.pdf.

Bennett, J. and H. Barrett (2003). Rangeland management in communal areas of the central Eastern Cape Province, South Africa. A case study of two contrasting cases. African Journal of Range and Forage Science 20, 197-201.

Brauteseth, N.L. (2000). Environmental law. Available at: http://www.lawinfo.org.za/general/lawinfo.asp.

Breedlove, G. (2002). South African conservation policies. Available at: http://upetd.up.ac.za/thesis/available/etd-08062003122913/unrestricted/02.pdf.

Buckles, D. and G. Rusnak (2006). Conflict and collaboration in natural resource management. Available at: http://unpan1.un.org/intradoc/groups/Public/ documents/APCITY/UNPAN122242.pdf.

Chapeyama, O. (2004). Rangeland management in Lesotho: Report on assessment of needs for reintroduction of grazing fees. Available at: http://www. sarpn.org.za/documents/d0001094/P1207-Lesotho_June2004.pdf.

Cocks, A., M. Dold and I. Grundy (2006). Challenges facing a community structure to implement CBNRM in the Eastern Cape, South Africa. Available at: http://www.africa.ufl.edu/asq/v5/v5i3a4.htm.

Coetzee, L., R. Manshwe and A. Jooste (2005). The marketing of livestock on communal lands in the Eastern Cape Province: constraints, challenges and implications for extension services. South African Journal of Agricultural Extension 34, 81-101.

Commark Trust, (2003). Red meat – Eastern Cape. Available at: http://www.commark.org/pages/Default.asp?SectionID=191.

Corbett, A. and B. Daniels (1996). The legal aspects of Governance in CBNRM in Namibia. Available at: http://www.cbnrm.net/pdf/corbett-001.pdf.

Cousins, B. (1995). Common property institutions and land reform in South Africa. Development Southern Africa 12, 583-493.

Cousins, B. (1999). The contribution of communal rangelands to the rural livelihoods in South Africa. Development Southern Africa 16, 499-510.

CSIR (2004). 2004 Eastern Cape State of the Environment Report. CSIR, Division of Water, Environment and Forestry Technology. Durban, South Africa. Available at: http://www.environment.gov.za/soer/ecape/download_report.htm.

De Bruyn, T., P. Lent, and N. Van Averbeke (2000). Policies on natural resource management. Available at: http://les.man.ac/ses/research/CAFRE/ indicators/wp4.pdf.

Department of Agriculture, (2003). Department of Agriculture Strategic plan 2003-2006. Available at: http://www.info.gov.za/view/DownloadFileAction?id=70175.

Deshingkar, P. and S. Cinderby (1999). Renewable natural resource use in post-apartheid South Africa: Summary of key issues and findings. Stockholm Environment Institute, Stockholm, Sweden.

De Wet, C. (1995). Moving together drifting apart: betterment planning and villagisation in South African homeland. Witwatersrand University Press, Johannesburg, RSA.

Duvel, C.H. and D.B Afful (1997). Towards understanding of stocking rate decisions for commercial rangelands in South Africa: The role of culture. South African Journal of Agricultural Extension 26, 216-227.

Eastern Cape Parks (2007). Overview of east London coast nature reserve. Available at: http://www.ecparks.co.za/.

Eastern Cape Provincial Government (2006). The Eastern Cape Province. Available at: http://www.ecprov.gov.za/index.php?module=press_releases.

Engel, A. and B. Korf (2005). Negotiation and mediation techniques for natural resource management. Available at: http://www.fao.org/docrep/008/a0032e/a0032e00.htm.

Environmental Law Alliance Worldwide (1999). South Africa – National Environmental Management Act 107 of 1998. Available at: http://www.environment.gov.za/polleg/legislation/natenvmgmtact/natenvmgmtact.htm.

Food and Agricultural Organisation (1991). Sustainable agriculture and development in Asia and the Pacific. Regional document No. 2 FAO/Netherlands Conference on Agriculture and Environment, S-Hertogenbosch, the Netherlands.

Ghee, T. and Valencia, M. J., (1990). Conflict over natural resources in South-East Asia and the Pacific. Available at: http://www.unu.edu/unupress/unupbooks/80a04e/80A04E00.htm.

Gopalakrishnan, S. (2005). Collective action in the management of common resource: is there an alternative to the rationale choice model. Unpublished MSc thesis, Department of Agricultural Economics, Michigan State University, East Lansing, USA.

Hoare, D. (2002). Biodiversity and performance of grassland ecosystems in communal and commercial farming systems in South Africa. Proceedings of the FAO's Biodiversity and Ecosystem Approach in Agriculture, Forestry and Fisheries Event: 12-13 October, 2002. Food and Agriculture Organisation of the United Nations, Rome, Italy, pp. 10-27.

Hoffman, M.T., S. Todd, Z. Ntshona and S. Turner (eds.) (1999). Land degradation in South Africa. Department of Environment Affairs and Tourism, Pretoria, RSA.

International Center for Living Aquatic Resource Management (1996). Analysis of fisheries co-management arrangements: a research framework. Available at: http://www.ifm.dk/reports/21.pdf.

Kariuki, S.M. (2004). Failing to learn from failed programmes? South Africa's communal land rights act (ClaRA 2004). Available at: https://openaccess.leidenuniv.nl/dspace/bitstream/1887/488/1/02PUB0000001753.pdf.

Kirkland, T., M.L. Hunter, and W. Twine (2005). The bush is no more: insights on institutional change and natural resource availability in rural South Africa. Society & Natural Resources 20, 337-350.

Kowero, G., M. Bruce, U. Campell, and R. Sumaila (2003). Policies and governance structures in woodlands of Southern Africa. Available at: http://www.cifor.cgiar.org/publications/pdf_files/Books/policies/PoliciesSouthern Africa-1.pdf.

Letsoala, R. (1987). The betterment planning. Available at: http://www. york.ac.uk/inst/sei/projects/booklet.pdf.

Makhanya, M.E. (1999). The sustainability of rural systems under the communal property association act in South Africa. Available at: http://www.iucn.org/ themes/ceesp/publication/Makhanya.pdf.

Masiteng, T.J., C. Van De Westhuizen and M. Matli (2003). Aspiration and needs of farmers on communal grazing areas in the Free State. South African Journal of Agricultural Extension 32, 85-91.

Matiru, V., N. Hart and P. Castro (eds.) (2000). Conflict and natural resource management. FAO, Department of Forestry, Rome, Italy.

McCall, J.C. (1997). Rethinking ancestors in Africa. Available at: http://www.jstor.org/view/00019720/ap040255/04a00060/0.

McCarthy, N., B.M. Swallow, T. Kirk. and P. Hazell (2000). Property rights, risk, and livestock development in Africa. International Livestock Research Institute Nairobi, Kenya.

Meadows, M. and M.T. Hoffman (2002). The nature, extent and causes of land degradation in South Africa: legacy of the past, lessons for the future? Available at: http://www.blackwell-synergy.com/doi/pdf/10.1111/1475-4762.00100?cookieSet=1.

Messer, N.M. (2001). Mapping traditional structures in decentralisation policies: illustrations from three countries in Sub-Saharan Africa and the Near East. Available at: http://www.fao.org/docrep/006/ad721e/ad721e00.htm#Contents.

Multi Purpose Community Centers (2004). Available at: http://www.gcis.gov.za/.

Ngaido, T. and N. Mccarthy, (2004). Collective action and property rights for sustainable development: institutional options for managing rangelands. Available at: http://www.ifpri.org/2020/focus/focus11/focus11_09.pdf.

Nilsson, T. (2001). Management of communal grazing land: A case study on institutions for collective action in Endabeg village, Tanzania. Available at: http://dlc.dlib.indiana.edu/archive/00000961/00/WebThesis.pdf.

Nkosi, Z. (2005). Spirituality, land and land reform in South Africa. Available at: http://www.wcc-coe.org/wcc/what/jpc/echoes-16-05.html.

Nothard, B.W., G.F. Ortmann and E. Meyer (2005). Institutional and resource constraints performance in the small-scale sugar cane industry in Kwa-Zulu Natal. South African Journal of Agricultural Extension 34, 65-75.

Omamo, S.W. (2005). Institutional costs as a theoretical framework for transformation in agriculture. Agrekon 45, 21-26.

Ortmann, G.F. (2000). Promoting competitiveness in South African agriculture and agribusiness. The role of institutions. Agrekon 39, 367-387.

Ostrom, E. (1990). Governing the commons. The evolution of institutions for collective action. Cambridge University Press, Cambridge, UK.

Paudel, G.S. and G.B. Thapa (2003). Impact of social, institutional and ecological factors on land management practices in mountain watersheds of Nepal. Applied Geography 24, 35-55.

Payne, G. (1997). Urban land and property rights in developing countries. Intermediate technology publications: London, UK.

Philip, K. (2003). Co-operatives in South Africa: their role in job creation and poverty reduction. Available at: http://www.sarpn.org.za/documents/d0000786/P872-Coops_October2003.pdf.

Peden, M.I. (2005). Tackling the most avoided issue: communal rangeland management in KwaZulu-Natal, South Africa. African Journal of Range and Forage Science 22, 167-175.

Quail Report (1980). Ciskei commission report. Conference Associates (Ltd), Pretoria, RSA.

Quinn, C.H., H. Huby, C. Kiwasila and J.C. Lovett (2007). Design principles and common pool management: An institutional approach to evaluating community management in semi-arid Tanzania. Journal of Environmental Management 84, 100-113.

Ramirez, R. (2002). A conceptual map of land conflict management: Organizing the parts of two puzzles. Land Tenure Service, FAO Rural Development Division, Rome, Italy. Available at: http://www.fao.org/sd/2002/IN0301a_en.htm.

Roy, D. (2005). Traditional customary laws and indigenous peoples in Asia. Minority Rights Group International. Available at: http://www.minorityrights.org/download.php?id=131.

Sanginga, P.C., R.N. Kamugisha and A.M. Martin (2007). The dynamics of social capital and conflict management in multiple resource regimes: a case of the southwestern highlands of Uganda. Ecology and Society 12 (1), 6.

Scogings, P., P. Lent and R. Bally (2000). Policy and institutional frameworks for natural resource management. Available at: http://les.man.ac/ses/ research/CAFRE/indicators/wp4.pdf.

Shackleton, C. (1993). Are communal rangelands in need of saving? Development Southern Africa 10, 65-67.

Shackleton, C. and B. Cousins (2000). Re-valuing the communal rangelands of South Africa: new understanding of rural livelihoods. Available at: http://www.odi.org.uk/nrp/62.html.

Smajgl, A., K. Vella and R. Greiner (2003). Frameworks and models for analysis and design of property rights in Outback regions. Scoping report Desert Knowledge CRC.

Soga, J.H. (1937). The Ama-xosa life and customs. Lovedale press: South Africa.

Stevens, J. and S. Treurnicht (2001). Culture as a resource for development: some challenges for extension management. South African Journal of Agricultural Extension 30, 104-110.

Swatuk, L.A. (1996). Environmental issues and prospects for South African regional co-operation. Available at: http://www.iss.co.za/Monographs/No4/ Swatuk.html.

Terblanche, S.E. (2005). Participation and linkages for improved extension delivery: The role of the extension worker. South African Journal of Agricultural Extension 34, 122-134.

Timothy, O. (1998). Multiple uses of common pool resources in semi-Arid West Africa: a survey of existing practices and options for sustainable resource management. Available at: http://www.Odi.org.uk/nrp/38.html.

Tonah, S. (2002). Migrant Fulani herdsmen, indigenous farmers and the contest for land in Northern Ghana. Available at: http://www.vad-ev.de/papers/tonah.pdf.

Tshabalala, M.T. (1995). The social organization of range utilization and livestock production in Lesotho. Grassland Society of South Africa. Pietermaritzburg, RSA.

Tucker, C. (2000). Common property design principles and development in a Honduran community. Available at: http://fletcher.tufts.edu/praxis/archives/xv/Tucker.pdf.

Twine, W.C. (2005). Socio-economic transactions influence vegetation change in the communal rangelands of South African lowvelds. African Journal of Range and Forage Science 22, 93-99.

Uma, T., G. Gebru, A. Aboud, and D. Little (2003). Borana pastoralists of Southern Ethiopia: the role of indigenous institutions in resource and risk management. African Journal of Range and Forage Science 20, 200-201.

Van Averbeke, W. (2000). Policy and institutional frameworks for natural resource management. Available at: http://les.man.ac/ses/research/CAFRE/ indicators/wp4.pdf.

Vink, N. (1989). An institutional approach to livestock development in South Africa. Unpublished PhD thesis, Department of Agricultural Economics and Extension. University of Stellenbosch, Cape Town, RSA.

Warner, M. (2000).Conflict management in community-based natural resource projects: experiences from Fiji and Papua New Guinea. Available at: http://www.odi.org.uk/publications/working_papers/wp135.pdf.

Wendel, K. (2004). Institutions and common pool resource: a case study of grazing land in the Matsheng area, Kgalgadi District, Botswana. Available at: http://epubl.luth.se/1404-5508/2004/233/LTU-SHU-EX-04233-SE.pdf.

World Bank (2004). Rural poverty in Africa. Available at: http://www.ruralpovertyportal.org/english/regions/africa/index.htm.

Zenani, V. and A. Mistri (2003). A desktop study on the cultural and religious uses of water using regional case studies from South Africa. Available at: http://www.dwaf.gov.za/Documents/Other/RMP/SAADFCulturalWater UseJun05.pdf.

Part VI
Alternative trade and support arrangements to enhance livelihoods and welfare of small farmers

11. Recent changes in markets and market relationships and lessons for the design of effective support programmes[2]

Binganidzo Muchara and Ajuruchukwu Obi

Abstract

As a result of poor planning and inappropriate procedures for beneficiary selection, the land reform programme in Zimbabwe resulted in a drastic decline in production and productivity of the agricultural sector. At the same time, the perceived political agenda of the process sparked off a reaction in the investment climate that manifested in worsening balance of payments, reduction in industrial production, and growing unemployment. It is safe to conclude that the Zimbabwean land reform programme resulted in an economy-wide decline in effective demand occurring simultaneously with a sharp fall in physical agricultural output as appropriated farms lay idle and poor farm practices depressed productivity levels. A low-equilibrium trap thus ensued, culminating in the deterioration of livelihoods across the broad spectrum of society as a hyper-inflationary situation developed and assumed scandalous proportions. A situation such as this set off chain reactions that touched all segments of the economy and produced diverse effects, including the disappearance of markets, the emergence of informal exchange arrangements that represent adjustments to the aberrant phenomenon of hyper-inflation emerging in the absence of both cash and goods/services. To date, there has been no systematic assessment of this dimension of the land reform programme with the aim of drawing lessons that can provide some basis for developing strategies for revamping the economy now that clear signs are emerging that the long-running political crisis might soon end. To understand what might have happened, an initial broad appraisal of specific indicators of agricultural output and prices for the livestock and maize products has been conducted in one district and formed the basis for assessing any spatial and temporal patterns in markets and marketing relationships.

11.1 Overview of agricultural commodity markets in developing countries

Low production levels, poor product quality and thin markets characterise the smallholder sector in Sub-Saharan Africa. Countries in Southern Africa are generally regarded as food deficit areas whose food requirements are often met through supplementation with donor hand-outs. Within the same region however, some countries like South Africa, Botswana and Namibia are performing reasonably well in terms of basic food provision, while the majority of the countries have been in food deficits for the better half of the last decade.

[2] This chapter is in part based on a larger study funded under the 'Small Grants Scheme on Livelihoods after Land Reform in Zimbabwe' of the Institute of Development Studies of the University of Sussex, and the comments of Professor Ian Scoones on the draft of the main report are gratefully acknowledged.

Holistic strategies that seek to boost food production and efficient distribution through well planned marketing systems remain the cornerstone of efforts to eradicate poverty in developing countries. Policy has tended to put emphasis on maximising food production as a way of ensuring food self-sufficiency among the poor, while less focus is given to distributional functions like product marketing. It cannot go unnoticed that effort by the International Monetary Fund and the World Bank in the 1980s and 1990s to increase market participation by developing countries, through Structural Adjustment Programmes (SAPs) for instance, had wide ranging impacts. Markets in third world economies are generally controlled and hence Structural Adjustment Programmes recommended trade liberalisation, reduction of government subsidies/expenditure and adoption of free exchange rate policies. Though a few countries like Uganda and Mozambique experienced positive improvements due to these Structural Adjustment Programmes (SAPs), some countries like Zambia, Zimbabwe and Kenya had to battle to reverse the negative effects of the SAPs. The food security situation in developing countries has not changed significantly inspite of the WB and IMF policy prescriptions, instead some countries have worsened. Policy makers believe that liberalisation has led to substantial increase in prices of food crops, associated with high volatility of prices, which introduced some level of uncertainty in farm planning and management (Chirwa, 2006).

Malawi's experience can be used to illustrate some of the interesting outcomes of the SAP and how land ownership patterns play a role. The liberalisation of agricultural produce marketing in Malawi resulted in the declining importance of Agricultural Development and Marketing Corporation (ADMARC), which is a government agency (Chirwa, 2006), and attracted private traders in the marketing of smallholder agricultural produce in the sector (Fafchamps and Gabre-Madhin, 2001; Chirwa, 2006). The Malawian maize case is very important in explaining smallholder market dynamics within staple food markets. Chirwa (2006) also observed that the extent of maize sales in Malawi is positively associated with increasing land holding sizes, and that those whose maize failed had the lowest land holding size. The significance of landownership is emphasized, where land hunger has caused instability in Zimbabwe and still remains a hot issue in most other countries like Namibia and South Africa.

In recent years, and especially after the inception of political instability and agricultural reform programme in Zimbabwe in 2000, South Africa became the only reliable exporter of white maize in the Southern African region (Jayne *et al.*, 2006). Although Mozambique, Zambia, and Malawi produce white maize surpluses, these surpluses are usually depleted halfway through the marketing year (Jayne *et al.*, 2006), leaving both urban and rural maize deficit regions increasingly dependent on South Africa for their residual national white maize requirements. Drawing on FAOSTAT data, Jayne *et al.* (2006) indicated that 92% of the white maize imports to Malawi, Zambia, and Mozambique have been coming from South Africa since 1998. While formal trading of grain is through marketing boards in most

Southern African countries, bureaucratic impediments and poor administrative structures that fail to accurately anticipate demand and supply and price trends have tended to increase the cost of grain to the disadvantage of the poor consumer, both rural and urban based. This results in inefficient grain supplies to the deficit areas. Informal trading by small dealers and millers naturally develop to fill this gap but they work best when distances between surplus and deficit zones are small and road and travel conditions allow for supplies to be easily accessed. Where movement of grain involves transportation across two country boarders, the role of informal traders in grain provision tends to dampen.

11.2 Maize commodity marketing and the emergence of informal markets in Zimbabwe

The fast track land reform programme (FTLRP) of 2000 in Zimbabwe featured ad-hoc policies and accelerated institutional reforms within the agricultural sector. As such, agricultural markets for most products were affected. Although technically efficient, state marketing systems which used to be dominant in Zimbabwe proved to be economically inefficient and regressive, thus having negative effects on both growth and equity (Rukuni, 2006). These inefficiencies were brought about by monopoly power and pricing inefficiencies, inappropriate bureaucratic procedures and the restrictions on trade between communities in different areas (Jayne *et al.*, 2006). At the same time, the ensuing marketing controls implicitly taxed the poorest rural people some 20-30 percent of potential producer prices in most years.

Cereal marketing in the formal markets of Zimbabwe has for the recent past (2000-2008) been constrained by low production and market irregularities. The fast track land reform programme (FTLRP) in Zimbabwe coincided with frequent droughts and hyper-inflationary environment in Zimbabwe, which caused uncertainties in the agricultural sector. According to a study conducted in Mwenezi District of Zimbabwe, the district is inherently dry, with rainfall averaging 500 mm/year. As such, cattle ranching have been the dominant commercial activity among large scale farmers, augmented by game farming. Mwenezi District post fast track land reform programme saw the subdivision of the cattle ranches into small and medium sized farms occupied by beneficiaries of the land reform programme. This saw the introduction of crop farming in these areas together with small scale cattle rearing at household levels.

Traditionally, the Grain Marketing Board (GMB) of Zimbabwe served as a government strategic grain reserve for the country. Its main responsibility is to buy all cereal produce from farmers and subsequently sell the cereals to millers, institutions and individual consumers at affordable prices. GMB is the sole importer and exporter of government cereal products. Since 2000, GMB has been affected by cash shortages, weak pricing structures, transport and fuel shortages. The effectiveness and efficiency of the Grain

Marketing Board (GMB) following the fast track land reform was at stake given the hyper inflationary environment under which the parastatal was operating. From 2000 to 2008, Zimbabwe experienced grain shortages and importation of maize through the GMB was one of the strategies adopted by the Zimbabwean government to cope with the ever souring shortages. Commercial imports were augmented by food handouts from donor agencies. However, Zimbabwe remained constrained with foreign currency shortages as a result of reduced exports by Zimbabwean producers, and an ever increasing need for imports in both agricultural and manufacturing sectors. Table 11.1 gives an insight into how Zimbabwe managed its food sector by highlighting domestic maize procurement and importation by the Grain Marketing Board of Zimbabwe.

Though Table 11.1, represents a trend where maize imports were greater than locally procured maize nationally, except for the 2004/05 and 2006/07 seasons. The role of informal market in augmenting government procurement efforts is evident from Table 11.1. However, the question of how much grain the informal sector imported into the country during this shortage periods is not obvious. Some informally imported maize ended up on the formal markets like retail shops and supermarkets at exorbitant prices, while some continued to be channelled to final consumers through backyard trading and road side marketing. The failure by the Zimbabwean government to suppress the mushrooming of informal markets in the post Fast Track Land Reform Programme period can partly be blamed for the low levels of maize supplies to the Grain Marketing Board (GMB) from local producers. This was strongly influenced by the scandalous inflationary levels and high volatility of the Zimbabwean dollar, which crippled the capacity of state procurement agencies in favour of profiteering informal traders who had a lot cash to throw around. Though Zimbabwe's

Table 11.1. National maize procurement statistics for Grain Marketing Board (GMB, Zimbabwe).

Marketing year	Local intake/procurement (tonnes)	Imports (tonnes)
2001/02	154,847	88,656
2002/03	49,418	763,594
2003/04	244,187	375,198
2004/05	186,661	75,609
2005/06	181,219	685,983
2006/07	543,725	161,235
2007/08	194,331	401,285
2008/09	35,593	417,825
2009/10[1]	6,062	3,580

[1] Figures for the 2009/10 marketing year are as at week ending Friday July 10, 2009

2000-2008 case represents an abnormal situation, it remains important in calling attention to the consequences of bad policy on markets and welfare of the people. By 2009, Zimbabwe witnessed an adoption of multiple currencies (US dollar and the South African Rand) in the economy as a way to curb hyperinflation and price stabilisation. This had a positive effect since inflation dropped from 2.31 million per cent, in July 2008 (GOZ, 2009) to -3% after the dollarisation of the economy in the first quarter of 2009.

Land redistribution in Zimbabwe led to a wide dispersion of people in rural areas as farmers occupied land in former cattle ranches. Since 2004, Mwenezi District had its GMB depot relocated from Sarahuru business centre in the old communal area to Rutenga Business Centre in the former large scale commercial area. Though such a move was believed to be politically motivated, it could be justified by the location of Rutenga Business Center at a well serviced location along the Beitbridge-Harare highway, together with a rail network that links the district to main towns like Bulawayo, Chiredzi and Beitbridge as well as direct link to Mozambique, Botswana and South Africa. Such links are important to a district like Mwenezi that has always benefited from food acquired from outside the district boundaries. The issue of high transactions costs, which is normally associated with poor road infrastructure, poor information flow and the role of middlemen in agricultural product marketing, was partly addressed by the relocation of the Grain Marketing Board (GMB). The current GMB location is however too far for most residents, with most of the households in the newly resettled areas preferring to acquire their maize from informal sources (farmer to farmer trading), rather than travelling for long distances to the GMB, where they were not sure to get the maize.

Another consequence of the FTLRP associated with the emergence of the informal trade in maize has been on widening the gap between the haves and the have-nots due to differential access to sources of foreign currency for the non-working population who normally received remittances from relatives abroad or through cross border trading (FAO, 2009). There is also the emergence of small scale millers among entrepreneurs with the means to directly import maize from neighbouring countries, especially South Africa. This development ensured continuous supply of maize and maize meal on the informal markets. In 2008, a 50 kg bag of maize was being sold by small scale millers for between R300 and R350, which was more than four times the price of maize on the international market and consequently out of reach of many consumers.

Farmer to farmer trading also represented another form of informal marketing of maize in Zimbabwe and specifically Mwenezi district. Due to inflation, cash trading of maize was limited among farmers except where such transactions were denominated in the South African rand. To overcome the cash constraint, most farmers preferred barter-trading with livestock (goats, sheep, cattle, chicken, and donkeys), clothes, household utensils, farm equipment (ploughs, scotch carts, hoes) or building materials. Such trade was also meant to bypass the government's price control measures and was an opportunity for farmers to bargain for better returns. It is possible that the emergence of barter trading was another reason the supply of local grains to the GMB fell drastically during this time (Table 11.1).

The government of Zimbabwe responded predictably to these developments. For the 2008/09 cropping season, the Zimbabwean government set maize floor price at US$265/ton whilst private buyers pegged their price at US$400/ton (Figure 11.1). As is well known, agricultural commodity pricing has always been at the centre of most debates in developing countries. In order to motivate farmers to produce, governments in developing countries artificially maintain prices above equilibrium as incentives to producers to deliver increased output. However, this contradicts the broader development goal to promote food security part of which is achieved by making foods affordable to the populace. The ultimate effect of this floor price policy is a powerful lesson on how not to intervene in commodity markets when a government is confronted with a situation similar to what prevailed in Zimbabwe in the post FTLRP era.

Having gazetted the producer price at US$265/ton, it became obvious that the government could not pay this price, given its circumstances, to domestic producers to deliver maize

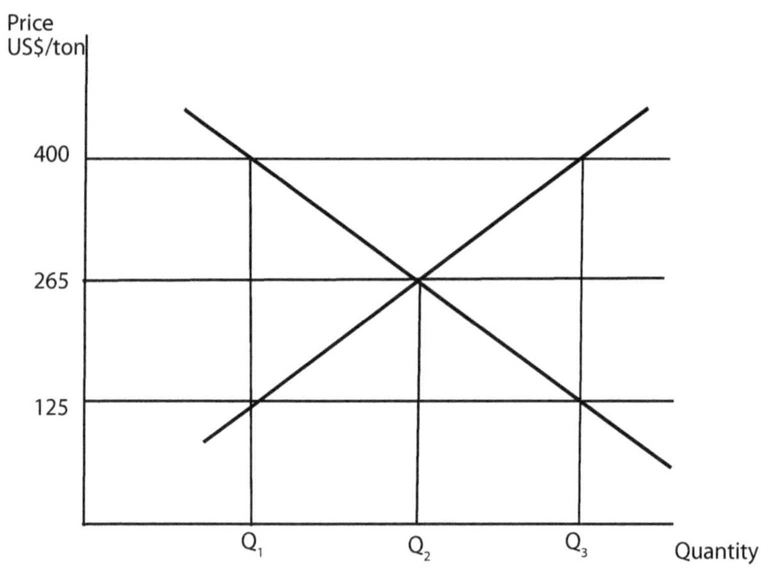

GMB Price = US$265/ton
Private buyer's starting price = US$400/ton
Final private price = US$125/ton

Figure 11.1. Maize price structure in Zimbabwe for 2008/09 season (Survey data, July 2009).

to the GMB. Private buyers and millers immediately caught on to this fact and saw it as an opportunity to edge off the Government's procurement agent (GMB) (Figure 11.1). A situation had arisen where farmers who had been encouraged by the price floor to produce could not find a buy since the government had no cash to pay. Private buyers took advantage of this and underpriced the maize to as low as R20-R30/20 kg bucket which worked out to US$125-US$188/ton, allowing them to realize margins as large as US$100/ton in some instances. Of course, this became a disincentive to producers and further exacerbated the supply bottlenecks. As the government of Zimbabwe continues to be cash-strapped and hence unable to buy maize from domestic producers, uncertainties among farmers about delayed payments for their maize deliveries to the Grain Marketing Board remain a deterrent to channelling maize to the institution. The few who deliver maize to GMB are owed for protracted periods.

The marketing problems of maize farmers are complicated by two other fallouts of the FTLRP. One is the weakened role of the Zimbabwe Farmers Union (ZFU) and the other is the capacity constraints that have made the extension service ineffectual. Prior to 2000, the ZFU was active in price setting debates. However, as the economy crashed between 2000 and 2008, its involvement dwindled due to its shrinking revenue base, which was mainly through deduction of 1% levy from farmers' sales, annual subscriptions and card selling to new farmers. Where farmers were hardly selling anything, their subscriptions fell off and representation became almost non-existent. Farmers' representation is an integral part in product marketing in developing countries where the majority of the smallholder farmers are either illiterate or semi-illiterate.

The other source of price/marketing information is the agricultural extension service. But due to lack of resources, their mobility has become highly restricted, with the result that farmers have been deprived of vital marketing information. More so, where such information could be easily accessible, extension personnel especially in drought prone regions indicated their deliberate omission in educating farmers about maize marketing. Somehow, the extension service has written off the smallholder farming in the face of dwindling productivity and production. An interview with an extension officer on their role in advising farmers on how to market their produce revealed that they deliberately omit marketing information in their messages to smallholder farmers. One extension agent in the Mwenezi District had this to say:

'We rarely talk about maize marketing with farmers, except those who do market gardening. How can you tell someone who is always begging how to market his insufficient food? We always advise them to keep their food and not to market' (Interview, Agritex Extension Worker, Mwenezi District, 22 July 2009).

There is no written policy pertaining to what the extension officer say to farmers, but his sentiments were based on the fact that Mwenezi District of Zimbabwe has always relied on food hand outs from the donor community. He also emphasised that although beneficiaries of the Fast Track Land Reform Programme (FTLRP) in Mwenezi District seem to produce more food than their communal counter parts, there is no clear evidence of surplus to warrant any formal marketing of basic cereals like maize, sorghum and finger millet. Productivity on per hectare basis is estimated to be high for resettled farmers (A1) than communal farmers (CA) as shown in Table 11.2 at provincial level.

A closer look at the provincial statistics (Table 11.2) indicate a fairly higher average yield from Fast Track Land Reform beneficiaries under the small scale A1 schemes (1.79 t/ha) as compared to 0.50 t/ha from the communal areas (CA). However, this has not translated into a noticeable participation of farmers on the formal market given that both district and provincial averages (0.55 t/ha and 0.68 t/ha respectively) are below 1999 national average of 1,516 kg. This is a clear indication of the negative productivity effect of the Fast Track Land Reform Programme (FTLRP) in Zimbabwe.

Table 11.2. Maize production in the districts of Masvingo Province of Zimbabwe (tonnes/ha) (GOZ, 2009).

District	LSCF[1]	SSCA[2]	A2[3]	A1[4]	OR[5]	CA[6]	District average
Bikita	0	0.95	0	0	0.71	0.57	0.60
Chiredzi	1.0	0.73	1.46	0.74	0.50	0.45	0.64
Chivi	0	0.76	0	0	0.93	0.61	0.60
Gutu	3.78	1.32	2.62	1.06	1.32	0.70	0.70
Masvingo	2.01	0.82	1.30	0.94	0.74	0.61	0.82
Mwenezi	0	0	0	0.51	0.71	0.60	0.55
Zaka	0	0	0	2.13	2.00	0.60	0.65
Average	1.94	1.01	1.47	0.79	0.87	0.50	0.68

[1] LSCF refers to large scale commercial farms which are not part of the Fast Track Land Reform Programme.
[2] SSCA refers to Small Scale Commercial Areas which are not part of the Fast Track Land Reform Programme.
[3] A2 Farmers are medium-large scale commercial farmers who benefited in Zimbabwe's Fast Track Land Reform Programme between 2000 and 2009.
[4] A1 Farmers are small scale farmers who benefited in Zimbabwe's Fast Track Land Reform Programme between 2000 and 2009.
[5] OR-refers to Old Resettlement areas that were formed from 1980 -1998 under the government's Land Reform and Redistribution Programme.
[6] CA refers to Communal Areas that have never been reformed by any of Zimbabwean government's land reform programmes.

11.3 Impact of FTLRP on agricultural input markets

Zimbabwe's input markets following fast track land reform was marred by irregularities, where major crop and livestock inputs were generally in critical short supply and direct purchases were very limited as there were inadequate stocks on the open market. Most input and products were sold on the informal markets at exorbitant prices beyond the reach of most farmers. However, Zimbabwean farmers also benefited from several agricultural input programmes. The effectiveness and scale of the input supply programmes still need to be looked into given the unsatisfactory production levels of farmers after the fast track land reform programme. Some of the input schemes included:

- The Presidential Programme, which was for targeted farmers especially the Champion farmers.
- Reserve Bank Programme.
- Social Welfare Programme for vulnerable households.
- SADC Agricultural Assistance Programme, supporting communal, old resettlement and small scale commercial farming.
- Contract Schemes, which are mainly found in the cotton sector.
- NGO Programmes (Red Cross, Care International, and Plan International). NGOs mainly targeted vulnerable persons/households such as those affected by HIV/AIDS either as patients or as care-givers, the old and the poor. NGOs did not offer much help in newly resettled areas.

Smallholder farming sectors (communal areas – CA), Old Resettlement (OR), A1 and small scale commercial areas (SSCA) contributed more than 90% of the area planted to maize with communal sector contributing about 64% (GOZ, 2009). Due to persistent input shortages and rampant market failures following the fast track land reform programme (FTLRP) and the hyperinflationary pressure until end of 2008, bulk of the maize inputs were from NGOs especially to cater for the old communal areas, whose people did not directly benefit from the fast track land reform programme. Open Pollinated varieties (OPVs) also dominated the maize input market in the period 2000-2008. These included Red Cork, Red Cob, Hickory King and ZM 521. The traditional high breed Seed Co-op[3] maize seed varieties (SC 401, SC 501, etc.) were in short supply on the formal market. Shortages of seed on the formal market led to unscrupulous tendencies by informal market dealers to sell fake seed, packed in Seed Co-op packages. Some seeds that were planted by farmers were informally sourced from South Africa and other neighbouring countries, and as such, their viability under Zimbabwean conditions requires further scrutiny.

The fertilizer market was also not spared the unstable economic conditions in Zimbabwe. Low production by major fertiliser companies (Windmill, Sable Chemicals, ZFC) in

[3] Seed Coop: a seed house company that specialises in maize breeding and trading of high breed seed in Zimbabwe.

Zimbabwe culminated in severe shortages of fertilizer. Major nitrogen, phosphorous and calcium fertilizers were imported and this affected small scale agricultural producers who were cash-constrained. The large scale farmers equally faced challenges in accessing foreign currency from the Reserve Bank of Zimbabwe for purposes of importing fertilizers from outside Zimbabwe. This situation had a negative impact on the productivity and total output from the large scale commercial sector with the result that even the formal output markets did not escape the scourge of the FTLRP.

11.3.1 The model

Although the Zimbabwean FTLRP has affected almost all segments of the economy, it is possible to isolate the impact on agriculture in general and specifically on the marketing activities as they have been affected by this programme. On the basis of the foregoing discussion, it is safe to say that with respect to the agricultural sector, an important effect has been on market participation of the domestic producers, be they small-scale or large commercial producers. It is therefore possible to obtain some policy-relevant insights by modelling this variable and examining what factors have been central to the observed trends.

Economic theory predicts direct relationships between a vast array of socio-economic and community variables and the willingness or otherwise of economic actors to participate in the process of exchange. It is therefore possible to fit a simple linear model of the form:

$$Y = f(x_1, x_2, ..., x_n) \qquad (1)$$

where:
Y is the dependent variable representing some measure of market participation for the particular enterprise, while the x's are the explanatory variables.

Following convention, the model can be specified as:

$$Y = \beta_0 + \beta_1 X_1 + \beta_2 X_2 + \beta_3 X_3 + ... + \beta_n X_n + \mu_i \qquad (2)$$

where:
β_0 = the intercept or constant term;
$\beta_1, \beta_2, \beta_3, ..., \beta_n$ = slope or regression coefficient;
$X_1, X_2, X_3, ..., X_n$ = explanatory or independent variables;
μ_i = error or disturbance term.

The model was estimated to identify factors affecting the participation of new farmers in formal and informal market channels.

Given the rather large number of variables enumerated, the likelihood of correlation among independent or predictor variables is high. For this reason, the test of multi-colinearity was applied. Assuming two variables, X_1 and $X_{2,}$ co-linearity is suggested if:

$$X_1 = \lambda X_2 \tag{3}$$

However, Equation 2 demands that a more robust function be developed to cater for the several predictor variables in the model. This can be presented as:

$$\lambda_1 X_{1i} + \lambda_2 X_{2i} + \ldots + \lambda_k X_{ki} = 0 \tag{4}$$

where λ_1 are constants and X_1 are the exploratory variables that might be linearly correlated.

The speed with which variances and covariances increase can be seen with the variance-inflating factors (VIF), which shows how the variance of an estimator is inflated by the presence of multi-colinearity. A formal detection tolerance or the variance inflation factor (VIF) for multi-colinearity as illustrated by Gujarati (2003) can be used as follows:

$$VIF = 1 \, / \, tolerance \tag{5}$$

where tolerance = $1\text{-}R^2$

Tolerance of less than 0.21 or 0.10 and/or VIF of 5 or 10 and above indicates multi-colinearity of variables. Where multi-colinearity was detected on the basis of the value of the VIF, the highly collinear variable, that is those with very high VIF, were deleted from the model.

Finally, a test was conducted to detect any possible serial correlation indicated by the size of the Durbin-Watson (DW) statistic by establishing that:

$$\mu_t = \rho \mu_{t\text{-}1} + \varepsilon_t \tag{6}$$

Or that the error terms are not correlated.

11.3.2 Model variables

Socio-economic and technical variables that hinder efficiency in trading of agricultural produce by newly resettled farmers have been analysed. Table 11.3 presents a summary of these variables, their units of measurements, types, and hypothesized relationships with the dependent variable.

Table 11.3. Definition and description of variables employed in analysis.[1]

Variables	Unit	Type of variable	Expected sign(+/-)
Age	actual in years	continuous	-
Gender	male or female	categorical	-
Household size	actual number	continuous	+/-
Educational level	attended formal school or not	categorical	+
Marital status	have family or none	categorical	+
Farm labour	actual number	continuous	-
Land ownership	actual size (ha)	continuous	+
Land satisfaction	satisfied or not	categorical	+
Land fertility	fertile or not	categorical	+
Need extra land	actual ha	continuous	-
Market participation	formal or informal	categorical	-
Market access support	available or not	categorical	-
Market distance	far or near	categorical	-
Marketing revenue	exact amount	continuous	-
Output	actual number	continuous	+
Plough ownership	have or do not have	categorical	-
Scotch cart ownership	have or do not have	categorical	-
Granary ownership	have or do not have	categorical	-
Planter ownership	have or do not have	categorical	-
Road condition	good or bad	categorical	-
Area accessibility	accessible or not accessible	categorical	-
Extension availability	available or not	categorical	+
Source of market information	formal or informal	categorical	+
Market participation before FTLRP	yes or no	categorical	+
Market participation after FTLRP	yes or no	categorical	+

[1] Developed from survey data.

Variables in Table 11.3 were systematically regressed to identify their respective effects on maize market participation by Fast Track Land Reform beneficiaries in Zimbabwe.

The measure of the maize market participation was the extent to which the household head sold produce in the market during the relevant period namely the season preceding the survey season (2007/2008 season). In order to establish the extent of maize market participation post fast track land reform programme in Zimbabwe, a regression analysis was performed with Maize Market Participation (INPMAIZE) as the dependent variable. The results are presented in Table 11.4.

The regression results indicate that age of the household head, land ownership satisfaction, farm labour availability, plough ownership and maize storage method had a strong influence on farmers' participation in formal maize markets, at 1% significance level (Table 11.4). Land fertility and satisfaction with land holding also showed high significance, as a factor affecting market participation by farmers. These results are consistent with the hypothesized expectations. Chirwa (2006) also observed that the extent of maize sales in Malawi is positively associated with increasing land holding sizes, and that those whose maize failed had the lowest land holding size. This is an important finding given the perception that land reform programmes in developing countries are more instruments of political appeasement than addressing economic and market goals.

Table 11.4. Factor variables affecting maize market participation (Survey data, 2009).

Variables	Coefficient (B)	t-value	Significance
Age	-0.28	-9.161	0.000*
Marital status	-1.07	-1.667	0.104
Farm labour	0.19	7.812	0.000*
Employment status	0.032	0.345	0.732
Plough	0.332	2.789	0.008*
Planter	-0.02	-0.207	0.837
Land ownership	-0.039	-0.386	0.702
Land satisfaction	0.798	6.81	0.000*
Need for extra land	0.003	3.56	0.001*
Maize revenue	2.166	0.759	0.453
Maize storage method	-0.728	-7.136	0.000*
Maize price setting Authority	0.55	1.126	0.268
Source of marketing Information	-0.579	-8.427	0.000*
Market participation Post FTLRP	-0.147	-2.254	0.031**
Market participation Before FTLRP	-0.191	-2.525	0.016**
Market distance	-0.075	-0.739	0.465
Market access Support	0.332	3.448	0.001*
Model summary	R Square = 0.933; Adjusted R Square = 0.896; Durbin-Watson = 2.173; F =25.534		

* Shows significance at 1% level; ** shows significance at 5% level.

Another important finding that merits some discussion is the weak effect of market distance in the regression results. The initial assumption was that land reform resulted in a wider dispersion of settlements away from their established markets and business centres. Though this dispersion is real, and most farmers indicated that maize markets were far from their homesteads, they still sacrificed to travel long distances to buy the staple products, including maize. However, the issue of transaction costs becomes important if smallholder farmers are to remain efficient and viable in the market. Whilst consumers travel long distance to procure maize and maize meal for domestic consumption, it might be difficult for farmers to make comparable sacrifice in ferrying their produce to these formal markets for sale, unless strong market incentives like higher prices and cash payment of products on delivery are offered. This gap has been identified and manipulated by informal traders who establish relations with villagers and entrust them with cash to procure maize at low prices, which will later be collected for resale in other lucrative markets especially in urban areas. Farmers are therefore trapped between low prices within their localities amid higher prices on very distant markets associated with higher transactions costs.

Market access, which also goes hand in glove with source of market information by farmers are some factors that the government of Zimbabwe have to consider to boost formal marketing of maize by new farmers. Most farmers indicated that they relied on informal channels to access both the information and the markets. Information is usually obtained through informal conversations among farmers or through national radio stations. Timing of advertising messages of critical farming information is normally a challenge to smallholder farmers, who normally come across such information by chance upon switching on their radios.

11.4 The livestock production and marketing in Zimbabwe

Market liberalization in 1991 and the subsequent growth in private abattoir participation in the domestic market in direct competition with the Cold Storage Company, the successor of the Cold Storage Commission was regarded as a major development in the marketing of beef in Zimbabwe (Sibanda and Khombe, 2006 in Rukuni *et al.*, 2006).The Cold Storage Company, which was set up to cater for the European Union beef market, failed to meet its annual quota despite its monopolistic advantage. Notably, it failed to offer attractive prices to beef producers due to its cost structure. The drastic reduction of the commercial beef herd has resulted in the shortage of beef in the domestic market and insufficient volumes of export quality beef (Sibanda and Khombe, 2006 in Rukuni *et al.*, 2006). Most high-class abattoirs, including the Cold Storage Company, were operating at about 20 percent of their installed capacity during the fast track land reform period (2000-2008) (Sibanda and Khombe 2006 in Rukuni *et al.*, 2006).

By 2003, only 5% of formal beef sales went through CSC and by mid-2007, the government of Zimbabwe announced price controls on beef, and the closing of private abattoirs, with the requirement that all meat be marketed through the CSC (Mavedzenge *et al.*, 2008). Beef price controls resulted in beef shortages on the formal market, while informal market prospered at premium prices (Mavedzenge *et al.*, 2008)

Sibanda and Khombe (2006) also noted that veterinary control systems were weakened because the country lacked capacity to import vaccines, while uncontrolled movement of animals which was allowed when new farmers took up their farms also compromised animal disease control. Between 2001 and 2002, a total of 26 outbreaks of foot and mouth disease were reported, mostly resulting from cattle coming into contact with wildlife (Sibanda and Khombe, 2006). A decrease in volume of beef from the quasi government 'Cold Storage Company' shows effects of government policies on overall performance of the agricultural sector.

11.5 The role of local authorities in livestock markets

Local government authorities have since emerged as the major coordinators of the cattle markets in Mwenezi District of Zimbabwe. It is apparent that the participation of council authorities in commodity marketing has sustained the operations of the council as a revenue base through taxes. As such the local district council plays major roles such as:

- Scheduling cattle market dates by producing annual calendar and circulating it to farmers and potential buyers.
- Contacting potential market participants such as private companies, abattoirs and individual cattle buyers.
- Contacting cattle market stakeholders such as the veterinary department and the Zimbabwe Republic Police Anti Stock Theft unit to attend the auctions for clearance of all stock sales in the district.
- Alerting all stakeholders before auction dates, in case auctions are cancelled by the veterinary department due to notifiable disease outbreak in the area or surrounding regions.
- Registration of all buyers who intend to do business in the district.
- Regulating the buying and selling of cattle in the district.

Mwenezi Rural District council is responsible for organizing both public and private cattle sales. However, for private sales to take place the private players need to notify relevant ward councillors through village headmen, who will in turn notify the council, veterinary and ZRP anti stock theft unit. This should be done two weeks before the private auction takes place. This arrangement is done to curb cattle theft in the District. Table 11.5 summarises the various stakeholders that are active in cattle marketing.

Cattle trading in Zimbabwe is regarded as a lucrative venture by both private and public traders, hence an increase in the number of stakeholders to monitor the process (Table11.5).

Table 11.5. Major participants/stakeholders in the cattle industry (survey data, 2009).

Stakeholder	Reasons for market participation
Resettled farmers (A1 & A2 farmers)	Participate in public markets as buyers and sellers of heifers and steers for restocking purposes
Private abattoirs	Buy livestock for resale to their clients. Some abattoirs do export the meat products
Private butcheries	Most butcheries buy cattle for slaughter for their local market
Supermarkets	Normally buy cattle products from abattoirs or slaughtered livestock directly from farmers
Zimbabwe Republic Police (ZRP)	For verification and clearance of cattle sold at the market to avoid trading of stolen livestock
Veterinary department	Check the health status of all livestock at the auction. Works with the police in clearing all sold cattle
District council	Coordinates public auctions through scheduling dates, registering market participants and contacting market participants before auction dates
Resettled small scale farmers (A1 farmers) and communal farmers	Attend auctions as sellers of cattle to various buyers at the auction. Few of them buy cattle from auction, but instead prefer farmer to farmer trading for restocking purposes

Local district governing authorities were not active in cattle trading, whose involvement early 2000, as a way to generate revenue for the cash-strapped councils. Before 2000, Mwenezi Rural District Councils only received cattle levies annually from farmers within the district, but of late a levy of 7.5% of gross purchases is collected from traders at auctions. The impact of a multiple stakeholder involvement (Table 11.5) on market bureaucracy, efficiency and transactions costs needs further scrutiny. In general, such a structure has a direct negative or positive impact on market prices (Table 11.6), depending on the policies and coordination mechanisms governing such a market structure.

An overview of cattle prices in Mwenezi District showed that farmers were getting less than the prices indicated in Table 11.6. Figures of the public cattle auction conducted on 2nd April, 2009 indicated that the average weight of marketed cattle was 350kg for which an average price of R2 630 was paid. This translates to an average live weight price of R7.51/kg (US$1.00/kg). It is however important to note that the auction was purely dominated by cattle traders and middlemen who procured cattle for sale at the prices depicted on Table 11.6. Retail prices are expected to be considerably higher that the figures in Table 11.6. Farmers were found to lose considerably by not directly sending their livestock to abattoirs.

Table 11.6. Beef prices as at July 2009.

Cattle grades	Bulawayo auctions (US$)	Mwenezi RDC (US$ averages)
Super	3.00	2.80
Choice	2.80	2.60
Commercial	2.60	2.20
Economy	2.30	2.20
Manufacture	2.00	1.10
Average prices (dressed)	2.54/kg	2.18/kg

The belt weighing method for livestock is now dominant on cattle auctions in Mwenezi District of Zimbabwe. This replaced the use of spring scales that were becoming insensitive due to aging and wear and tear. More so, the scales were too heavy to be carried from one auction to the other and as such have always been an inconvenience to the council, which had no transport facilities to move these scales across the district.

The need to understand the real market dynamics and farmers' selection of specific market led to numerous interviews with farmers and market participants (Box 11.1).

Though farmers are not normally encouraged to sell heifers, economic pressure has resulted in farmers selling them at public auctions leaving oxen for their draft power. The economic value of heifers by small scale farmers have always been outweighed by the traditional lifestyles.

Box 1. Farmers' selection of cattle market channel.

An in-depth interview with Mr John (not real name), an officer at Mwenezi Rural District Council in Zimbabwe who has the responsibility to conduct all district cattle auctions revealed important features in the marketing of cattle. His experience with farmers is that, if they could, farmers would prefer not selling their cattle directly to abattoirs despite the high cash returns associated with that marketing channel. This is because most abattoirs pay farmers using cold dressed mass, after their beasts have been slaughtered and assessed for quality and health status. This process leaves farmers with uncertainties that might even result in rejection of their cattle if the carcass is found to be of poor quality or diseased. To avoid such risks, farmers end up selling to middle men who pay them based on visual assessment of animal and live mass for guaranteed income. Selling cattle directly to big abattoirs also means that farmers incur high transport costs which they try to lower by combining loads to do collective transportation of cattle. But this causes problems among small scale farmers who find it difficult to distribute costs or losses due to unforeseen mortalities of cattle on transit due to poor health, accidents and/or theft of cattle on transit.
Interview, Mwenezi District Council, 24 July 2009.

Public auctions are encouraged by major stakeholders to minimize cattle thefts through public exposure of animals, sellers and buyers. Farmers have used auctions as sources of market information by bringing their cattle for weighing and have a feel of their market values without any intentions to dispose them. This prepares such farmers for participation on upcoming auctions.

Cattle market regulation has always been at the centre of both public and private players in the industry. According to government regulation, no cattle slaughtering takes place without clearing from police and veterinary services department, working in conjunction with local authorities such as council, village headmen, and chiefs, relying as well on farmers' dipping records.

11.5.1 Cattle market channels

The cattle market channels in the survey area were assessed. Figure 11.2 sketches the intricate relationships and networks among the different market participants, institutions and outlets involved in cattle trade in the district.

Farmers are exposed to different market channels that include both formal and informal channels (Figure 11.2), where several factors affect channel selection by the farmer. Inefficient pricing system, information asymmetry and poor infrastructure in new settlements post Zimbabwe's fast track land reform programme promoted the emergence of the informal

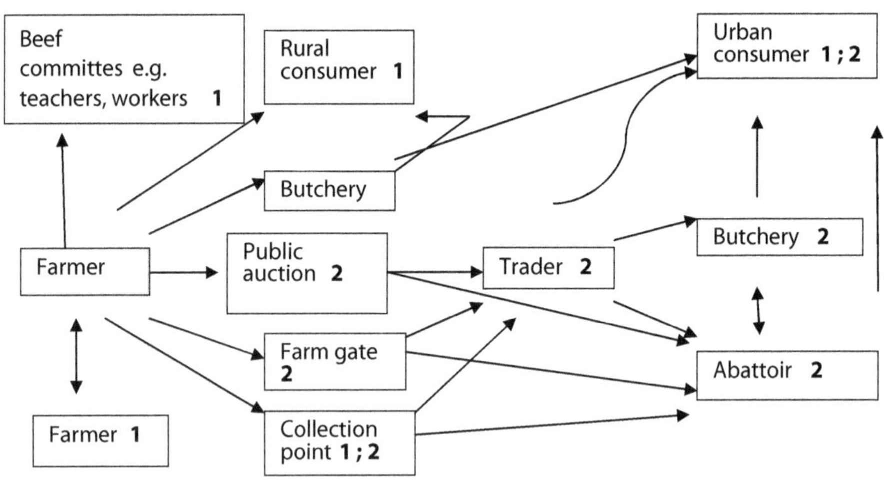

1 = informal market, 2 = formal market

Figure 11.2. Cattle market channels (Developed from survey data, 2009).

cattle sector in favour of the formal sector (Figure 11.2). As in the maize marketing system, the informal cattle trading became increasingly dominated by barter trading and use of foreign currency.

11.5.2 Public auctions

Public auctions remain one of the formal market channels in the cattle industry. Before land reform, farmers used to sell their cattle to abattoirs and companies like Agri-auctions, CC-Sales which are now inactive in the Mwenezi area. The old arrangement saw butcheries buying their cattle products from the abattoirs. Commercial farmers also used to contact their auctions at Bubi Cattle Pens being co-ordinated by CC sales and Agri-Auctions. These seized to participate after the FTLRP. According to one newly resettled commercial farmer:

> *'Some buyers don't participate in auctions for fear of being labelled as supporters of the land reform' (Interview, New Large Scale Farmer, Mwenezi District, 22 July 2009).*

Currently, new buyers have emerged in the market. Prominent among these are Montana meats, Cold Storage Company, Circle Y, Bulawayo Abattoirs, Carswell Meats and small unregistered dealers. Big abattoirs have not been very active in direct cattle purchases in Mwenezi District, but have relied heavily on cattle traders and middlemen.

The FTLRP period has also witnessed the closure of most farm-based small butcheries. At the height of economic downturn in Zimbabwe, Colcom meat processors closed and it's now trading as Montana meats. After the land reform and the decline of Cold Storage Company in Zimbabwe, an estimated sixteen (16) indigenous abattoirs emerged and mostly based in Bulawayo. These have played a significant role in cattle marketing in Mwenezi District, which falls within the procurement zones of Bulawayo.

Sources from AGRITEX and Mwenezi RDC confirmed that no effective cattle sales were contracted in newly resettled farms until 2004. Between 2000 and 2004, there was a lot of uncertainty about the future of the new farmers in their newly acquired farms. Most farmers had therefore not moved their livestock from the communal homelands until 2004, when the Local government authorities (Mwenezi District Council) established buying points and started conducting public cattle auctions in the new farming areas. Though this was a positive move, cattle sales were however halted at certain intervals due to the outbreak of cattle diseases like foot and mouth and anthrax. Uncontrolled cattle movement from old communal areas to new areas was blamed for the increase in disease outbreaks. Longer periods would lapse before auctions could resume due to short supply of cattle vaccines and veterinary personnel to curb the devastating diseases. Although, formal markets were regularly interrupted, informal and illegal cattle trading continued. These were from farmer to farmer or farmer to traders, who would illegally smuggle the cattle to urban areas especially at night for slaughter. Such practices deprived the local authorities' revenue in form of taxes, but more serious was the noncompliance with government guidelines on animal disease control. Participation of farmers in cattle markets post fast track land reform (2008) in Mwenezi district of Zimbabwe has been summarised on Table 11.7, showing both new resettlement and communal areas.

The data displayed in Table 11.7 show that more auction points (totalling nine) are located in new farms as compared to four in the old communal lands of Mwenezi district. Dispersion of the new settlements as a result of Fast Track Land Reform Programme could be the reason for this distribution. More so, more cattle were purchased from the new farmers than the old communal areas in 2008. Various factors including good pastures

Table 11.7. Cattle market statistics in Mwenezi district for 2008 (Compiled from Mwenezi district council auction records for 2008).

Auction points	Jan	Feb	Mar	Apr	May	Jun	Jul	Aug	Sep	Oct	Nov	Dec	Total
New resettlement areas (post fast track areas)													
Chamakudo	0	150	213	250	206	270	381	114	290	287	541	308	3,010
Chovelele	0	270	167	333	478	0	255	201	144	460	616	523	3,447
Chizenge	0	0	0	152	200	269	247	216	250	315	443	561	2,653
Mananga	0	218	396	98	0	244	289	116	261	267	350	693	2,932
Welkom	0	161	0	300	216	0	107	246	0	251	361	469	2,111
Chingwizi	0	300	0	114	200	0	136	160	222	465	728	386	2,711
Stelmaco	0	243	416	213	0	0	239	111	210	444	587	401	2,864
Mucheni	0	197	127	304	0	234	284	126	197	366	405	819	3,059
Muzhanjire	0	113	289	166	259	0	278	323	0	248	293	427	2,396
Total	0	1,652	1,608	1,930	1,559	1,017	2,216	1,613	1,574	3,103	4,324	4,587	25,183
Average	0	184	179	214	173	113	246	179	175	345	480	510	2,798
Old communal areas													
Maranda	0	215	188	200	257	288	315	173	188	516	0	956	3,296
Chemvana	0	489	256	280	0	267	140	154	264	311	613	752	3,526
Chitemere	0	0	250	293	197	0	160	211	257	198	356	280	2,202
Neshuro	0	173	0	278	167	153	127	96	226	0	186	439	1,845
Total	0	877	694	1,051	621	708	742	634	935	1,025	1,155	2,427	10,869
Average	0	219	174	263	155	177	186	159	234	256	289	607	2,717
Combined marketing figures													
Total	0	2,529	2,302	2,981	2,180	1,725	2,958	2,247	2,509	4,128	5,479	7,014	36,052
Average	0	201	176	239	164	145	216	169	204	301	385	558	2,758

in new areas, contributed to this situation. One communal farmer indicated that he is currently left with only six heads of cattle for draught power, and the bulk of his cattle were given to relatives and friends who benefited from the FTLRP for safe keeping and good grazing. In return such recipients would benefit through draught power, milk and manure for the fields. Such relationships between the fast track land reform beneficiaries and their communal counterparts make it difficult to conclude on the independent dominance of land reform beneficiaries' on productivity and marketing of produce. A comparative study of the two land holdings, that is communal and resettlement, will provide more insights into such movement of cattle between the two farming communities.

As noted by Mavedzenge *et al.* (2008) in Marongwe, (2008), households owning cattle averaged 49.2% across communal sites and 52.1% across the A1 sites of Southern Zimbabwe. The study also noted that restocking in the new A1 resettlement schemes was continuing, with more cattle being transferred from communal to resettlement areas. However, the study notes that, as is the situation in communal areas, many herds in the A1 resettlement schemes were still low enough not to allow the putting together of a span of cattle for cultivation. As such, many of the households still depended on hired draught power, like those without cattle at all.

11.5.3 Farmer to farmer trading

This is dominant among small scale farmers, both newly resettled and communal farmers. The channel has been dominated by barter trading especially between 2000 and 2008, where farmers experienced erratic rains and poor harvests and continuing cash constraints. Farmers used their livestock to exchange for food. Some farmers have also used this channel to maintain their breeds in the area by selling their cattle to neighbours.

11.5.4 Cattle rustling

Cattle rustling, or the act of stealing cattle as part of organized raids has been on the increase in Mwenezi District. This practice which has been going on for centuries among pastoralists without any connotation of criminality is now seen as destructive of local economies and has become quite dangerous as cattle owners have devised less traditional means of protecting themselves and their livestock (Eavis, 2002). But it is a sign of how desperate the food situation is that cattle rustling is growing nonetheless. In the Mwenezi District, as in other parts of Zimbabwe, his has resulted in a thriving informal market for cattle. Unofficial sources claim that cattle from Matebeleland North and Bulawayo went to Zambia and Botswana, while those from Mwenezi District were destined for Mozambique. More so, unverified information also alleges that former large scale white farmers were paying people to move their cattle to neighbouring countries soon after the onset of Fast Track land reform programme, which has a negative effect on the size of the national herd.

11.5.5 Cattle levies

All cattle buyers in Mwenezi District are charged 7.5% levy of gross purchases, which is payable to the council. This amount is normally paid on the auction date, but where buyers convene a private auction without the knowledge of the district council, a flat fee of R150/animal is charged to the buyers. An understanding of the impact of the cattle levy will be helpful for formulating policies on cattle marketing. Though cattle levies are paid by the buyer at auction level, the ultimate offer price is affected by the levy. Buyers bid less for cattle so as to reduce the total levies to be paid to the council after the purchases. This directly affects the farmers' revenue by depriving him/her a potentially high bidding price if the buyers were not directly taxed.

11.6 Policy focus

The empirical evidence of the impact of Zimbabwe's Fast Track Land Reform Programme (FTLRP) on maize, cotton and cattle markets and market relationships gives important ground work for future policy formulation in Zimbabwe and other developing countries in similar circumstances. In order to achieve this objective, it became necessary to broaden the analysis to cover variables that could have directly been affected by FTLRP such as market participation, extension availability, market availability, market access support, road condition, market distance and several other socio-economic variables implicated in market access and channel choice. It is clear from the results that Zimbabwe's FTLRP was characterized by high volatility and uncertainties from which immense lessons can be drawn to ensure successful future land reform programs in the country and for countries in the process of reforming their agricultural and economic sectors.

Without a doubt, the FTLRP in Zimbabwe has led to far-reaching changes in markets and market relationships, and has exerted considerable influence on the levels and patterns of agricultural production especially on commercial products like cotton and cattle. Land variables such as ownership, fertility and satisfaction with existing land have been shown to be important in maize, cotton and cattle marketing. It therefore can be argued that land reform, when properly organized, can be an effective instrument for incorporating emerging farmers into profitable commodity markets. But the evidence from the analysis is that land-related assets need to be augmented by infrastructural investments even at household levels. This probably explains the significant effects predicted for such variables as ownership of plough, planters, scotch carts and granaries in terms of farmers' market participation. FTLR beneficiaries among small-scale farmers have significantly invested in these structures but it is clear that government and institutional support is required to boost their already noticeable efforts. Such assets improve productivity and efficiency which leads ultimately to an increase in market participation by farmers.

As Groenwald (2003) has noted, successful land reform does not depend only on principles or special conditions, but depends on the execution and delivery of land reform policies, where people and institutions have to formulate policies, devise means, procedures and administrative bodies to perform the agreed tasks. In as much as land reform is aimed at increasing productivity and people welfare, this goal must not be achieved in isolation to major variables like markets and economic performance of the country. Much of the worldwide criticism of the Zimbabwean FTLRP has been based on such convictions.

The Zimbabwean market post FTLRP was dualistic in nature, with the formal market trailing behind the thriving informal market. Market participation by farmers required extra effort and a certain level of market intelligence/awareness. Expectedly, educational level and age were shown to play a major role in determining farmers' market participation. Though these are critical factors that can even be considered during land reform beneficiary selection, if seriously considered, they distort the essence of the program, which is poverty alleviation and livelihood improvement among the poor and the disadvantaged. Training FTLR beneficiaries through increasing extension visits can be a great step towards integrating small scale farmers in the mainstream economy.

In the light of the foregoing, it is clear that the process of land reform is a complex, if not also complicated, process, and everything cannot be achieved simultaneously, where successful management and execution of the programme is dependent on the clear identification of priorities. Success of land reform programmes will logically be maximised if early efforts are expended on those activities and locations offering the highest probability of success. The land reform process must take into account various institutional, technical, financial and time constraints, in view of the need to achieve the broad social and economic goals of such a program within the shortest desirable period.

References

Chirwa, E.W. (2006). Commercialisation of food crops in rural Malawi: insights from the household survey. Working Paper No. 2006/04.EU. University of Malawi, Chancellor College, Department of Economics.

Eavis, P. (2002). Small arms and light weapons (SALW) in the Horn of Africa and the Great Lakes region: challenges and ways forward. The Brown Journal of World Affairs 9, 251-261.

Fafchamps, M. and E. Gabre-Madhin (2001). Agricultural markets in Benin and Malawi: the operations and performance of traders. The World Bank, Washington, DC, USA.

FAO (2009). FAO Zimbabwe newsletter, February-March 2009. Harare, Zimbabwe.

GOZ, (2009). Second Round Crop and Livestock Assessment report. Ministry of Agriculture, Mechanisation and Irrigation Department. Government of Zimbabwe. 12 February 2009.

Gujarati, D.N. (2003). Basic econometrics. Fourth edition. McGraw-Hill: New York, USA.

Groenwald, J.A. (2003). Conditions for successful land reform in Africa. Paper presented at Pre-IAAE Conference on African Agricultural Economics, Bloemfontein, South Africa.

Jayne, T.S., M. Chisvo., M. Rukuni and P. Masanganise (2006). Zimbabwe's food insecurity paradox: hunger amid potential. In: Rukuni, M., Tawonezvi, P. and Eicher, C. (eds.), Zimbabwe's agricultural revolution revisited. University of Zimbabwe Publications, Harare, Zimbabwe.

Marongwe, N. (2008). Redistributive land reform and poverty reduction in Zimbabwe: a working paper for the research project on 'livelihoods after land reform. Available at: www.larl.org.za/zimbabwe.

Mavedzenge, B.Z., J. Mahanehene, F. Murimbarimba, I. Scoones and W. Wolmer (2008). The dynamics of real markets. cattle in Southern Zimbabwe following land reform. Development Change 39, 611-639.

Rukuni, M. (2006). Revisiting Zimbabwe's agricultural revolution. In: Rukuni, M., Tawonezvi, P., Eicher, C. (eds.), Zimbabwe's agricultural revolution revisited. University of Zimbabwe Publications: Harare, Zimbabwe.

Rukuni, M., P. Tawonezvi and C. Eicher (2006). Zimbabwe's agricultural revolution revisited. University of Zimbabwe Publications: Harare, Zimbabwe.

Sibanda, S. and C.T. Khombe (2006). Livestock research and development. In: Rukuni, M., Tawonezvi, P., Eicher, C. (eds.), Zimbabwe's agricultural revolution revisited. University of Zimbabwe Publications: Harare, Zimbabwe.

12. Livelihoods, institutions and the small farmer

Ajuruchukwu Obi

Abstract

Without a question, it can be concluded that institutions still matter although there is need for restraint to avoid a mindset that institutions are the answer to all the development gaps in developing countries. Nonetheless, the case made for undertaking a comprehensive analysis of the institutional constraints faced by the small farmers in Southern Africa has been justified. The case was made on the grounds that a combination of institutional, climatic and macro-economic constraints seemed to have led to some of the worst food and humanitarian crisis in the region, culminating in drastic food shortages and famines and deepening of poverty levels. It is expected that this will form a basis for recommending alternative approaches to farmer support and other interventions to improve livelihoods and incomes and contribute to poverty alleviation. The collective innovation models that are suited to dealing with the inherent complexities and uncertainties in both the smallholder farming systems of Southern Africa and the policy/development environment are recommended as one way to ensure that increased knowledge about rural conditions as revealed by more robust institutional analysis, translate into practical actions acceptable to all concerned.

12.1 Introduction

The central question posed by this book at the outset was: what are the factors holding back smallholders from producing at levels that would make farming so attractive that they would continue to farm, or at least make a sustained dent on the extreme poverty that characterizes this sector? The foregoing chapters have shown that there are a multitude of factors in the production and marketing environments of the small farmer that can be considered under the broad theme of institutions. The standard definition of 'institutions' which see them as the over-arching environment within which economic actors operate to create wealth, whether private or public, was adopted and their diverse intellectual origins were explored. Whether we see them as 'rules of the game', formal or informal, including norms, or as the enforcement mechanisms and skills that allow for the optimal extraction of benefits from property rights, institutions are an omnibus phenomenon and their effects are felt in intra-household bargaining, farm-level production, in marketing and distribution, as well as in the consumption of farm produce. A useful analysis of institutional constraints will have to be quite broad since all of the elements enumerated above are themselves influenced by a host of other variables which are economic, social, religious, etc.

But understanding what needs to be analyzed does not answer the question as to what needs to be done to develop the smallholder sector and improve the livelihoods. The focus of this chapter will be on that: taking stock of what has been examined in the previous chapters to provide some sort of summary and then drawing a few general conclusions that will help point in the direction of change where needed. The first three sections of this chapter will address the question as to whether institutions are relevant as a summing up of the literature reviews and document analysis that are a feature of studies of this nature. In answering that question, some of the shortcomings as well as strengths of existing approaches to both analyse institutions, and deploy them, for reform will be examined. So, the questions will be whether institutions still matter, the scope of the problem to be addressed by institutional change, and what is being done and how that is being done. These sections will be followed by a presentation of the chapter summaries with analytical discussions of the key findings. The final section takes a look at the future directions of institutional analysis in relation to the implications for smallholder development and provides an opportunity to proffer some recommendations for policy as promised at the beginning of the book.

12.2 Do institutions still matter?

It is appropriate at this point to attempt to draw together all the strands of the foregoing chapters that deliberately expanded the theme of institutional analysis to capture the multiplicity of the arrangements and issues that define the environment of the smallholder farmer in Southern Africa. The initial observations that smallholder farming in Southern Africa has been on the decline remain valid today as they were several years ago when many of the studies reported in this book were designed. The food security situation in the region today is still desperate and many of the countries of the region still have a large part of their populations listed as vulnerable and in need of humanitarian assistance. According to the Southern Africa Regional Food Security Update issued by the Joint United Nations Mission (2010) in January, 2010, Lesotho, Swaziland, Namibia, Mozambique and Zimbabwe seem to be the most at risk. The problem of food insecurity in these countries has been blamed on many factors over the years, including the policy framework in respect to the allocation of resources to the farm sector, how much support small farmers receive from the extension services, the institutional capacity of the extension services themselves in terms of the size of their workforce and how much they know about what needs to be done to support the small farmers, the condition of the infrastructure available, such as the road condition and the facilities available to the farmers. These border on formal rules and the associated characteristics for their enforcement which constitute 'institutions' in the way they are defined (North, 1990, 1994). To the extent that the food security situation remains dire, it is important to analyse these issues.

But there is nothing in the descriptions of these 'institutions' that provides adequate guidelines on how to make a judgement about whether or not they matter. That insight can only come from knowledge about their efficacy in transforming the smallholder sector and causing growth to happen. In order to be in a position to pronounce on the precise role of institutions in the development process and which institutional variables work, to what extent and in what circumstances, it is necessary to assemble sizeable amounts of data over time and analyze them with appropriate methods. The studies reported in this volume made an attempt at filling this gap but they are only limited in terms of their temporal coverage. The variables need to be tracked over time to ascertain their patterns of action and how they can be incorporated into the policy process. Such information exists for developed countries. For now, it suffices to fall back on general theories and draw lessons from developed country experiences and our limited studies and attempt to incorporate these in simulation models as part of policy experiments.

Development practitioners and theorists have particularly found that good policies alone, like good intentions, are hardly sufficient to bring about sustainable and positive change in rural livelihoods. No where is this truer than in South Africa today where the plethora of policies and programmes put in place since 1994 to bring about rapid and drastic transformation of the rural sector have produced little or no improvement in the conditions of the under-resourced black farmers who reside and eke out a hard existence in those areas. Academic researchers and commentators have often argued that this is because rural people need to be assisted and facilitated to access both private and public resources and services that lead to wealth creation. There are issues with service delivery all across the country today and it is convenient to blame the service delivery shortcomings on the lack of implementation capacity of personnel employed in the local municipalities. Quite rightly, rural people, whether farming or otherwise, need institutions that promote transparent governance so that conflicts arising from unclear rules are minimized. But this may be only part of the problem and possibly not the most crucial.

The key to resolving this development impasse possibly lies elsewhere. North (1990) was indeed categorical about the primacy of institutions when he said: 'that institutions affect the performance of economies is hardly controversial. That the differential performance of economies over time is fundamentally influenced by the way institutions evolve is also not controversial'. This seems to leave little room for dissenting views but it still leaves the obligation to find out exactly where that link lies and how it operates. An important question posed by Bardhan (2005) also deserves a definitive answer: 'which ones?' Even if it is accepted as an article of faith that institutions work, given that there are many, one should be interested in identifying which ones work best and when, etc., more so as resources are scarce and there is need for finer targeting and rational use of available resources.

North (1994) himself has thrown what might be seen as a lifeline in this regard. In a short note entitled 'institutions matter', North (1994) set off to disaggregate 'institutions' into its two main components namely, formal rules and informal norms. He then looked at the unique characteristics of each of these different components to see if there is anything in the way they present that can form the basis for linking institutions to national economic development in general and to smallholder development in particular. North (1994) makes the following key points in relation to the two components:

a. While the formal rules can be changed legislatively and instantaneously, the informal norms (which include belief systems, culturally-determined ways of doing things, and habits based on experience) take a long time to change.
b. Both formal rules and informal norms must change for sustained economic performance in the desired direction.
c. When only formal rules change, the resulting system is unstable and meaningful change cannot be realized.

The South African example is perhaps useful at this stage. In 1994, the political system changed in the country as legislative apartheid was overturned to usher in a democratic dispensation. This represented a change in the formal rules of the game. Having waited for a very long time for this change, people are naturally anxious to see change happen in the form of improved living conditions for the majority of the black population whose socioeconomic exclusion was the reason for the long-drawn out struggle. There is a feeling that 16 years is too long a time to wait for these formal rules to have made poverty history. Across the country, local people are rising up in violent protest against 'poor service delivery' and cannot understand why they still live in the shacks, lack electricity and water, and cannot even find enough food to avert starvation.

Nothing makes North's point clearer and more powerful as this situation. While it required only the launch of the constitution of a New South Africa for apartheid to 'disappear', there are still norms and habits and belief systems about which it is difficult to pass a legislation and therefore impossible to outlaw in the short-term. Given the link that has been made earlier between human actions and institutions, it is possible to attempt an explanation of the situation by examining briefly the aspect of the dynamics of inter-racial relationships at the human level in South Africa. White and black South Africans became accustomed to certain unwritten codes-of-conduct in their interactions with one another and those are persisting in much the same way as they were practised during the apartheid era. As is now becoming quite clear, the end of apartheid would not necessarily end that mindset overnight. The 'mental models' which are determined by norms and belief systems shape the choices that people will make. Black parents still send their children to schools where mathematics and science are not taught because all they aspire to is to obtain the 'Matric' certificate or the terminal high school qualification which was recognized from apartheid days as the ultimate 'poverty fighter'. The certificate would enable the recipient to land a supervisor job in a furniture store, at Woolworth or Shoprite. This is what is observed but the underlying dynamics need to be assessed more systematically before definitive statements can be made about the pattern of response of the black population to the incentives currently on offer. Among those who go to University, the vast majority is satisfied with terminating their university education at the Junior Degree level and do not see a need for an Honours degree. This is seems to be the norm, the exception being those who go on to obtain Masters and PhDs and aspire to, and actually do, rise to the top. But this is a frustratingly small minority. A number of questions immediately come to mind, principally: how can mindsets change fast enough for the majority of high school leavers to move into the science and technology disciplines and commit to a long-term involvement with academia that is needed to develop strong skills and competence to bridge the capacity gaps? There is no shortage of scholarships and bursaries and international cooperation agreements in pursuit of this goal, but it may well be that the fundamental problems are not yet properly defined. Again, it is not know for certain if this observation represents the product of free choice or rather some inevitability as a consequence of conditions that are deliberately manipulated to achieve those outcomes. Whichever way, these are institutional questions that call for deeper scrutiny.

This works much the same way in farming when a smallholder who has access to a large piece of land actually settles for a small garden around the homestead. Orthodox neo-classical economics teaches that a farm unit will continue to deploy variable productive resources to fixed factors such as land as long as the marginal product is rising. This relates to the concept of allocative efficiency. But this does not happen in environments where the 'mental model' does not make provision for surplus production. The farmer would invariably stop producing even when it would have made sense to continue. Researchers and policy makers do not understand why the poor rural people cannot bring more land under cultivation to expand their agricultural production. Even with numerous credit schemes available, the majority of the resource-poor people do not consider that they qualify and do not even apply. The former Minister of Agriculture, Ms Lulu Xingwana, once decried the existence of uncultivated land which she described as 'dead assets in the hands of the poor'. In the Eastern Cape Province, government is trying to change this by implementing the Massive Food Production Programme, but both the interest and the coverage are limited (Dirwayi, 2010). Though only anecdotal at this stage, it has been said that a sizeable number of rural people prefer the social grants they receive for old age, child, disability, etc. Researchers often try to explain the low demand for institutional credit and uptake of improved technology in terms of insufficient information available to remotely-located rural people. But there does not seem to be any difficulty for the rural people in obtaining information about the various social grants. How do they know about the social grants and yet lack information about production credits and farm support schemes. Could it be in the way the message is packaged that makes them feel excluded from productive credits? There are obviously more to this than meets the eyes and we need more information before we can conclude one way or another. These are clearly institutional questions that need to be understood and addressed.

In attempting an explanation of the sizeable welfare and income differences between rich and poor countries, Rodrik and Subramanian (2003) short-listed three key determinants, namely geography, integration, and institutions. On the basis of an innovative empirical assessment involving the use of instrumental variables approach, they confirmed 'the primacy of institutions' and showed that the quality of institutions is consistently a positive and significant determinant of income levels (Rodrik and Subramanian, 2003). In his own contribution to the debate, Acemoglu (2003) employed a historical approach to isolate all but two 'candidates', namely geography and institutions, as the main reasons countries differ in terms of the average incomes commanded by their citizens. Roe (2003) and Dorward *et al.* (2005) focused attention on markets as institutions that contribute to economic development through the facilitation of exchange and coordination functions in a market economy. More recently, Van Huylenbroeck and Espinel (2007) have confirmed the importance of institutions and governance structures for market access, drawing from African examples, among others. It is indisputable that institutions matter. There is also hardly any doubt that differences between countries in terms of growth experience have a lot to do with the quality of the institutions that have emerged in response to new economic opportunities. With the awarding of the 2009 Nobel Economics prizes to works in economic governance and two of its most avid apostles in the persons of Elinor Ostrom and Oliver Williamson (Nobel Memorial Prize in Economic Sciences, 2009), there is no question that the endorsement of institutions and their role in economic development is universal.

12.3 How much is too much?

But, as with many other effective remedies, it is easy to build up dependence and this is already happening in respect to the appeal to institutions to explain anything that poses some confusion in the development context. Today, a number of analysts and commentators on development have come to see a growing trend towards what is disparagingly termed 'institutional fundamentalism' which can be interpreted as a tendency to see the fingers of 'institutions' in every development 'pie' (Sachs, 2003; Rodrik, 2006; Bulte, 2008). But, as the chapters in this book show, there is a need to understand the specific role of institutions in any particular context and their multi-dimensional character and how they affect different aspects of household and communal life, be it in social, economic, or political terms.

This chapter will re-visit the basic problem context and the justification for engaging in these enquiries and then summarize the key results in respect to the effect of institutional factors on the operations and development of the smallholder sector. The different dimensions of these effects will be examined and discussed and indications of future directions of institutional analysis of smallholder agriculture presented.

12.4 How big is the problem and where are we towards addressing it?

On the specific question of the development challenges that this book confronted, there is no question that all relevant bodies and practitioners within the region and beyond continue to agonize about the serious problem of agricultural underdevelopment, especially in respect to the circumstances of the small-scale farmer. The paradox that this class of farmers who constitute about 70% of the population, maintain households that grow at 3% per annum, and account for the bulk of rural employment, are unable to generate enough surplus produce to be competitive, is of immense concern. Even if dependence on food aid and emergency relief has lessened somewhat as a generalized response, all the countries in the region still contain pockets of extreme destitution which belie the progress many are making on the level of macro-economic management. Although this situation may seem more desperate in some countries and much more dramatic, no country in the region is completely free from the scourge of rural blighting that is associated with the high rates of poverty, both relative and absolute, that has become endemic. Even in relatively affluent South Africa, many villages present a picture of areas through which a violent hurricane has just passed.

These pressures have obviously been exacerbated by the dramatic developments that have taken place in the agro-food industry in the past few decades. Market reforms that were embarked upon by several governments in the region, as elsewhere in the developing world, have joined forces with globalization to create new relationships in the production and marketing of basic commodities with far-reaching consequences for household welfare. Some of the key challenges that have been identified as being linked to these developments include: changing roles of smallholder sector, declining fortunes of traditional agriculture in the face of stiffer competition both internally and externally, emergence of unorthodox but innovative exchange arrangements, new relationships between the private and public sectors, and different negotiating platforms and procedures at the global scene.

If at least one example can be given from contemporary experience, the recent conflictual debates on climate change that have drawn in new players and new issues as water security is identified as the nexus between energy security and food security and is manifested most glaringly in the desperation exhibited by the poor who is disadvantaged on both fronts (United Nations -COP15, 2009 and North Atlantic Treaty Organization, 2009). There are also much broader implications for national and international security if differential impacts of climate change on contiguous communities and territories give rise to responses and coping strategies that lead to conflict. In at least one case in northern Kenya, prolonged droughts that led to high stock mortality has been associated with growing gun running among local communities and incessant outbreaks of violent conflicts (Conservation Development Centre-International Institute for Sustainable Development-Saferworld, 2009). According to the Secretary-General of the North Atlantic Treaty Organization (NATO), such a scenario can easily be sketched in respect to national concerns over resource constraints with significant cross-border implications. It is not difficult to understand this in the context of a region such as southern Africa where ethnic groupings have rarely been discouraged by national boundaries to claim identities and shared ancestry. We see a lot of these cross-border relationships in southern Africa, as in other parts of Africa (Brown and McLeman, 2009; Cilliers, 2009). Among the actions prescribed by Conservation Development Centre-International Institute for Sustainable Development-Saferworld (2009) to avoid or minimize these, the strengthening of customary governance institutions rank high. And it is not only in the resource-based conflicts

When the poor is in the developing country, especially one in Africa, the matter becomes even more complicated because in many cases the institutions are either non-existent or are poorly developed and therefore ineffective. From concern over the possibility (and indeed, reality) of market failures which became ready justification for launching the structural adjustment programmes (SAPs) in the 1980s, many of these countries began to experience missing markets. The situation described by Stiglitz (2002) in relation to reform era Russia, in which the lack of the necessary 'institutional infrastructure' compromised the effectiveness of existing institutions, is also true of contemporary Africa; calling something an 'institution' does not automatically imbue it with the capabilities that similarly named arrangements in the developed world have. Whether or not the 'institutional infrastructure' exists makes all the difference. These and other obstacles remain palpable in Africa, and more so in parts of the continent such as Southern Africa where remote and recent history combine to complicate the picture even more.

Expectedly for post-apartheid South Africa, dissatisfaction is growing as the populace confronts a different reality from the lofty expectations at the dawn of democracy in 1994. The problem has become so large that even the Government has acknowledged the emergence of a '...*second economy* characterized by poverty, inadequate shelter, uncertain incomes and the despair of joblessness...' in which many South Africans are 'trapped' (Government of South Africa, 2004). The service delivery protests that rocked South Africa in recent years bear testimony to loss of patience among the people. In the former 'independent homelands' where services had been rather basic in the best of times, things have become worse in the face of a clear lack of capacity to deliver the simplest development outcomes to a people that have suffered deprivations for too long.

The new administration that came into power in 2009 in South Africa has gone a step further than its predecessors in recognizing the role of institutional innovations in addressing this undoubtedly potentially explosive situation. New government departments that more explicitly focus on human settlements, rural development, land reform, and governance have been created as more immediate and innovative approaches to capacitate the local population, create jobs, expand income-earning opportunities, and generally stimulate the local economy. Given these clear indications of the diversity and multi-disciplinary nature of the problems of the Southern African agricultural scene, the focus of this book was deliberately expanded to capture the multiplicity of the concerns of a diverse stakeholder community seeking deeper understanding of the complexities involved and how best to address them.

12.5 Rural livelihoods dynamics

As indicated in Chapter 1, the intention was to present the smallholder setting as it currently exists to provide a baseline picture against which the subsequent analyses can be evaluated. This process led to a review of the current pattern of livelihoods pursuits and official engagement. It is clear from the analysis that South Africa's post-apartheid economy has maintained the same mindset that agriculture is the dominant occupation in the rural areas and therefore continues to invest in agricultural development while all available evidence points in a different direction. The persistence of poverty among households despite the considerable expansion of investments in agricultural development might suggest that the core livelihoods options have been misspecified by officialdom. There is increasing realization among researchers that agriculture is only one part of what rural people do for a living. Chapters 3 and 4 pick up this question by means of household and community level data based on which simultaneous equations were developed to estimate the determinants of both poverty and income from activities practised in representative rural areas. The observation that wage employment and agricultural employment seemed equally matched is an important one from the point of view of resource allocation for rural development. The estimation of inequality carried out by means of Lorenz curves and Gini decomposition technique revealed the same highly skewed income distribution observed elsewhere in South Africa and the rest of the region. It is important to pay attention to the key determinants of activity choice such as level of education, labour availability, social capital, physical capital and geographical location. To improve rural income and reduce poverty, urgent improvements in education, capital resources, infrastructure and communication are necessary.

12.6 Dynamics of marketing channel choice

Chapter 5 focused on the central theme of the book and examined the range of institutional challenges smallholders confront in produce marketing. In turn, these challenges influence their selection of marketing channels. By means of a multinomial logistic regression model, the influence of these factors was tested and it was clear that access to market information, expertise on grades and standards, availability of contractual agreements, existence of extensive social capital, group participation and reliance on traditions remain key considerations. The Kat River Valley exemplifies other communal areas of South Africa and the region where interactions between modern systems and traditional practices work to the disadvantage of the smallholder with limited means to face up to the challenges of modern market participation.

12.7 Institutional innovations in research and community outreach

The University of Fort Hare is among the 'black institutions...in outlying, often rural areas, 'disarticulated' from the economic heartland of industrial South Africa' (Swartz, 2005). For most of these institutions, linkage arrangements that enhance their relevance to their catchment areas are quite limited if not non-existent. To that extent, the innovative activities related to the development of the community outreach programme around the indigenous Nguni cattle represents a significant break with tradition. The efforts around the Nguni cattle also endorse contemporary intellectual views regarding the importance of market access. As Jooste and Van Rooyen (1996) concluded, the transition of the small scale sector towards commercial production will ultimately be determined by its access to markets where produce can be marketed profitably to serve as incentives for output expansion. According to Stroebel (2004), several constraints affect the efficient marketing of livestock in the Eastern Cape Province of South Africa, chief of which are related to market availability and high transactions costs. Currently there is poor response of consistent cattle supplies to existing market outlets by small-scale farmers on communal lands (USAID, 2003). It is often convenient to explain low off-take rates in traditional livestock production in terms of customs and traditions, but this is clearly simplistic and tends to uphold the stereotype of the tradition-bound peasant who is unresponsive to economic incentives.

The empirical tests carried out in Chapter 6 suggest that the severe marketing constraints faced by small-scale cattle farmers are the reasons they sell few of their cattle. Factors such as market availability, market distance, transport availability, transport affordability, information access, current market price and condition of cattle, herd size, method of payment, household size, gender, age and level of education of head of household are all linked to the institutional architecture and were examined. The indication is that addressing marketing constraints will enhance off-take rates thereby increasing household income earnings and standards of living of the beneficiaries of the University of Fort Hare livestock improvement programme. There is immense scope for a long-term adaptive and collaborative research on these issues to integrate natural science know-how with socio-economic analysis to gain deeper insights into the dynamics of the traditional livestock economy. It is also important to identify areas of flexibility in the system where science can help to enhance livelihoods in an environment that still remains depressed many years after democratic rule was enthroned in South Africa. Designing such a collaboration and ensuring that it works effectively is an important step towards improving rural livelihoods.

12.8 Impact of trade regulation

Maize production in Swaziland has been falling in recent years, forcing the country to increasingly rely on imports from neighbouring South Africa. Weather patterns that have featured unrelenting droughts in the country, as in the other Southern African countries, have contributed in no small way to this situation. But it is also considered that excessive regulation of the maize market by the statutory body, the National Maize Corporation (NMC) has played a major role as an institutional constraint to the development of the smallholder sector in that land-locked country. For instance, the NMC has the sole responsibility to anticipate domestic maize demand based on which imports are programmed. In turn, importation of maize can only be done upon receipt of an import license which is not only costly but also involves very cumbersome processes.

The study established that the maize industry of Swaziland is uncompetitive and that this situation gives rise to the serious distortions in the maize market. It was also observed that the system features high degree of efficiency losses as a result of the misallocation of resources in the economy. Prospects for any improvement look quite bleak. In 2009, the maize production deficit remains high despite improvements in the rainfall levels and timing. The chapter concludes by calling for a full deregulation of the maize market and the dismantling of the statutory structures that currently manage the system. Implications for effective food chain management are highlighted.

12.9 Land and natural resource management

The sub-theme of land and natural resource management is a crucial one in a discussion of institutions and the African farmer. Land tenure systems relating to ownership patterns and whether or not land can be alienated and under what conditions merely reflect the significance attached to this resource by different societies. As can be seen quite easily, the more land serves other than productive purposes, the less easily is alienation or sale permitted. It is often the case that elaborate systems and mechanisms are created for controlling the asset and ensuring that it remains within the society that owns it. It is a question of group identity and survival. In fact, in certain societies, it is so important that the traditional power élite has to hold it in trust for the rest of the community and no person can have other than usufructuary rights of an asset that virtually represents the soul of the society.

In South Africa, land remains an emotive issue. Commentators and researchers attribute this to the history of land as a tool in the hands of settlers for creating economic and social dichotomies in the country rather than its contribution to gross national output (Bundy, 1987; Mabin, 1991; Adams *et al.*, 2000, among others). Frost (1998) observes that to the South African black, land has both territorial significance as well as symbolic power that is intimately linked to their very identity as a people. Hence, the bitterness about forced removals from their land, an action that was seen as a symbolic erasure of black identity and insights (Frost, 1998). According to the Reconstruction and Development Programme document (RDP), land is a 'basic need' of the people of South Africa (African National Congress, 1994; Machete, 1995). The discussions about how to develop the smallholder in the communal areas will therefore always start and end around land issues. To a large extent, such discussions are important in other traditional societies in the region such as Lesotho and Swaziland. Quite recently, such issues about the ownership of land were at the centre of a protracted legal battle between the Government of Botswana and the indigenous peoples of that country.

Chapter 9 examines the institutional factors governing the existing patterns of communal rangelands utilization in the Eastern Cape Province of South Africa. The discussion covered various institutional factors particularly those relevant to natural resource management at the communal level. Informal institutions such as customary laws and traditional institutions seem to have gone out of fashion in many places while in a number of communities indigenous institutions still operate but competitions between them and State structures have resulted in conflicts and confusion over definition, responsibilities, management procedures and enforcement mechanisms. It was necessary therefore to include in this book a substantive discussion on these various dimensions to provide a basis for future work on the role of collective action and how collective innovation can contribute in alleviating poverty within the communal areas of South Africa and elsewhere. A key finding is that lack of government support to communities has resulted in the poor management of resources. Elements of the classical tragedy of the commons are embedded in the three cases examined and these will be explored in greater depth on follow-up projects. The consequence of this has been deterioration of the rangelands and impoverishment of the communities. The indication is that transaction costs and good governance structures are important in the common property institution and high levels of trust and cooperation between members are necessary to reduce these transaction costs. Furthermore, collective action on communal lands enables communities to share ideas on how best to manage their resources to ensure sustainability.

12.10 Changes in markets and marketing under institutional pressures

As a result of poor planning and inappropriate procedures for beneficiary selection, the land reform programme in Zimbabwe resulted in drastic decline in production and productivity in the agricultural sector. At the same time, the perceived political agenda of the process sparked off a reaction in the investment climate that manifested in worsening balance of payments, reduction in industrial production, and growing unemployment. It is safe to conclude that the Zimbabwean land reform programme resulted in an economy-wide decline in effective demand occurring simultaneously with a sharp fall in physical agricultural output as appropriated farms lay idle and poor farm practices depressed productivity levels. A low-equilibrium trap thus ensued, culminating in the deterioration of livelihoods across the broad spectrum of society as a hyper-inflationary situation developed and assumed scandalous proportions.

A situation such as this set off chain reactions that touched all segments of the economy, including the disappearance of markets, the emergence of informal exchange arrangements that represent adjustments to the aberrant phenomenon of hyper-inflation emerging in the absence of both cash and products. To date, there has been systematic assessment of this dimension of the land reform programme with the aim of drawing lessons that can provide some basis for developing strategies for revamping the economy now that clear signs are emerging that the long-running political crisis might soon end. To understand what might have happened an initial broad appraisal of specific indicators of agricultural output and prices for the livestock and maize products has been conducted in one district.

12.11 Future directions for institutional analysis

As the World Bank (2008) noted in its *World Development Report 2008*, institutional innovation is still a 'work-in-progress'. This has long been recognized as the profession grapples with the challenge of dealing decisively with developing country hunger. But there are still many disagreements about definition and methodology which may stand in the way of gaining deeper understanding about the workings of institutions and how they influence income growth among smallholder farmers. There is obviously a lot more empirical work done around the conception of institutions as market-enhancing arrangements and mechanisms than in their other roles and functions (Rodrik, 2003).

But Rodrik's (2003) suggestions to address this short-coming ends up recommending even more market-related enquiries. This is probably because traditional academic literature and the contemporary thematic concerns of development organisations have reached a new consensus on the efficacy of market access which is often seen as being more about marketing and prices than anything else. This is also probably why it is fashionable to focus recommendations for improving market access around the need to remove trade barriers such as tariffs and quotas, etc. (Partnerships for Sustainable Development, 2004). In reality, because product quality has become an important issue these days, especially with growing concerns about food safety and the role food plays in human health, market access is as much a production issue as it is a marketing one and both production and marketing environments need to be considered in an analysis of market access. The crucial role that the market access concept plays in the extent to which small producers are able to sell produce at a profit is therefore worth examining in a systematic manner as a guide to more viable policy responses. Over and above all these, it is also important that institutions are in place to enhance the capacity of the beneficiaries of the increased income to cope effectively with change, often through collective action and other mechanisms that have evolved in response to the increasing complexity of society (Eigen-Zucchi *et al.*, 2003).

On the basis of the work of Ruttan (1984), Hayami and Ruttan (1985) and several others, it is clear that countries that fail to undertake institutional innovation are eventually penalized heavily in terms of forgone growth opportunities. The consensus is that institutions make it possible for new knowledge to be deployed towards influencing the pace and direction of technical change needed for agriculture to grow and enhance the welfare of those employed in it. It is therefore considered politically expedient to adopt policies that create opportunities to establish and sustain institutions that support rural development. To this end, four groups of end-uses for institutions and information about have been identified by researchers and other practitioners, have been guided by four principles as follows:

1. Using institutions to implement a net transfer of assets to small producers that provides an incentive for increased productivity (Delgado, 1999).
2. Using institutions to deliver relevant technical, management and market information that improve the knowledge and skills available to producers. Many economists have advocated the convenient route of collective action to enhance social learning as a means for empowering small producers in terms of knowledge, skills and assets and dealing with market failures (Hagedorn, 1992; Becker, 1983). In that regard, the benefits of farmers' organizations have been given a lot of attention in the literature (Carney and Van Rooyen, 1996).
3. Employing institutions for purposes of risk sharing and provision of agricultural services and inputs to small farmers.
4. Using institutions as a means to enhance access to profitable markets for small holders.

An important future direction for institutional analysis in southern Africa has to do with the changing environment in which development takes place today. Taking South Africa as a test case, there is no denying the significant progress that has been made at the political level and how far the country has come from its apartheid past. The development scene of today is characterized by increased complexity and much greater openness than before. While South Africa's agriculture continues to feature high levels of unemployment and underemployment, poverty and inequitable distribution of assets, the relevant institutions have been undergoing constant and rapid change as new political priorities call for adaptation in established institutions (ICRA, 2007).

Without question, this is a carryover from past approaches which did not put adequate emphasis on the strengthening of institutions that serve smallholders in South Africa. The present regime in the country has to contend with agricultural support institutions that lack the personnel with the appropriate rural development perspective to be of much use to resource-poor farmers. For instance, many Local Municipalities do not have qualified agricultural specialists on their staff. In many cases, rural and agricultural development matters are bunched under a single portfolio managed by a Local Economic Development (LED) unit that lacks the necessary rural orientation to tackle small farmer problems. Thus, the important policy shifts of the past few years and the substantial investments to tackle rural development and poverty reduction issues are not achieving the desired results.

The foregoing realities call for long-term engagement of research, extension and training expertise. A systems approach is needed which will allow for investments in all the relevant aspects of agricultural development to be coordinated simultaneously. This approach recognizes that higher education can play a strong role in agricultural development if it is linked to well-focused, development-oriented research and extension. One of the most significant shortcomings of the educational system is that the current curriculum for training of agricultural specialists does not equip them adequately to function effectively in the kind of participatory contexts that are necessary to engage the local farmers and tap the rich indigenous knowledge they possess (Chitsike *et al.*, 2009). This is what collective innovation entails. The current rural development context of Southern Africa, or at least South Africa, makes this imperative. Since the mid-1980s, agricultural development literature has given prominence to the idea of integrating research, extension and training within a function system that ensures that the curriculum is made more relevant to the needs of the beneficiary communities. Such an idea has been promoted under various labels such as 'Agricultural Knowledge System', 'Agricultural-Knowledge Information System', and 'Agricultural Knowledge Triangle' (Eicher, 1999).

For many years, the International Centre for development-oriented Research in Agriculture (ICRA) based in the Netherlands, has promoted the idea of collective innovation as part of the concept of 'Agricultural Research for Development' (ARD). In 2004, the South African government invited the International Centre for development-oriented Research in Agriculture (ICRA), to provide capacity building through the Agricultural Research Council of South Africa (ARC) to institutionalize the concept through simultaneously developing curricula and training teaching staff to deliver more appropriate material in 5 participating Agricultural Faculties in 5 Universities, namely the University of Fort Hare, the University of Limpopo, University of Venda, the University of Free State, and the University of Kwazulu-Natal. The first 4-year phase of this support has ended in 2009 and a new phase focusing on curriculum development will commence in 2010.

The questions that could be posed in this respect are: 'Why collective innovation as part of Agricultural Education and Training?' and 'How does it contribute to speeding up the pace of agrarian reform in the region, or specifically South Africa?'

Some of the most frequent arguments around which the ARD proponents have tried to market it include the fact that it is crucial to:

- Integrating the perspectives, knowledge and actions of different stakeholders around a common theme.
- Integrating the learning by stakeholders from working together.
- Integrating analysis, action and change across the different (environmental, social, economic) 'dimensions' of development.

- Integrating analysis, action and change at different levels of spatial, economic and social organization.
- Replacing patron-client relationship with cooperation and collaborative partnership.

In the light of the foregoing, applying the ARD approach to institutional analysis in South Africa's second economy, namely the rural and under-developed parts of the country where conditions are similar to what obtains elsewhere in the region and the continent, promises to provide a much stronger platform on which to build the institutional base that will serve the needs of the resource-poor smallholder. But that is not the end of the matter. Analysis will increase the stock of information available to us about the enormity of the problem and how best to tackle it, but will not do the job for us. When we know enough about the root causes of the slow progress towards transformation in society, the challenge will be to identify appropriate strategies to realize sustained change and be able to secure the required buy-in to ensure that we succeed. It must be possible to implement the needed practical actions in an atmosphere that effectively responds to real grassroots concerns. Finding a way to do this will provide the answer to the second question as to how collective action can help to speed up the pace of agrarian reform in South Africa.

References

Acemoglu, D. (2003). Root causes – a historical approach to assessing the role of institutions in economic development. Finance and Development 40, 27-30.

Adams, M., B. Cousins and M. Siyabulela (2000). Land tenure and economic development in rural South Africa: constraints and opportunities. In: B. Cousins (ed.) At the crossroads: land and agrarian reform in South Africa into the 21st Century. Braamfontein, RSA: PLAAS and NLC, pp. 111-126.

African National Congress (1994). The reconstruction and development programme. Pretoria, RSA: the African National Congress (ANC).

Bardhan, P.K. (2005). Institutions matter, but which ones? Economics of Transition 13, 499-532.

Becker, G.S. (1983). A theory of competition among pressure groups for political influence. Quarterly Journal of Economics 98, 371-400.

Brown, O. and R. McLeman (2009). A recurring anarchy? The emergence of climate change as a threat to international peace and security. Conflict, Security & Development 9, 289-305.

Bulte, E. (2008). Institutions fundamentalism – In-sti-tu-tions. The Broker: connecting worlds of knowledge, http://www.thebrokeronline.eu/en/layout/set/print/articles/In-sti-tu-tions.

Bundy, C. (1987). Land and liberation, popular rural protest, and the national liberation movement. In: S. Marks and S. Trepido (eds.) The politics of race, class and nationalism in the twentieth century South Africa. Essex: Longman Group.

Carney, D. and C.J. Van Rooyen (1996). Empowering small farmers through collective action: the case of technology development and transfer. Agrekon 35, 211-219.

CDC/IISD/SAFERWORLD (2009). Climate change and conflict – lessons from community conservancies in northern Kenya. Conservation development centre-international institute for sustainable development-saferworld. Available at: http://www.iisd.org/pdf/2009/climate_change_conflict_kenya.pdf.

Chitsike, C., T. Ngcobo, A. Obi, S. Dube, D. Norris, N. Raidimi, and D. Enserink (2007). Aligning teaching & learning to development challenges: institutionalization of ard in agricultural schools of five South African universities. Paper presented at the ARD Symposium on 'Enabling collective innovation in agrarian research, development and education – lessons from the last decade' 28-29 October 2009, Pretoria, RSA.

Cilliers, J. (2009). Climate change, population pressure and conflict in Africa. ISS Paper 178, Institute for Security Studies, Pretoria, RSA.

Delgado, C. (1999). Sources of growth in smallholder agriculture in sub-Saharan Africa: the role of vertical integration of smallholders with processors and marketers of high-value added items. Agrekon 38 (Special Issue), 165-189.

Dirwayi, T.P. (2010). Application of the sustainable livelihoods framework to the analysis of the provincial growth and development plan of the Eastern Cape Province – A case study of the massive food production programme in Nkonkobe and Buffalo City municipalities. Unpublished MSc Agric Dissertation. University of Fort Hare, RSA.

Dorward, A., J. Kydd, J. Morrison and C. Poulton (2005). Institutions, markets, and economic development: linking development policy to theory and praxis. Development and Change 36, 1-25.

Eicher, C.K. (1999). Institutions and the African farmer. Third Economist Distinguished Lecture. CIMMYT, Mexico D.F. 60 pp.

Eigen-Zucchi, C., G.S. Eskeland, A and Z. Shalizi (2003). Institutions needed for more than growth. Finance and Development 40 (2), 42-43.

Frost, B. (1998). Struggling to forgive – Nelson Mandela and South Africa's search for reconciliation. London, UK: HarperCollins Publishers, pp. 173-206.

Government of South Africa (2004). Budget speech to parliament by Mr Trevor Manuel, Minister of Finance. Cape Town, South Africa.

Hagedorn, K. (1992). The impact of institutional particularities on agricultural policy. In: C. Csaki, T.J. Dams, D. Metzger and J. Van Zyl (eds.) Agricultural restructuring in Southern Africa, IAAE/Agrecona. Windhoek, Namibia.

Hayami, Y. and V. Ruttan (1985). Agricultural development: an international perspective. Baltimore, USA: Johns Hopkins University Press.

ICRA (2007). Agro-enterprises and the small farmer. ARD Learning Cycle, International Centre for development-oriented Research in Agriculture, Wageningen, the Netherlands.

Joint United Nations Mission (2010). Southern Africa regional food security update. United Nations Resident Coordinator's Office, Pretoria, RSA.

Jooste, A. and C.J. Van Rooyen (1996). Constraints to small farmer development in the red meat industry. Red Meat Producers Conference, Roodevallei, Pretoria, 27 March 1996.

Mabin, A. (1991). The impact of apartheid on rural areas of South Africa. Antipode 23, 33-46.

Machethe, C.L. (1995). Approaches to rural development in the Third World: lessons for South Africa', MSc dissertation submitted to the Department of Agricultural Economics, Michigan State University, USA.

Nobel Memorial Prize in Economic Sciences (2009). Laureates. Available at: http://nobelprize.org.

North Atlantic Treaty Organisation (2009). NATO Chief warns of climate change security risks. News Briefing.

North, D. (1990). Institutions, institutional change, and economic performance. Cambridge, UK: Cambridge University Press.

North, D. (1994). Institutions matter. Economics history working paper 9411004. Economics working paper archive, Washington University, St. Louis, USA.

Partnerships for Sustainable Development (2004). Market access through meeting quality standards for food and agricultural products. Information sheet, Available at: http://webapps01.un.org/dsd/partnerships/public/partnerships/248.html.

Rodrik, D. and A. Subramanian (2003). The primacy of institutions – and what this does and does not mean. Finance and Development 40 (2), 31-34.

Rodrik, D. (2006). Goodbye Washington consensus, hello Washington confusion? Journal of Economic Literature 44, 973-987.

Roe, T. (2003). Markets, trade and the role of institutions in African development. Paper Presented at the Pre-IAAE Conference, 13-14 August, 2003, at the President Hotel, Bloemfontein, South Africa.

Ruttan, V. (1984). Models of agricultural development. In: C.K. Eicher and J.M. Staatz (eds.) Agricultural development in the Third World. Baltimore, USA: The Johns Hopkins University Press.

Sachs, J.D. (2003). Institutions matter, but not for everything. Finance and Development, Volume 40 (2), 38-41.

Stiglitz, J. (2002). Globalization and its discontents. London, UK: Penguin Books.

Stroebel, A. (2004). Socio-economic complexities of smallholder resource-poor ruminant livestock production systems in Sub-Saharan Africa. PhD Thesis, University of Free State, RSA.

Swartz, D. (2005). New pathways to sustainability: African universities in search of meaning in a globalizing world. University of Fort Hare Professorial Lecture Series 26 July 2005.

United Nations (2009). COP15 – United Nations conference on climate change. Copenhagen, Denmark.

USAID (2003). Agri-link II Project 2003. Monthly progress report #22, South Africa.

Van Huylenbroeck, G. and R.L. Espinel (2007). Importance of institutions and governance structures for market access and protection of property rights of small farmers in developing countries. In: E. Bulte and R. Ruben (eds.) Development economics between markets and institutions – incentives for growth, food security and sustainable use of the environment. Wageningen, Wageningen Academic Publishers, the Netherlands.

World Bank (2008). Supporting smallholder competitiveness through institutional innovation. In: The World Development Report 2008, The International Bank for Reconstruction and Development/The World Bank, Washington, DC, USA.

Appendix - The methodologies of the studies

A.1 Introduction

The three component studies implemented under the FIRCOP Project No. 6 were framed around the central goals of defining and understanding the sampling frame and identifying the key institutional constraints to smallholder market access. One study was therefore designed in such a way as to determine the baseline status of the sample in terms of occupational and employment patterns and livelihoods strategies. The results of that study were expected to facilitate the determination of the design parameters for the constraints studies. The study on the welfare and incentive effects of maize market deregulation was conducted using the partial equilibrium model based on the Policy Analysis Matrix (PAM). The Natural Resource Management study employed the Institutional Analysis and Development (IAD) framework and this has already been discussed in the relevant chapter of the book. These various methods relevant to the other chapters are described in this section.

A.2 Model for baseline-related study

One organizing conceptual principle for understanding household livelihoods strategies is based on the household model which relates poverty measurements to household resource use patterns. The simplest starting point is the utility maximizing scenario which can be depicted by a general participation function of the form:

$$Y^j = f(u, k, i) \tag{A.1}$$

where:
Y^j = household participation in activity j;
u = capacities;
k = incentives;
i = institutions.

This function is derived from similar analysis by De Janvry and Sadoulet (2001) and Sanchez (2005). According to De Janvry *et al.* (2005), two major categories of factors determine a rural household's economic participation decision: first, the factors that affect the relative returns and risks of participation (incentives); second, the factors that determine the capacity to economic participation, namely, capacities, such as education, asset endowments, etc. Any factor that positions agriculture as a risky or low return activity increases participation in non-farm activities, and vice versa. Factors increasing risk or reducing returns in non-farm enterprises increase households' participation in farming. This can only be possible if the required capacities to participate in these activities are available to a household.

In a similar analysis, Sanchez (2005) proposes that labour supply is a function of returns and risks in an activity, capital available to the household and household characteristics. Returns and risks explain the incentives to participate in a given activity and both capital available to a household and household characteristics reflect the capacities to participate in a certain activity.

The household utility function can be solved by specifying a model of the form:

$$U_{ij} = \beta_j X_i + \mu_{ij} \qquad \text{(A.2)}$$

where:
U_{ij} = the utility associated with the ith household choice of activity j;
X_i = a vector of parameters of the ith household;
μ_{ij} = the error term.

But utility U_{ij} is not observable (Alvarez and Nagler, 1995). What we observe is decision to participate, which is represented by a dummy and can be expressed as a probability function employing the binary linear model (BLM) that allows for choices between a set of two alternatives to be analyzed:

$$\rho\,(Y=1) = e^{\beta'\chi} / 1 + e^{\beta'\chi} \qquad \text{(A.3)}$$

$$\rho\,(Y=0) = 1 - e^{\beta'\chi} / 1 + e^{\beta'\chi} = 1 / 1 + e^{\beta'\chi} \qquad \text{(A.4)}$$

which for market participation choices can be expressed as a logit transformation of the type:

$$\text{Logit}\,(P_i) = \ln\,(P_i / 1 - P_i) = \alpha + \beta_1 X_1 + \ldots + \beta_n X_n + U_t \qquad \text{(A.5)}$$

Where:
$\ln\,(P_i / 1 - P_i)$ = logit for market participation choices;
P_i = not participating in markets;
$1\text{-}P_i$ = participating in markets;
β = coefficient;
X represents covariates;
U_t = error term.

For the baseline status study, the dominant model applied sequentially introduced dependent variables that represented the probability of participating in own business, own agriculture, wage employment and remittances. This decision is explained by a set of socio-economic factors that distinguished the households into categories.

In the context of poverty measurements, the concept of income inequality decomposition was applied to calculate Gini Coefficients which allowed for the extent of inequality between household categories to be assessed. Consideration was also given to the fact that the Gini index satisfies the following five properties:

- Transfer sensitivity which allows the measure to respond to the relative positions of the source and target of an income transfer – transferring income from a rich person to a poorer person results in a more equal distribution and hence a lower Gini measure than otherwise. This is also referred to as the 'transfer principle' (Shorrocks and Foster, 1987; Debraj, 1998).
- Symmetry – which ensures that the relative position of individuals in the income order does not nothing to alter the absolute size of the inequality measure. Debraj (1998) describes this as 'anonymity'. When individuals switch places in the income order through some shuffling that did not affect the distribution, the inequality measure will not be affected.
- Mean/Scale neutrality which ensures that when all incomes change by the same proportion, there is no effect on the size of the measure of inequality.
- Population independence/homogeneity which ensures that the size of the population does not affect the measure of inequality – it is the distribution of income across the population that matters rather than the absolute size of the economy itself.
- Decomposability – the measure should be able to capture the contributions of sub-groups and activities (income sources) to the total inequality (Foster, 1985; Adams, 1999).

Assuming n households engaged in k diverse economic activities for purposes of income generation, the total household income of the ith household can be depicted as y_{ik}. It is also possible to represent the mean household income for the sample of households by the standard notation μ, while the cumulative distribution of the total household income would be shown as F(Y). A generalized function of the following form was defined:

$$G = \frac{(2\,\mathrm{cov}[Y, F(Y)])}{\mu} \tag{A.6}$$

where:

$\mathrm{cov}[Y, F(Y)]$ is the covariance of total income, Y, with mean μ and its cumulative distribution, F. By decomposing the total household income into k sources to take account of the contribution of each component of income to the overall inequality, equation A.6 can be rewritten and expanded so that the overall Gini Coefficient can be expressed as:

$$G = \sum_{k=1}^{k} \left(\frac{\mathrm{cov}[Y_k, F(Y)]}{\mathrm{cov}[Y_k, F(Y)]} \right) \left(\frac{2}{\mu_k} \mathrm{cov}[Y_k, F(Y_k)] \right) \left(\frac{\mu_k}{\mu} \right) \tag{A.7}$$

which can be shortened further to:

$$G = \Sigma^{k}_{k=1}\, G_k\, R_k\, S_k \tag{A.8}$$

where:
S_k is the share of income source k on overall income (i.e. $S_k = \mu_k/\mu$));
G_k is the Gini coefficient measuring the inequality in the distribution of income from activity k within the group, and R_k is the Gini correlation of income from source k with total income which can be expressed separately as:

$$R_k = \frac{\text{cov}[Y_k, F(Y)]}{\text{cov}[Y_k, F(Y_k)]} \tag{A.9}$$

Contribution of source k to overall income inequality (P_k) is given by the product of $G_{k,}$ R_k and S_k

$$P_k = G_k R_k S_k / \text{G} \tag{A.10}$$

Adams (1999) pointed out that the larger the product of G_k R_k S_k, the greater the contribution of income from source k to overall income inequality. S_k is the fraction of income from source k to total income, it should be noted that it is always positive and less than one, G_k is always positive and may exceed one, and R_k is between -1 and 1.

Source elasticity on total inequality P_k measures the response of inequality to changes in income from source k. It assesses whether an income source is inequality-increasing or inequality-decreasing on the basis of whether or not marginal increase on share of that income source leads to an increase or decrease in overall inequality. This can only be valid under the assumption that additional increments in income are distributed in the same manner as the original units since anything that disrupts the pattern of increase also affects overall inequality (Van de Berg and Kumbi, 2006).

$$P_k = S_k (R_k G_k - \text{G}) / \text{G} \tag{A.11}$$

The above procedures were used to establish the extent of rural poverty, the pattern of rural activity choices, and extent of inequality which have been reported in Chapter 4 of this book.

A.3 Models for the component studies assessing institutional constraints

The generalized choice model was employed to estimate the probability that a smallholder farmer in the enumerated localities faced or did not face a specific constraint. Since only two options are available, namely 'all produce sold' or 'not all produce sold', a binary model is set up which defines Y = 1 for situations where the farmer sold all produce, and Y = 0 for situations where some or all produce was not sold. Assuming that x is a vector of explanatory variables and ρ is the probability that Y = 1, it is valid to proceed in the same way as suggested by Equations A.3 – A.5 above.

In line with Hosmer and Lemeshow (2000), Agresti (1990), Gujarati (1992), Liao, (1994), and Harrell (2001), the right-hand term in Equation A.3 is the natural logarithm of the modelled variables. A goodness-of-fit test following Hosmer-Lemeshow, can be carried out by examining the Pearson Chi-square outcomes calculated from the table of observed and expected frequencies as follows:

$$X^2_{HL} = \sum_{i=1}^{g} \frac{(Oi - Ni\pi i)^2}{Ni\pi i(1 - \pi)} \tag{A.12}$$

where:
N_i = the total frequency of the items in the ith group;
O_i = the total frequency of obtaining particular event outcomes in the ith group;
π_i = the average estimate of the probability that a particular event outcome in the ith group would be realized.

In addition to the binary and multi-nomial linear models, tests of association and analysis of variance (ANOVA) were carried out. In the case of market access constraints for small-scale cattle producers, market off-take rates were calculated using the formula:

$$\textit{Market offtake} = \frac{\textit{cattle sold in each municipality in the last 12 months}}{\textit{municipality herd size} \times 100\%}$$

According to Ba *et al.* (1996), off-take rate is the number of animals sold, donated, slaughtered, or loaned, as a percentage of the adjusted number of animals. In this project the prime focus was on *market off-take* rather than *general off-take rate* as the one used by Ba *et al.* (1996). Market off-take rate is essential in this study, since many development projects try to encourage small scale farmers to sell more livestock. It is commonly thought that communal areas are over-stocked and that higher rates of commercial off-take would be good for the national economy and for the environment (Sieff, 1999; Hendricks *et al.,* 2007).

Farmers in communal areas differ on how they market their cattle. Others sell while others do not. Therefore, this implies that the problem that needs to be solved needed a method that was able to explain a binary endogenous variable (yes / no) by a set of covariates that determine the outcome of the decision. A typical method used to solve such dichotomous variables is the logistic regression (Hosmer and Lemeshow, 2000). According to Kleinbaum (1994), there are two main reasons for using logistic regression in economics research. Firstly, the logistic function is extremely flexible and easily applicable, and secondly the interpretation of the results is straight forward and meaningful. The logistic model also imposes for threshold and interaction effects and allows for examination of social interaction.

An important concern when categorical and continuous variables are included together in the same model is the presence or otherwise of heteroscedasticity or unequal variance of the disturbance term, μ. According to Gujarati (2003), one source of heteroscedasticity which is relevant to this study is skewedness in the distribution of income, wealth, education, etc. As this study examined these variables, such a problem is worth noting. In such a case, the classical linear regression model (CLRM) assumption that $E(\mu^2_i) = \sigma^2$, for all $i = 1, 2, ..., n$, is violated. In place of that, a situation exists where $E(\mu^2_i) = \sigma^2_i$, implying that the variance is no longer constant but varies as the number of observations varies. The consequences of this problem are well-documented, including the fact that estimates obtained by the regression are not efficient and best estimators (Gujarati, 2003), which can lead to wrong conclusions, and the 'usual hypothesis-testing routine is not reliable ...'. Detection of heteroscedasticity by means of the White's General Heteroscedasticity Test is a handy procedure which proceeds by conducting an auxiliary regression in which the square of the residual term is the dependent variable which is regressed on the original variables or predictor variables as follows:

$$\hat{}\mu^2_i = \alpha_1 + \alpha_2 X_{2i} + \alpha_3 X_{3i} + \alpha_4 X^2_{2i} + \alpha_5 X^2_{3i} + \alpha_6 X_{21} X_{3i} + v_i \quad \text{(A.13)}$$

where $\hat{}\mu^2_i$ is the squared residuals which is regressed on the predictor variables, X_i to X_{ni}, the squared values of these original predictor variables, and their cross-products, and an error term v_i. When the sample size is multiplied by the coefficient of multiple determination, R^2, resulting from this regression, a value is obtained which follows a chi-square distribution with degrees of freedom equal to the number of regressors (less the constant term) in the auxiliary regression. That is:

$$n \times R^2 \sim X^2_{df} \quad \text{(A.14)}$$

Setting the null hypothesis that there is no heteroscedasticity, the observed chi-square value obtained in relation (A.14) above is compared with the tabulated chi-square and the null hypothesis is accepted if the observed chi-square value exceeds the tabulated or critical chi-square value. This would mean that heteroscedasticity is confirmed. The White's Heteroscedasticity-Constant Variances and Standard Errors procedure is available for correcting this problem where it is detected.

A.4 Methods for the study on welfare and incentive effects of maize market deregulation in Swaziland

This section is devoted to a description of the theoretical underpinnings of the Standard Partial Equilibrium model and its application in analysing the welfare effects of trade restrictions. The procedure for analysing the differential impact on consumers, producers and government revenue are specified. *Ex ante* procedures for analysing markets are employed in the study to determine the normative scenarios for a possible deregulation of the maize market of Swaziland. To that extent the key assumptions for such a normative analysis are detailed. Furthermore, the Policy Analysis Matrix (PAM) is described as it is applied in assessing the viability of markets in private and social terms. The section concludes by describing the nature of the data collected as well as the data collection procedure, including the sources and the transformations necessary to render the data useable within the model.

A.4.1 Theoretical framework

In this section, the theory behind the Partial Equilibrium model is described. In addition, the steps followed in applying the framework will be clarified. This will be followed by a description of the PAM, how it is constructed and the way it is interpreted in the study.

A.4.1.1 The partial equilibrium model

The focus of this study was markets as institutions for mediating exchange and instruments for enhancing the economic participation of smallholders in order to promote their welfare. Markets will differ in the extent to which they perform this function and it is possible to measure this performance. In general, economists employ two types of models to analyse markets for purposes of assessing various policy scenarios and problems. These models or frameworks are the Partial Equilibrium model and the General Equilibrium model. The difference between these two approaches lies in the fundamental assumptions made about the interrelationship between the variables studied and the rest of the economy. The Partial Equilibrium analysis uses a few variables in isolation and which are allowed to vary while it assumes that all other variables remain constant. In short, as the name implies, the Partial Equilibrium framework is a model structure most frequently used to understand one-sector policies or circumstances, without having to deal with the corresponding changes in other policies/circumstances and other sectors of the economy (Hieller, 1973; Reed, 2001). Of course, there is no suggestion that the problems at hand are any less complex than they really are.

On the other hand, the General Equilibrium approach makes effort to accommodate the inherent complexity in the system. In that regard, it continues to be used for economy-wide policy analysis, for instance to examine the effects of change on the economy of an economic policy such as the imposition of tariffs or quotas on imported goods. It can also be used to explore the effects of an increase in price or reduction in the supply of imported goods. How these measures impact on the welfare of the average citizen of the country can be deduced from the results of such an analysis. One of the major qualities of the General Equilibrium model is its ability to trace effects of large changes in specific sectors throughout the entire economy. All economic variables and their interactions with the rest of the economy can be analysed today with the Computable General Equilibrium Models for which several computer softwares are available. This means that the General Equilibrium framework tries to approximate the real world in which all the variables are constantly changing. This makes the General Equilibrium modelling process to be a dynamic modelling process (Hieller, 1973; Scarf and Shoven, 1984). The short comings of the General Equilibrium models are its inadequacy to treat money and financial institutions. The General Equilibrium model has great difficulty in allowing for unemployed resources and is unable to cope with large-scale industrial enterprises that are capable of exerting significant influences on prices (Scarf and Shoven, 1984).

The basic assumption of the Partial Equilibrium model in analysing trade policy issues is that when one commodity is traded between two countries or markets they are essentially similar to the rest of the world market (Söderstem and Reed, 1994). The Partial Equilibrium framework uses the conventional supply and demand curves to depict the price effects of policies and then the concept of consumer and producer surpluses. Revenue collected and costs incurred are also determined and their impact on government annual budget is assumed to determine the welfare and distributional effects of policies. This methodology is used to analyse the effects of policy actions examined only in the markets that are directly affected by the policy under consideration. The analytical structure makes a distinction between the welfare of consumers who purchase a product and that of the producers. Consumer welfare is measured using consumer surplus, while producer welfare is measured using producer surplus. It is assumed that the revenue collected by government is redistributed in the economy and is used up either on public goods or to raise economic welfare (Söderstem and Reed, 1994; Reed, 2001).

A.4.1.2 Welfare effects of a tariff case of a small importing country

Welfare economics is concerned with comparing different economic situations in terms of social welfare, including issues related to efficiency, equality and public economic policy (Just *et al.*, 1982). Welfare economics is the fundamental principle of using resources optimally so as to achieve maximum well-being for all individuals in a society. One difficulty with welfare economics assessment is the fact that welfare is not an observable variable (Just *et al.*, 1982). According to Fraser (1992), assessing the welfare effects of government policies has a long history in agricultural economics. Traditional welfare assessment methodologies use concepts like economic rent, producer and consumer surplus in identifying the transfer (or expected transfer) between producers and consumers or taxpayers (Just *et al.*, 1982; Fraser, 1992).

According to Just *et al.* (1982), welfare economics rests on the principle that social gains are maximised at competitive markets and where non-competitive interferences exist, the economist are justified to recommend policy measures that eliminate the distortions. In recent years, numerous commercial policies affecting economic welfare have proliferated, partly because of the many trade theory innovations and policy instruments that have been introduced into the world market. These international theoretical developments have helped improve the economist's ability to measure the potential costs and benefits of trade policy interventions.

For a long time, welfare analysis has been largely qualitative with most research merely focusing on likely effects of liberalisation (Greenaway, 1983). To fill this gap, in 1990, Isabella Tsakok developed the Partial Equilibrium framework to quantify the welfare effects of a given liberalisation process (Greenaway, 1983). In the last decade, several empirical policy analysis models have been developed to analyse the impact of trade policies on welfare. Most of these modes have the ability to go beyond the traditional border analysis. Faced with all these diversity in empirical analysis tools, to date, limited research work has been undertaken to analysis the welfare effects of trade policies in the Swaziland maize sector, not even the Partial Equilibrium model. It is on this premise that this study will utilise the Partial Equilibrium model to quantify the welfare effects of the current maize market regulations in Swaziland. The Partial Equilibrium analysis is preferred for its ability to analyses single-sector market while assuming that the effects in the other markets are negligible or remain constant.

In this study we will consider a small importing country case with the assumption that the country's imports are a small share of the world market, and complete elimination of imports from the world market have no noticeable effects on product quantity and world price. Even if a small country restricts trade by implementing a tariff barrier on imports, no much effects are felt in the world market (Lindert and Kindleberger, 1982; Reed, 2001).

When a small country faces international or world price P_w in free trade, then the free trade equilibrium is depicted in Figure A4.1. P_w is the free trade equilibrium price. At that price, domestic demand is given by $D_{1,}$ domestic supply by S_1 and imports are the difference D_1-S_1 (the line DH in the figure). When a small country implements a specific tariff it will raise the domestic price by the full value of the tariff. Suppose the price in the importing country rises to P_d because of the tariff. In this case the tariff rate would be $t = P_d - P_w$, which is equal to the length of the line segment BF in Figure A4.1. As a result of the implemented tariff barrier, domestic demand decrease from D_1 to D_2 and supply increase from S_1 to S_2 (Cramer and Jenses, 1994.).

The tariff barrier brings up welfare changes whose direction and magnitude are summarised in Table A4.1 below. The notations used in Table A4.1 are based on the information already presented in Figure A4.1. The magnitude of the welfare changes can be calculated using the formulations in Table A4.2 that is adapted from Tsakok (1990). Both Tables A4.1 and A4.2 present the aggregated national welfare effects of the tariff to the importing country.

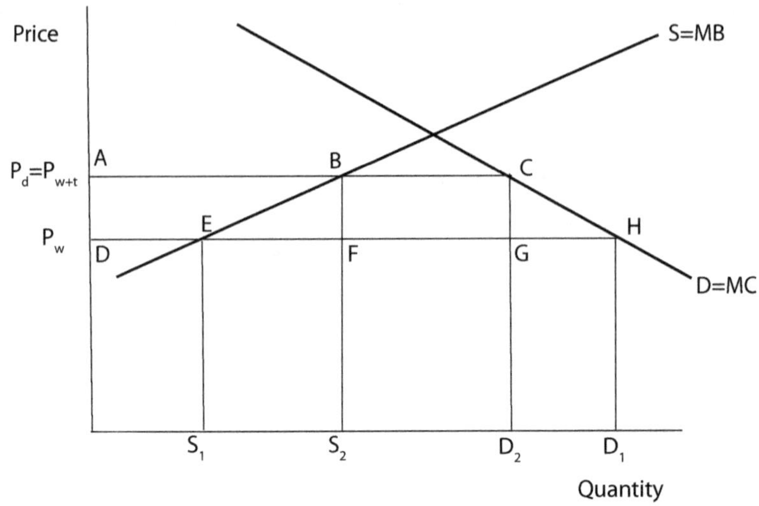

Figure A4.1. Welfare effects of tariffs on small country

Table A4.1. Summary of welfare effects import tariff for a small importing country (Reed, 2001).

	Graphically
Consumer surplus (negative)	ACHD
Producer surplus	ABED
Government revenues	BCGF
National welfare (negative)	BFE and CGH

Table A4.2. Methodological framework to calculate the welfare effects of a tariff (Tsakok, 1990).

NEL_p	Deadweight loss in production	$0.5 \times e_s \times t'^2 \times V'$
NEL_c	Deadweight in consumption	$0.5 \times n_d \times t'^2 \times W'$
WG_p	Change in producer surplus	$t'V' - NEL_p$
WG_c	Change in consumer surplus	$-(t'W' + NEL_c)$
ΔGR	Change in revenue	$t'(W' - V')$
Loss/capita	Loss per capita	WG_c / total population

A.4.1.3 Consumer surplus

A French engineer, Jules Dupuit first formulated the concept of consumer surplus in 1850. His major concern was the amount of worthwhile subsidy given towards constructing a bridge. Dupuit knew that consumers would be willing to pay more for the bridge than they actually paid, because the consumers obtained more satisfaction or surplus from using the bridge than they were being charged (Ng, 1979). The consumer surplus concept gained popularity after the Scottish Alfred Marshall first introduced the concept to the English-speaking world, defining consumer surplus as the excess price which consumers are willing to pay rather than go without the consumption commodity (Ng, 1979). Winters (1985) said consumer surplus represents the utility that consumers derive from consuming a good over and above what they had to sure up. In simple terms, consumer surplus is the difference between what consumers are willing to pay for a unit of the good and the amount they actually pay for the product at the market (Nicholson, 1978). It is the welfare measure of a group of consumers who purchase a particular product at a particular price (Nicholson, 1978). The Marshallian framework used the triangular area under the demand curve to quantify consumer surplus similar to Figure A4.2 sum of the area of the rectangle. Therefore, total consumer surplus is the amount of extra utility that consumers derive from consumption that is reasonably measured as the area between the demand curve and the horizontal line drawn at the equilibrium market price as shown in the Figure A4.2 (Södersten, 1970).

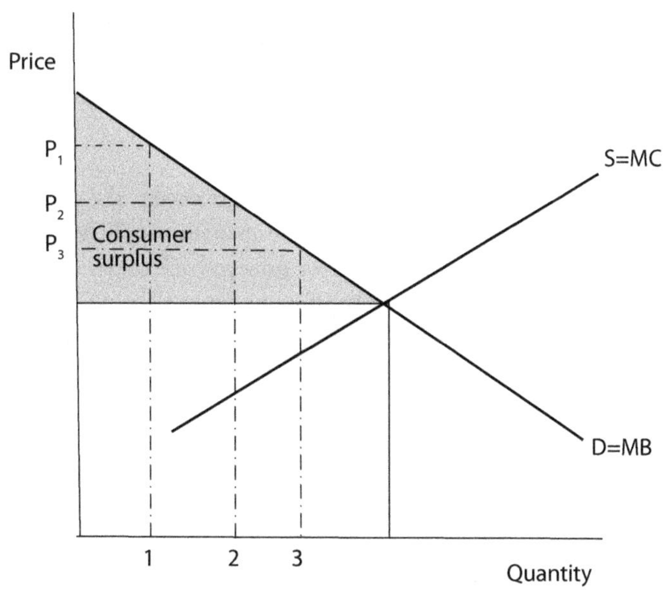

Figure A4.2. Graphical representation of consumer surplus (Nicholson, 1978).

When a tariff barrier is imposed on imports, consumers in the importing country are made worse-off because the imposed tariff raises consumer prices. The price increase of both domestic and imported goods reduces the quantity of goods that consumers can buy. As a result, they suffer a loss of consumer surplus in the market. The lost surplus is redistributed to producers who are the winners of the tariff policy (Cramer and Jensen, 1994). The magnitude of the change in consumer surplus is the area ACHD as illustrated in Figure A4.1.

A.4.1.4 Producer surplus

Producer surplus is analogous to the consumer surplus concept. It is used to measure the welfare of a group of firms who sell a particular product at a particular price (Just *et al.*, 1982). By definition producer surplus is the difference between what producers actually receive when selling a product and the amount they would be willing to accept for a unit of the good. A firm's willingness to accept payments can be determined from a product supply curve. The market supply curve shows the quantity of the goods that a firm would supply at each and every price that might prevail. In other words, the supply curve tells us the minimum price that producers would be willing to accept for any quantity demanded by the market. Therefore, producer surplus is that triangular area under the demand curve and above the price (Price, 1977).

The procedure outlined earlier for calculating the consumer surplus can be adopted for calculating producer surplus. Suppose that only one unit of a good is demanded in a market, as shown in Figure A4.3. A firm would be willing to accept the price P_1 to supply that single unit it produced. If two units of the good were demanded in the market then the minimum price to induce two units to be supplied is P_2. A slightly higher price would induce another firm to supply an additional unit of the good. Three units of the good would be made available if the price were raised to P_3 (Just *et al.*, 1982).

According to Kreinin (1975), Suranovic (1999) amongst others, the price that ultimately prevails in a free market is that price which equalises market supply with market demand. In our example (Figure A4.3) that price is P_e if we now go back to the first unit demanded. Assuming there is a firm that is willing to supply one unit at the price P_1 but ultimately receives P_e for the commodity sold. The difference between the two prices represents the amount of producer surplus that accrues to the firm. For two units another firm would be willing to supply at price P_2 but ultimately receives P_e, in which case the producers' surplus is a smaller amount as shown in Figure A4.3. The total producer surplus in the market is given by the sum of the areas of the rectangles formed by the intersections of thee equilibrium price line and the various quantities supplied.

We therefore conclude that total producer surplus can be practically measured as the area between the supply curve and the horizontal line drawn at the equilibrium market price.

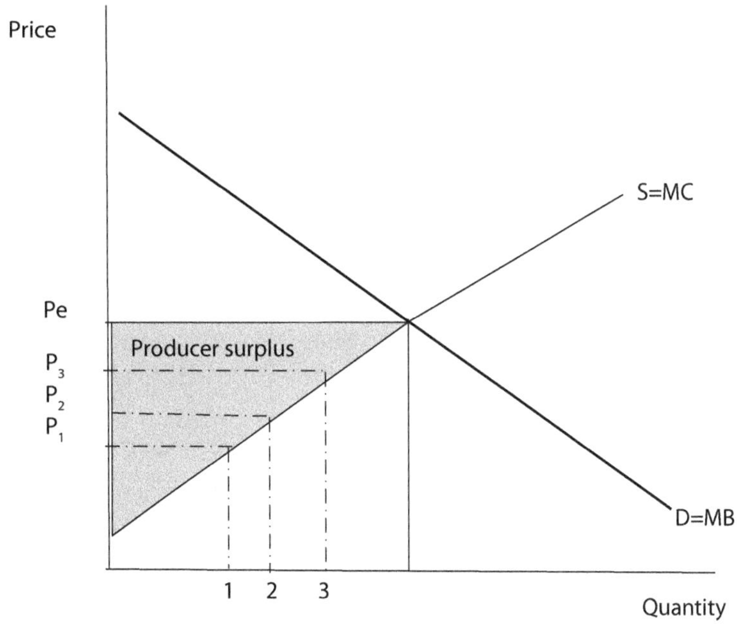

Figure A4.3. Graphical representation of producer surplus (Kreinin, 1971).

From the foregoing illustration producers in the importing country are made better off by the new tariff regime. A stylised outcome of the tariff regime can be suggested. The increase in the price of their product resulting from the tariff increases producer surplus. At the same time, this price increase induces an increase in output of existing farm firms and attracts new firms into production. This will result in an increase in employment and an increase in profit and or payments to fixed costs. These stylised outcomes are shown in Figure A4.1 as the area under ABED (Kreinin, 1975).

A.4.1.5 Government revenue

The government earns incremental revenue by imposing a tariff regime. This revenue equals the tariff rate per unit of imported goods times the quantity of goods imported. This government revenue is part of the consumers' surplus loss that comes with the tariff rule. Beneficiaries of the revenue collected vary depending on whom the government decides to spend the incremental earnings on. In most cases this revenue is used to finance a large portion of the governments' annual budget i.e. it is used to support diverse government spending programs. This way, the revenue benefits accrue to some sectors of the economy. Area BCGF in Figure A4.1 illustrates the amount of revenue collected by government with the imposition of a given tariff regime (Cramer and Jensen, 1994).

A.4.1.6 Net efficiency losses (total welfare effects)

The aggregate welfare effect for the country is found by summing the gains and losses to consumers, producers and the government. The net effect consists of two components: negative production efficiency loss (BFE), and negative consumption efficiency loss (CGH). The two losses together are typically referred to as 'deadweight losses' (Figure A4.3). The magnitude of the change in national welfare is computed as dead weight efficiency loss to consumption ($0.5 \times e_s \times t'^2 \times V'$) plus dead weight efficiency loss to production ($0.5 \times n_d \times t'^2 \times W'$). Since there are only negative elements in the national welfare change, the net national welfare effect of a tariff must be negative. This means that by imposing a tariff regime a 'small' importing country reduces national welfare (Tsakok, 1990).

Summary review of the welfare effects of imposing a tariff are as follows: whenever a 'small' country imposes a tariff rule national welfare drops to a level equal to area BFE and CGH in Figure A4.1. The higher tariff prices imposed the larger national welfare losses incurred and consumers are the worse affected because they pay for the producer and government gains. Since we are assuming that the country is 'small', the tariff has no effect on the rest of the world prices; thus no welfare changes for producers or consumers at world level. Even though imports are reduced, the related reduction in exports by the rest of the world is assumed to be too small to have a noticeable impact.

In conclusion, it is recommended that countries should open up their borders to trade, and concentrate their production activities on those commodities in which they have comparative advantage while importing the rest from the world market. This is said to boost trade for both trading parties. According to economic theory, inefficient producers are the ones likely to impose trade restriction policies to promote inefficient production systems at the expense of consumers' welfare.

A.4.2 Policy analysis matrix

Monke and Pearson (1989) developed the Policy Analysis Matrix (PAM) to analyse the effects of government policy's on efficiencies and competitiveness. The PAM is a Partial Equilibrium methodology that builds on the conventional social benefit-cost analysis. A PAM differentiates between private and social profits of a production system. Initially, a PAM was developed to analyse distorting policy effects on factor and product markets and the effects of the policy on new technology. Today, it has been extended to the analysis of the trade-off between efficiency and non efficiency objectives. The Policy Analysis Matrix results give an insight into the adverse effects of policy interventions. The general structure of a PAM is a 5 by 3 matrix with a two-way accounting identity similar to Table A4.3.

Table A4.3. Identities for calculating price effects of policy distortion (Monke and Pearson, 1989).

	Tradables		Domestic resources	Profits
	Output	Input		
Private	A	B	C	D=(A-B-C)
Social	E	F	G	H=(E-F-G)
Policy transfer	I=(A-E)	J=(B-F)	K=(C-G)	L=(I-J-K)=(D-H)

A.4.2.1 Welfare analysis

For this study, the static welfare effects related to Swaziland's maize market sector will be estimated using the traditional Partial Equilibrium model, as adopted from Tsakok (1990). The Partial Equilibrium model enables the estimation of the effects of price changes on the current market arrangement, ceteris paribus, and further looks at the economic welfare effects of the changes in price on the different economic groups. Here the welfare effects will be defined in terms of the amount of money that producers earn, consumers lose, government receives as revenue and the whole of society loses due to the rise in prices that reduces the demand for the commodity (deadweight losses).

The analytical formulation is presented by the equations below. A number of variables that form an integral part of this analysis are presented in the table and include the domestic maize grain price (P_d) being the price that large millers and other buyers pay the NMC for the maize that they require for milling, these prices were obtained from the NMC. The border price (P_b) is the import parity price for maize grain imported into Swaziland, i.e. the landing price for maize imports. These prices were also obtained from NMC offices. The Nominal Protection Coefficient (NPC) is the ratio of P_d to P_b (P_d/ P_b). This is a measure of protection received by either producers or consumers. An NPC value of greater than 1 (NPC>1) means that producers receive positive protection i.e. they received a higher price than they would have without the intervention. On the other hand, While an NPC<1 represents negative protection for producers suggesting that consumers receive the protection while producers are discriminated against by the policy intervention. The variable 't' is the implicit tariff (NPC-1), suggesting the level of efficiency losses to consumers as a result of domestic and border prices difference. The variable (V') is the value of domestic production at current domestic price, a product of P_d and domestic production, while W' is the value of domestic consumption at domestic prices and is the product of total supply and P_d. The last essential variable in these equations is the elasticities of supply and demand for maize in Swaziland, denoted as e_s and n_d respectively.

Equations

$$NPC = P_d/P_b \tag{A.15}$$

where:
NPC = nominal protection coefficient;
P_d = domestic price for maize;
P_b = border price for maize.

$$t = NPC-1 \tag{A.16}$$

where:
t = implicit tax;
NPC = nominal protection coefficient.

$$t' = tP_d/P_b \tag{A.17}$$

where:
t' = nominal tariff rate;
t = implicit tax;
P_d = domestic price for maize;
P_b = border price for maize.

$$V' = P_d \times \text{domestic production} \tag{A.18}$$

where:
V' = value of domestic production at domestic prices;
P_d = domestic price for maize;
Domestic production = domestic maize production.

$$W' = P_b \times \text{total supply} \tag{A.19}$$

where:
W' = value of domestic consumption at domestic prices;
P_b = border price for maize;
Total supply = total maize supply.

$$NEL_p = 0.5 \times e_s \times t'^2 \times V' \tag{A.20}$$

where:
NEL_p = deadweight loss in production;

e_s = price elasticity of supply;
t' = nominal tariff rate;
V' = value of domestic production at domestic prices.

$$NEL_c = 0.5 \times n_d \times t'^2 \times W' \qquad (A.21)$$

where:
NEL_c = deadweight loss in consumption;
n_d = price elasticity pf demand;
t' = nominal tariff rate;
W' = value of domestic consumption at domestic prices.

$$WG_p = t'V' - NEL_p \qquad (A.22)$$

where:
WG_p = change in producer surplus;
t' = nominal tariff rate;
V' = value of domestic production at domestic prices;
NEL_p = deadweight loss in production.

$$WG_c = -(t'W' + NEL_c) \qquad (A.23)$$

where:
WG_c = change in consumption surplus;
t' = nominal tariff rate;
W' = value of domestic consumption at domestic prices;
NEL_c = deadweight loss in consumption.

$$\Delta GR = t'(W' - V') \qquad (A.24)$$

where:
ΔGR = change in revenue;
t' = nominal tariff rate;
W' = value of domestic consumption at domestic prices;
V' = value of domestic production at domestic prices.

$$\text{Loss per capita} = WG_c/\text{total population} \qquad (A.25)$$

where:

WG_c = change in consumption surplus.

Currently, no study has been undertaken to specifically estimate the price elasticities of supply and demand for Swaziland maize. In this study, an effort was made to quantify these supply and demand elasticities although data availability presented some problems as the available time series data were not sufficiently long thus limited estimation possibilities.

Both supply and demand price elasticity estimates were insignificantly different from zero, implying that maize prices in Swaziland are inelastic, that is both supply and demand do not respond to small changes in prices. The supply elasticity is expected to be very low, almost zero especially when households produce their maize for household consumption in a resource poor production environment. This is the case in Swaziland where the majority (80%) of farmers on SNL produce the maize crop on marginal lands with limited crop production inputs and no irrigation facilities. The price elasticity of demand is also very low, approximately zero, but this is expected for a crop that is a dominant staple with comparatively no substitute. Even though rice is consumed in increasing amounts in Swaziland, maize remains the preferred crop even at higher consumer prices in the isolated consumer market.

Even though the maize price elasticities of supply and demand for Swaziland are negligible insensitive, economic theory maintains that the quantity of goods supplied and demanded should show some reaction no matter how small to price changes in a competitive market environment. But this is not the case in Swaziland. It has already been revealed above that the marketing of the staple food maize in Swaziland is not favourable for smallholders in a competitive market because of the failure of the market to react to market changes. For this reason, this study uses South African price elasticities that were estimated under similar market conditions (regulated markets). These elasticities were estimated by Wright and Nieuwoudt (1993). Although these elasticity figures are also very small, they have an advantage over the ones obtained for Swaziland in that they are statistically different from zero. This allows for the welfare analysis of the market organisation to be carried out, looking ahead to possible market alternatives that will enhance food security in the country.

The PAM identifies 13 different measures for policy effects and efficiency and they are as follows:

1. private profits;
2. social profits;
3. output transfer;
4. input transfer;
5. factor transfer;
6. net policy transfer;
7. private cost ratio (PCR);
8. domestic resource cost ratio (DRC);
9. nominal protection coefficient on tradable outputs (NPCO);

10. nominal protection coefficient on tradable inputs (NPCI);
11. effective protection coefficient (EPC);
12. profitability coefficient (PC);
13. subsidy ratio to producers (SRP).

The procedures for computing these measures are incorporated in the PAM framework and can be summarized in the following Equations:

Private revenue (A) = $P^D = Q_i P_i$ (A.26)

where:
^{D} = value of variable at private (market) price;
P^D = revenue at market price;
Q_i = output quantity;
P_i = output price.

Private revenue (E) = $P^S = Q_i P_i$ (A.27)

where:
^{S} = value of variable at social (shadow) price;
P^S = revenue at social price;
Q_i = output quantity;
P_i = output price.

Private tradable inputs (B) = $\Sigma_i p_i^D q_i^D$ (A.28)

where:
^{D} = value of variable at private (market) price;
p_i^D = price of tradable input i;
q_i^D = quantity of tradable input i per unit of output (Q).

Social tradable input (F) = $\Sigma_i p_i^S q_i^S$ (A.29)

where:
^{S} = value of variable at social (shadow) price;
p_i^S = price of tradable input i;
q_i^S = quantity of i per unit of output (Q).

Private domestic factors(C) = $\Sigma w_j^D l_j^D$ (A.30)

where:
^{D} = value of variable at private (market) price;
w_j^D = price of factor input j;
l_i^D = quantity of j per unit of output.

Social domestic factors = $\Sigma w_j^S l_j^S$ (A.31)

where:
^{D} = value of variable at private (market) price;
w_j^S = price of factor input j at social price;
l_i^S = quantity of j per unit of output at social price.

Private profit (D) = π^D (A.32)

where:
Private profit (D) = A-B-C;
^{D} = value of variable at private (market) price;
π^D = profit.

Social profit (H) = π^S (A.33)

where:
Social profit (H) = E-F-G;
π^S = profit;
^{S} = value of variable at social (shadow) price.

From the PAM, measures of efficiency and policy distortions can be calculated as follows:
1. private cost ratio (PCR) = C/(A-B);
2. domestic resource cost ratio DRC = G/(E-F);
3. nominal protection coefficient on tradable outputs (NPCO) = A/E;
4. nominal protection coefficient on tradable inputs (NPCI) = B/F;
5. effective protection coefficient (EPC) = (A-B)/(E-F);
6. profitability coefficient (PC) = D/H;
7. subsidy ratio to producers (SRP) = L/E.

Where:
- PCR show the extent to which a system can afford to pay domestic factor and still remain competitive.
- DRC ratio is an indication of whether domestic factors used are socially profitable (DRC<1) or not (DRC>1). It is a measure of comparative advantage. According to

economic theorym, a country is said to have comparative advantage if the DRC value is less than 1.

- NPCO measures the degree of output transfer to support a production system.
- NPCI allow analysts to contrast the effects of distorting polices on tradable-inputs cost in agricultural systems. An NPCI=1 indicate there is no transfer, an NPC>1 says there is a negative transfer because inputs costs are raised by policy and an NPCI<1 denotes a positive transfer because input costs are lowered by policy.
- EPC is a measure of incentive effects on commodity price policies but not as indicators of the total impact of policies that influence prices and costs. An EPC>1 suggest that government policies provide positive incentives to producers. While, an EPC<1 indicate the producers are no protected through policy interventions.
- PC is a measure of the degree to which net transfers have caused private profits to exceed social profits. It provides total incentive effects of policy including those influencing factor markets.
- SRP show the extent to which systems revenue have been increased or decreased because of policy. The smaller the SRP, the less distorted the agricultural system. A SRP value converted to percentage shows the output tariff equivalent required to maintaining existing private profits if all other policy distortions and market failures are eliminated. This indicates how much incentives disincentives the system receives from all the effects of divergences.

In addition, a sensitivity analysis was conducted to determine the possible effects of changes in pricing and yield on the key PAM ratios. The procedure employed for this analysis follows closely the recommendation made by Tsakok (1990) in the appendix to her book.

A.4.2.2 Data

Data on consumption, imports, production, prices and tariffs employed in this study were largely secondary data obtained from the databases of several organisations, particularly the National Maize Corporation (NMC), the National Early Warning Unit for food security and the Food and Agricultural Organisation (FAO) statistical database. In South Africa, such information is obtainable from the various publications of the Department of Agriculture, Forestry and Fisheries (formerly the National Department of Agriculture), and Statistics South Africa, among others. Published research reports and other material held at the Department of Agricultural Economics of the University of the Free State were also consulted for valuable information, especially on price elasticities measured in identical geographical and socio-economic settings. The welfare calculations were made for a six-year marketing period starting from 1998/1999 to 2003/2004.

References

Adams, R.H. Jr. (1999). Nonfarm income, inequality and land in rural Egypt. Working paper. World Bank, Washington, DC, USA. Available at: http://www.worldbank.org/html/dec/Publications/Workpapers/wps2000series/wps2178/wps2178.pdf.

Agresti, A. (1990). Categorical data analysis. New York, USA: John Wiley and Sons.

Alvarez, R.M. and J. Naglar (1995). The spatial model and specification of choice models. University of California, USA.

Ba, S.B., H.M.J. Udo and D. Zwart (1996). Impact of veterinary treatments on goat mortality and off take in the semi-arid area of Mali. Small Ruminant Research 19, 1-8.

Cramer, G.L. and C.W. Jenses (1994). Agricultural economics and agribusiness. John Wiley & Sons, Inc: New York, USA.

Debraj, R. (1998). Development economics. Princeton University Press: USA.

De Janvry, A. and E. Sadoulet (2001). Income strategies among rural households in Mexico: the role of off-farm activities. World Development 29, 467-480.

De-Janvry, A., E. Sadoulet and N. Zhu (2005). The role of non-farm incomes in reducing rural poverty and inequality in China. Johns Hopkins University Press: Baltimore, USA.

Foster, J. (1985). Inequality measurement. Proceedings of Symposia in Applied Mathematics 33, 31-68.

Fraser, R.W. (1992). The welfare effects of deregulating the producer prices. American Journal of Agricultural Economics 74, 21-26.

Greenaway, D. (1983). International trade policy – from tariffs to the new protectionism. Macmillan Press LTD: London, UK.

Gujarati, D. (1992). Essentials of econometrics. MacGraw Hill: New York, USA.

Gujarat, D.N. (2003). Basic econometrics (4th Ed). McGraw Hill. New York, USA.

Harrell, F.E. (2001). Regression modeling Strategies. New York, USA: Springer-Verlag.

Hendricks, H.H., W.J. Bond, J.J. Midgley and P.A. Novellie (2007). Biodiversity conservation and pastoralism – reducing herd size in a communal livestock production system in Richtersveld National Park. Journal of Arid Environments 70, 718-727.

Hieller, H.R. (1973). International trade: theory and empirical evidence. Prentice-Hall Inc.: Englewood Cliffs, NJ, USA.

Hosmer, D.W. and S. Lemeshow (2000). Applied logistic regression. 2nd edition. Wiley-Interscience Publication: New York, USA.

Just, R.E., D.L. Hueth, and A. Schmitz (1982). Applied welfare economics and public policy. Prentice-Hall Inc.: Englewood Cliffs, NJ, USA.

Kleinbaum, D.G. (1994). Logistic regression: a self learning text. Springer-Verlag: New York, USA.

Kreinin, M.E. (1975). International economics: A policy approach. 2nd edition. Harcourt Brace Jovanovich, Inc: Chicago, USA.

Liao, T.F. (1994). Interpreting probability models: logit, probit, and other generalized linear models. Sage Publications: Thousand Oaks, CA, USA.

Lindert, P.H. and C.O. Kindleberger (1982). International economics. New York, USA: Richard D. Irwin Inc.

Monke, E.A. and S.R. Pearson (1989). The policy analysis matrix for agricultural development. Cornell University Press: Ithaca, USA.

Ng, Y.K. (1979). Welfare economics: introduction and development of basic concepts. Macmillan Press Ltd.: London, UK.

Nicholson, W. (1978). Micro economic theory: basic principles and extensions. 2nd edition. The Dryden Press: Hindale, IL, USA.

Price, C.M. (1977). Welfare economics in theory and practice. The Macmillan Press Ltd: New York, USA.

Reed, M.R. (2001). International trade in agricultural products. Upper saddle river, NJ, USA: Prentice-Hall.

Sanchez, V. (2005). The determinants of rural non-farm employment and incomes in Bolivia. Masters of Agricultural Economics Thesis, Michigan State University, Michigan, USA.

Scarf, H.E. and J.B. Shoven (1984). Applied general equilibrium analysis. Cambridge University Press: London, UK.

Sieff, D.F. (1999). The effects of wealth on livestock dynamics among the Datoga pastoralists of Tanzania. Agricultural Systems 59, 1-25.

Shorrocks, A.F. and J.E. Foster (1987). Transfer sensitive inequality measures. Review of Economic Studies 54, 485-497.

Södersten, B.O. (1970). International economics. Macmillan and Co. Ltd.: New York, USA.

Södersten, B.O. and G. Reed (1994). International economics. Macmillan Press: Basingstoke, UK.

Suranovic, S.M. (1999). International trade and policy analysis. Available at: http://internationalecon.com/v1.0/ch80/ch80.html.

Tsakok, I. (1990). Agricultural price policy: A practitioner's guide to partial-equilibrium analysis. Cornel University Press: London, UK.

Van De Berg, M. and G.E. Kumbi (2006). Poverty and the rural non farm income in Oromia, Ethiopia. Agricultural Economics 35, 469-475.

Winters, L.A. (1985). International economics. George Allen & Unwin Ltd: London, UK.

Wright, P.D. and W.L. Nieuwoudt (1993). Price distortions in the South African maize economy: a comparative analysis. Agrekon 32, 51-59.

About the contributors

Dr. Abyssinia Mushunje holds a PhD in Agricultural Economics and lectures in the Department of Agricultural Economics and Extension of the University of Fort Hare. His research interests are in resource economics, natural resource management and land reform. He coordinates the seminar programme of the Department of Agricultural Economics and Extension and manages a programme for strengthening the teaching of economics at the high school level.

Dr. Ajuruchukwu Obi studied at the University of Nigeria, McGill University (Montreal, Canada), and the University of the Free State in Bloemfontein and holds a PhD in Agricultural Economics as well as an MA in Economic Policy Management. He is a Senior Lecturer in the Department of Agricultural Economics and Extension of the University of Fort Hare, and presents modules in agricultural marketing and development. Obi's current research interests include institutions, collective innovation and smallholder farming, particularly in relation to factors influencing market access and technology adoption.

Binganidzo Muchara was born in Zimbabwe. He obtained a BSc (Hons) in Agricultural Economics from the University of Zimbabwe and received a Cum Laude for the Honour degree in Agricultural Economics at University of Fort Hare. He is currently completing his dissertation for the award of the MSc in Agricultural Economics at University of Fort Hare, under which he is investigating the role of water use patterns in integrating smallholders to profitable value chains to enhance market access. He has utilized a small grant from the Institute of Development Studies of the University of Sussex to examine the impact of the Zimbabwe's Fast Track Land Reform Programme.

Bridget Jari is a PhD student in the department of Economics and Economic History, Rhodes University. She completed her BSc Agriculture (Agricultural Economics) and MSc Agriculture (Agricultural Economics) degrees at the University of Fort Hare in 2006 and 2008 respectively, receiving a Cum Laude for the MSc degree. Her MSc thesis investigated the institutional and technical factors that influence marketing among emerging and small-scale farmers. Her research interests are in developmental economics, especially on issues related to agricultural markets and trade.

Professor Gavin Fraser obtained a PhD in Agricultural Economics from the University of Stellenbosch and was Head of Department of Agricultural Economics at the University of Fort Hare for many years before moving to the Department of Economics at Rhodes University. His research has focused on marketing and institutional environment of smallholder farmers. For many years he has been on the Executive Committee of the Agricultural Economists Association of South Africa (AEASA) for which he was distinguished guest lecturer for the FR Tomlinson Commemorative Lecture in 2009.

Litsoanelo Mphahama obtained an Honours degree in Agricultural Economics and is currently completing her MSc in the Department of Agricultural Economics of the University of Fort Hare. Her research has focused on the impact of cooperatives on smallholder livelihoods and the role of institutions in agricultural development. She conducted the first systematic socioeconomic assessment of the flagship Agri-Parks Scheme of the Faculty of Science and Agriculture.

Lovemore Musemwa began his University education at the University of Zimbabwe and is a PhD candidate at the Department of Agricultural Economics and Extension of the University of Fort Hare. His research has focused on economics of livestock production and the role of institutions in agricultural and smallholder development. He has carried out several socioeconomic studies on the welfare impact of the Nguni cattle project of the Faculty of Science and Agriculture.

Mafabia Mokhahlane completed her MSc in Agricultural Economics at the University of Fort Hare and currently serves as lecturer at the College of Agriculture of the National University of Lesotho. Her research focuses on institutional issues and natural resource management. She lectures in Agricultural Marketing and Natural Resource Management.

Nkosazana Mashinini obtained an MSc in Agricultural Economics from the University of the Free State and worked as Research Fellow in the Faculty of Natural and Agricultural Sciences of the same University before moving to the Free State Department of Agriculture as Agricultural Economist/Planner. She is now pursuing her PhD in Agricultural Economics in Canada.

Simbarashe Ndhleve is a PhD candidate in Department of Agricultural Economics and Extension at the University of Fort Hare. His areas of research interest include rural livelihoods and poverty alleviation with special emphasis on the measurement of poverty and inequality. He also focuses on research design and data management for both quantitative and qualitative research. He has recently completed a study to quantify the extent of rural income inequality in South Africa which he is exploring further as part of his PhD research.

Dr. Tebogo Seleka studied at the Oklahoma State University and Cornell University in the United States and holds a PhD in Agricultural Economics. He was Senior Lecturer and Head of Department of Agricultural Economics at the Botswana College of Agriculture until 2004 before moving to the Botswana Institute for Development Policy Analysis where he currently serves in the capacity of Executive Director. He has published extensively on agricultural development and policy issues, and is focusing his research on Income disparities and welfare implications in developing areas in Botswana and the rest of the Southern African region.

Keyword index

www.ingramcontent.com/pod-product-compliance
Lightning Source LLC
Chambersburg PA
CBHW061157240926
51846CB00006B/1044